Mobile IPv6: Protocols and Implementation

Mobile IPv6: Protocols and Implementation

Qing Li
Blue Coat Systems, Inc.

Tatuya Jinmei
Toshiba Corporation

Keiichi Shima
Internet Initiative Japan, Inc.

AMSTERDAM • BOSTON • HEIDELBERG • LONDON
NEW YORK • OXFORD • PARIS • SAN DIEGO
SAN FRANCISCO • SINGAPORE • SYDNEY • TOKYO

Morgan Kaufmann Publishers is an imprint of Elsevier

Morgan Kaufmann Publishers is an imprint of Elsevier
30 Corporate Drive, Suite 400, Burlington, MA 01803, USA

This book is printed on acid-free paper.

Library of Congress Cataloging-in-Publication Data

Li, Qing, 1971-
 Mobile IPv6 : protocols and implementation / Qing Li, Tatuya Jinmei, Keiichi Shima.
   p. cm.
 Includes bibliographical references and index.
 ISBN 978-0-12-375075-4
 1. TCP/IP (Computer network protocol) 2. Mobile computing. 3. Mobile communication systems. I. Jinmei, Tatuya, 1971- II. Shima, Keiichi, 1970- III. Title.
 TK5105.585.L54 2009
 004.6'2--dc22
                    2009026151

British Library Cataloguing in Publication Data
A catalogue record for this book is available from the British Library

ISBN 13: 978-0-12-375075-4

For information on all Morgan Kaufmann publications,
visit our Web site at *www.elsevierdirect.com*

Typeset by: diacriTech, India

Printed in the United States of America
09 10 9 8 7 6 5 4 3 2 1

Contents

11 Macro and Type Definitions   113

12 Utility Functions   117

13 Common Mobility Header Processing   123

14 Home Agent and Correspondent Node   137

About the Authors

Li, Qing is a senior architect at Blue Coat Systems, Inc., leading the design and development efforts of the next-generation IPv6-enabled secure proxy appliances. Prior to joining Blue Coat Systems, Qing spent 8 years at Wind River Systems, Inc., as a senior architect in the networks business unit, where he was the lead architect of Wind River's embedded IPv6 products since the IPv6 program inception in 2000. Qing holds multiple U.S. patents. Qing is a contributing author of the book *Handbook of Networked and Embedded Control Systems* (2005, Springer-Verlag). He is also author of the embedded systems development book *Real-Time Concepts for Embedded Systems* (2003, CMP Books). Qing participates in open-source development projects and is an active FreeBSD src committer.

Jinmei, Tatuya, PhD, is a research scientist at Corporate Research & Development Center, Toshiba Corporation. (Jinmei is his family name, which he prefers be presented first according to the Japanese convention.) He was a core developer of the KAME project from the launch of the project to its conclusion. In 2003, he received a PhD degree from Keio University, Japan, based on his work at KAME. He also coauthored three RFCs on IPv6 through his activity in KAME. His research interests spread over various fields of the Internet and IPv6, including routing, DNS, and multicasting.

Shima, Keiichi is a senior researcher at Internet Initiative Japan, Inc. His research area is IPv6 and IPv6 mobility. He was a core developer of the KAME project from 2001 to the end of the project, and he developed Mobile IPv6/NEMO Basic Support protocol stack. He is currently working on the new mobility stack (the SHISA stack) for BSD operating systems, which is a completely restructured mobility stack.

Introduction

When communication resources were precious, it was natural to design a special method for better utilization of these resources. Thus, for a long time, many information network infrastructure providers developed their own network designs and protocols. The recent wide deployment of Internet Protocol (IP) technology provides a simple communication framework for any kind of information infrastructure, and it is integrating all information infrastructures into one protocol—IP.

The evolution first occurred for the wired infrastructure because the wired networks had a faster communication property than the wireless networks. Whereas many wired network infrastructures changed their dedicated network designs and protocols to the generic IP-based system, the wireless infrastructures kept their own designs. The wireless infrastructure, which had a slower communication property, could not accept the overhead of the generic protocol, even though having a common protocol had many benefits, such as interoperability, simplicity, and cost performance.

Recently, however, advances in wireless communication technology have resulted in much wider broadband infrastructures for the wireless environment than in the past. The IEEE 802.11-based technology will soon provide 600 Mbps communication speed, IEEE 802.16 (WiMAX) technology provides more than 70 Mbps with an approximately 50-km communication range, and IEEE 802.16e (Mobile WiMAX) provides more than 20 Mbps communication speed for moving nodes. There is no doubt that future wireless technology will provide much faster communication properties. They are still narrower than those of wired communication devices; however, the overhead of using IP over them is no longer a major issue.

Although the history of mobility technology research and development is quite long, the technology is still not widely deployed. We now have many mobile devices, such as laptop computers, PDAs, and mobile phones, but none of them currently use IP mobility technology. One of the reasons is that the wireless communication technology has not provided the required bandwidth and quality as described previously. Another reason is that we have not had a mobile-ready environment to apply IP mobility technology.

The situation is now drastically changing. In the past, we could not utilize the full advantages of IP mobility technology, even though we had the mechanism to do so. The goal of an unwired mobile Internet world will be achieved with the combination of recent advanced communication technology and the long-researched and -developed IP mobility framework.

1.1 History of IP Mobility

IP mobility protocol is not a special feature for Internet Protocol version 6 (IPv6). The mobility support protocol for IPv4 (Mobile IP) [RFC3344] also has a long history. The initial proposal of the mobility support protocol for IPv4 was presented in 1993. At that time, there were no real "mobile" computers. There were some small computers called laptops, but they were still relatively large and they were very expensive compared to desktop computers. Mobile phones were in use, but they were large and had poor computing resources. Even with this level of technology, IP engineers were trying to provide mobility support for computer devices as if they were foreseeing the future. Mobile IPv4 was finally standardized as RFC2002 (the latest revision is RFC3344) in 1996. In the late 1990s, the Internet era began. Some pioneers started commercial services to provide Internet connectivity. Many companies and universities started providing their information and services over the Internet. Individual users soon followed, and the Internet became the largest information network in the world. Unfortunately for Mobile IPv4, the communication technology, especially the wireless communication technology, was still poor at that time. Although the protocol could support handover from one network to another, we could not use networks in this manner. Mobile IPv4 is now used in the backend system of some mobile telephone service networks. In that sense, it is deployed in the real service network, but we still do not see Mobile IP devices near us and the benefit we receive is limited.

The discussion of Mobile IPv6 [RFC3775] started in 1996. The initial action of the standardization process of IPv6 mobility was very quick. Considering that the first draft of the IPv6 protocol specification was submitted in 1995, the discussion of IPv6 mobility was started almost at the same time as that of IPv6. However, the standardization of Mobile IPv6 was a thorny path. The final specification of Mobile IPv6 was published as RFC3775 in 2004. By contrast, from the first draft to publication as request for comment (RFC), Mobile IPv4 required only 3 years. Recently, the period required to publish a specification as an RFC has become increasingly longer, but 9 years is a surprisingly long time. The draft specification was revised 24 times before it was published as an RFC.

The first turning point of the Mobile IPv6 standardization process was its 13th draft in 2000. The Mobile IP working group reached a consensus on the specification and the 13th draft was submitted to the Internet Engineering Steering Group (IESG) for final review and publication as an RFC. However, the IESG rejected the specification.

Mobile IPv6 was trying to solve one major problem of Mobile IPv4—the path optimization mechanism between a mobile node and its communicating node. Mobile IP is a kind of automatic tunnel establishment protocol. The moving node registers its current location to the proxy node called the home agent. All the packets are forwarded once to the home agent and then sent to the final destination. Apparently, if the mobile node and its communicating node reside nearby and the home agent is located far away, the communication path becomes long and redundant. The Mobile IPv4 base protocol does not mention the optimization mechanism for this case. The Mobile IPv6 specification includes the optimization mechanism from the first draft of the specification. In the mechanism, the mobile node sends its current location to its communicating node. The problem concerns how the communicating node verifies the message sent from the mobile node. If there is no authorization mechanism of the message, any node can send a bogus message to the communicating node. If a malicious node sends such a request using the identifier of the mobile node, then all the data sent from the communicating node to the mobile node are stolen by the malicious node. The Mobile IPv6 specification before the 14th draft was using the IPsec mechanism to protect the message. However, it is usually difficult to set up IPsec parameters between two random nodes. IESG pointed out the difficulty of the IPsec setup process and judged that the specification was not feasible.

After receiving the rejection message from IESG, the working group invented a new mechanism to protect the message. The 14th and 15th drafts proposed a shared secret-based authorization mechanism. It was simple and easy to understand; however, the problem of how the two nodes share the secret remained. In 2002, the 16th draft introduced a completely new mechanism called the return routability mechanism to authorize the message. By using the mechanism, a mobile node and its communicating node can generate secret information with several messages exchanged between them before sending the notification message from the mobile node to register its current location. The detailed procedure of the return routability mechanism is explained later.

Finalizing the specification required only 2 years after the 16th draft. The final draft was published in 2003, and it became RFC3775 in 2004.

1.2 Benefit of IP Mobility

Mobile IP provides a mobility function to IP devices. But what is mobility? When we say "mobility," it implies that there are many different levels of mobility. For example, a cellular network can provide a mobility function to cellular phones. We can use our cellular phone almost everywhere with the same communication identifier—phone numbers. We can even establish an IP connection over the cellular network by using the dial-up connection function. Isn't this mobility? Another example is the e-mail system. We send an e-mail using a fixed identifier, such as bob@example.com. Wherever Bob is, the message will be delivered to the mailbox associated with the mail address bob@example.com, and Bob can retrieve the message independent of his location. It is a kind of mobility.

Figure 1-1 shows various levels of mobility support. SIP (Session Initiation Protocol) [RFC3261] is a session-layer protocol that establishes an application session between two application entities. Because it is independent of the actual location of the terminal on which the application is running, it can be considered as a mobile protocol in the session layer. SCTP (Stream Control Transport Protocol) [RFC2960] is a new transport protocol aiming to replace TCP (Transmission Control Protocol). It is defined on top of the IP layer and supports the IP address migration function while keeping the transport connectivity. HIP (Host Identity Protocol) [RFC4423] is another IP-layer mobility protocol. Unlike Mobile IP-based protocols, HIP is a completely new protocol to pursue the ideal mobility support in the IP layer. The design is cleaner than Mobile IP-based protocols; however, it does not have compatibility with the existing IPv4/IPv6 stacks, whereas Mobile IP-based protocols do. As demonstrated in Figure 1-1, the

FIGURE 1-1

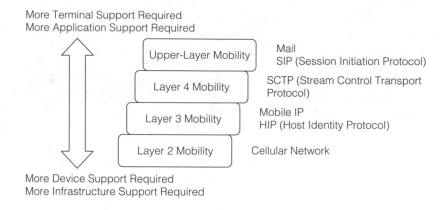

Different kinds of mobility technologies.

FIGURE 1-2

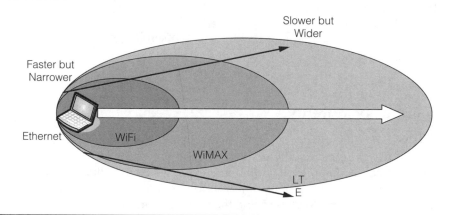

Combination of various kinds of communication media.

more we focus on the lower-layer technology, the more device and infrastructure support is required. For example, if we provide mobility support in cellular networks, we have to expand the same cellular-based network technology throughout the world. In contrast, the more we focus on upper-layer mobility technology, the more the application support is required. For example, if we want to use SCTP, all the transport stacks need to be upgraded to support SCTP.

The IP-layer mobility support, especially the Mobile IP-based protocol, provides a good solution. Because the protocol is located in the network layer, it can utilize various kinds of different data-link communication technologies. It can run over the cellular networks, WiFi networks, Ethernet networks, and many other future networks that support IP. The difference between media is abstracted by the IP layer. Recent advances in wireless communication technology strongly support the deployment of layer 3 mobility technology. As Figure 1-2 shows, the combination of multiple different communication media and Mobile IP enlarges the seamless communication area of the moving node. When a fast communication media such as Ethernet is available, the node can be wired to the network. If the node needs to move to other places, it can attach any kind of medium that is supported by the IP layer and seamlessly roam to the other networks using Mobile IP.

The upper-layer applications do not need to pay attention to the Mobile IP stack. Because Mobile IP provides a transparent IP layer to applications, the existing applications can be used without any modification.

1.3 Supplemental Technologies of Mobile IPv6

Although the function provided by Mobile IPv6 is excellent, it is difficult to convince people to use the technology only with the function provided by the basic Mobile IPv6 specification. There are two major concerns when we consider a real usage of the protocol. The first problem is that it is based on IPv6. The second problem is that it has to upgrade the protocol stack of moving nodes.

The title of the specification of Mobile IPv6 is "Mobility Support for IPv6." It clearly defines that the specification is dedicated to IPv6. This is natural because the mobility support protocol for IPv4 already existed as Mobile IPv4 when the specification was initially proposed. At that time, most people were not confident about IPv6 and there was no consensus by the Internet Engineering Task Force (IETF) that the specification should consider both IPv4 and IPv6. Recently, it has been suggested that protocol designers should make their protocols work on IP independent of its version number, which means that a new protocol should consider both IPv4 and IPv6.

The reason why we are focusing on both IPv4 and IPv6 is that we are seeing the IPv4 address exhaustion problem. Actually, the problem was raised a long time ago. One of the motivations of designing IPv6 was to provide a solution to the problem. However, all except a small number of IPv6 evangelists were not thinking about the problem seriously. Unfortunately, it is usually the case that we tend not to see the real problem until we must directly deal with the problem. Now that it is highly prospective that IPv4 addresses will be exhausted by 2011 if the addresses are used at the current pace [IPV4-ADDRESS-REPORT], we have to choose our future, but the options are limited. One option is to keep IPv4 and try to develop more efficient ways to utilize the existing address space, such as the multiple network address translation (NAT) technology. The other option is to switch to IPv6. There may be other options; however, many people believe that these two options are the most feasible. People on the IPv6 side believe that IPv6 is the final goal of the future Internet. However, they also understand that they cannot ignore the existing environment, IPv4. A few years ago, people were discussing how we could transit the IPv4 Internet to the IPv6 Internet. The discussion has changed. Many people believe that we will use IPv4 and IPv6 simultaneously for a long time. We may even use multiple NAT technologies in addition to IPv6 as an intermediate solution for the future Internet. Because IPv4 has been widely spread throughout the world, not only to researchers but also to the real world such as industrial networks, economic infrastructures, communication infrastructures, and even entertainment networks, it is not possible to replace it with other technology. Considering the situation, the Mobile IPv6 working group at IETF started the discussion of the specification of Dual-Stack Mobile IPv6 (DSMIPv6), which supports both IPv4 and IPv6. We provide a more detailed description of the technology in Section 1.3.1.

The other problem is an engineering problem. As discussed in this book, Mobile IPv6 requires modification of the core kernel function in many cases. It means that all the moving devices have to be upgraded to support Mobile IPv6. As can be easily understood, upgrading an operating system is tough work. Currently, it is very difficult to find such a system that supports Mobile IPv6 by default. If there are few users who can use Mobile IPv6, most service providers will hesitate to prepare a backend system for Mobile IPv6 service. As you know, it is not possible to use Mobile IPv6 just by upgrading local terminals. The protocol requires a home agent operation, which is a kind of proxy node for mobile terminals. This situation generates the chicken-and-egg problem. If there is no Mobile IPv6 service out there, who wants to sell the Mobile IPv6 stack for client nodes? To overcome this problem, IETF proposed two approaches, both of which try to reduce the large system upgrade that is the main cause of the problem.

The first approach is the network mobility function. This is standardized as RFC3963. This approach tries to reduce the modification of local moving terminals. The usage assumption of this protocol is that there are many terminals moving together as a large network. One of the nodes in the network becomes a mobile router and processes all the mobility-related protocol handing on behalf of the other nodes in the moving network. The nodes inside the network are not required to upgrade their operating system to support any mobility protocols.

The second approach is an operator-centric approach that removes all the upgrade burden from the mobile terminal side. In this approach, the mobile service operator upgrades its network infrastructure to accept normal terminals, which means that user terminals have no Mobile IPv6 capability as moving entities. To do that, all the attachment points to the mobile terminals of the service operator need to be upgraded. When a user terminal that does not speak Mobile IPv6 connects to the attachment point, the attachment point processes all the mobile-related signal processing on behalf of the user terminal. The mechanism is called Proxy Mobile IP and has also been standardized as RFC5213.

1.3.1 Dual-Stack Mobile IPv6

DSMIPv6 is a protocol currently discussed actively at IETF that provides IPv4 capability to Mobile IPv6 protocol. The specification is still in the draft phase [MEXT-NEMO-V4TRAVERSAL] and has not been published as an RFC. The idea of DSMIPv6 is simple and straightforward.

Before explaining how DSMIPv6 works, let's remember the base Mobile IPv6 protocol. Mobile IPv6 is a kind of tunneling protocol. When a mobile node moves from its home network to a foreign network, the mobile node registers its care-of address to the home agent that manages the binding information of the permanent home address of the mobile node and the temporal care-of address retrieved at the foreign network. Based on the binding information, the mobile node and the home agent create an IPv6-over-IPv6 tunnel between them. All the traffic sent from the mobile node is transferred once to the home agent using the IPv6-over-IPv6 tunnel. The source address of the packets is the home address of the mobile node. Because the home address is assigned from the home network, which is the logical attachment point of the mobile node, the mobile node cannot send a packet of which the source address is set to the home address from foreign networks. Such a packet may be dropped as an invalid packet of which the source address is forged. Mobile IPv6 solves this problem by utilizing the tunnel between the mobile node and the home agent. Because all the packets are transferred once to the home agent by the IP encapsulation mechanism from the care-of address of the mobile node to the home agent, the mobile node can safely transfer the outgoing packets to its home agent. The home agent forwards the packets after decapsulating the transferred packets. Because of this procedure, all the packets generated by the mobile node are sent from the home network. All other nodes communicating with the mobile node can act as if the mobile node is attached to its home network. When the mobile node receives packets from its communicating node, the opposite path is used. The communicating node sends a packet of which the destination address is the home address of the mobile node. Such a packet is routed to the home network based on the global Internet routing infrastructure. The home agent on the home network interrupts the packet on behalf of the mobile node and transfers the interrupted packet to the mobile node's care-of address using the IPv6-over-IPv6 tunnel created during the binding information exchange procedure.

The DSMIPv6 procedure extends the basic mechanism of Mobile IPv6. It is used for two different cases that the basic Mobile IPv6 protocol does not support: the IPv4 foreign network case and the IPv4 home address case.

When a mobile node moves to a foreign network and the foreign network supports only IPv4, the mobile node cannot keep connectivity as long as it is using Mobile IPv6 because Mobile IPv6 assumes all the foreign networks support IPv6. In DSMIPv6, the mobile node acquires the IPv4 address from the foreign network and uses it as a care-of address. Remember the fact that Mobile IPv6 is a kind of tunnel protocol. In the IPv4 care-of address case, the mobile node sends a registration message to its home agent to bind the IPv4 care-of address and the IPv6 home address of the mobile node. Once the registration message is accepted, an IPv6-over-IPv4 tunnel is created between the mobile node and the home agent. All the IPv6 packets sent from the mobile node are encapsulated and transferred to the home agent using the IPv6-over-IPv4 tunnel. The home agent decapsulates the packets and forwards them to the final destinations of the packets. Because the decapsulated packets are normal IPv6 packets, the communicating nodes do not notice the fact that the mobile node is in the IPv4 network. Figure 1-3 shows the mechanism.

On the reverse side, the same procedure happens as in the basic Mobile IPv6 case. The difference is that the tunnel is used to transfer packets sent to the mobile node. The home agent uses the IPv6-over-IPv4 tunnel to transfer packets sent to the mobile node from its communicating nodes. Apparently, to use this mechanism, the home network must be a dual-stack network to create an IPv4 tunnel between the home agent and the mobile node. This requirement may become a restriction when IPv4 addresses are actually exhausted; however, it seems a reasonable assumption when we see it as a tentative mechanism for IPv4 to IPv6 transition.

By using the previously discussed mechanism, the mobile node can now move to any IP network as long as it is using IPv6 as a network communication protocol. The next step is to provide an interaction mechanism to the legacy IPv4 world. Because there are many IPv4 applications and many people believe that we have to use both IPv4 and IPv6 for a long time to complete the transition from IPv4 to IPv6, it is important to provide a function to use IPv4 as a network communication protocol.

DSMIPv6 supports the IPv4 home address function. In this case, a mobile node can have a fixed IPv4 address assigned by the home network. As for the first mechanism, the home network must be a dual-stack network.

FIGURE 1-3

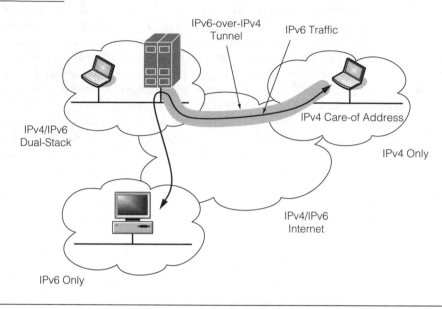

Dual-stack operation with an IPv4-only foreign network.

When the mobile node moves to a foreign network, it gets a care-of address and sends a registration message as usual. The different point is that the mobile node includes its fixed IPv4 address information in addition to the normal registration message, or the mobile node requests that it wants a fixed IPv4 address. The home agent receiving the registration message assigns an IPv4 address for the mobile node and starts the IPv4 home address processing. When the mobile node wants to use the IPv4 application, it generates an IPv4 packet of which the source address is the IPv4 home address registered to the home agent. The IPv4 packet is transferred to the home agent using the tunnel created between the mobile node and the home agent as a result of the registration process. The home agent decapsulates the packet and forwards the packet, which is a normal IPv4 packet generated by the mobile node, to the final destination node, which is also an IPv4 node. On the reverse side, the communicating IPv4 node sends an IPv4 packet to the IPv4 home address of the mobile node. In this case, the home agent is acting as a proxy ARP node for the mobile node's IPv4 home address. The home agent interrupts the IPv4 packet sent to the IPv4 home address of the mobile node and transfers it to the current location of the mobile node using the tunnel created between them.

As you may notice, there is no restriction against the IP version of the tunnel. It can be either an IPv4-over-IPv6 tunnel or an IPv4-over-IPv4 tunnel. For the latter case, the IPv4 home address mechanism is used in conjunction with the IPv4 foreign network support mechanism. This implies that it is possible to operate IPv4 mobility service using DSMIPv6, which is an extension protocol of Mobile IPv6. Some people may think it is too much to support all the cases by the DSMIPv6 protocol only. However, it is more important to provide a uniform protocol that supports all the cases. If we have to support IPv6 mobility by Mobile IPv6, IPv4 mobility by Mobile IPv4, and version-independent mobility by some other protocols, then the engineering task becomes more difficult and the deployment scenario becomes complex. If DSMIPv6 can support all four cases—IPv6 over IPv6 or IPv4, and IPv4 over IPv4 or IPv6—implementation and operation will be much easier.

Unfortunately, the protocol specification is still under discussion at IETF, even though it has long been the center of attention to help the deployment of both IPv6 and IP mobility technology. We believe that the quicker stabilization of the specification will accelerate the entire IP mobility deployment.

1.3.2 Network Mobility Basic Support

One of the reasons why it is difficult to deploy Mobile IPv6 is that to utilize the mobility function, the moving node must support the new protocol stack. Network Mobility Basic Support (NEMO BS) provides one of the solutions of the problem by providing a moving router.

There are two types of moving entities. One is a single moving entity such as a mobile phone. The other is a group of moving entities moving together. An example of the latter case is a transportation system such as trains and buses. In this case, many people move together as a single moving entity. If all the people want to use IP mobility service with Mobile IPv6, all the equipment they are using has to be upgraded. It is clear that there is an optimization in this case. If all the people are moving together, why do they need to manage their equipment individually?

NEMO BS introduces a notion of a moving network and adds a routing function to the basic Mobile IPv6. In NEMO BS, a mobile router has a fixed network prefix called a mobile network prefix (MNP). The MNP is assigned from the home network as shown in Figure 1-4.

The mobile network is allocated topologically naturally. When the mobile router is attached to the home network, the mobile network acts as a normal subnetwork. Once the mobile router moves to a foreign network, it registers the home address the same as Mobile IPv6, and it also registers the MNP associated with the mobile router. A node inside the mobile network, the mobile network node (MNN), can continue its communication the same as when the mobile router is attached to the home network. All the packets sent from MNNs go to the mobile router, which is the boundary node between the mobile network and the Internet. The mobile router then transfers the packets to the home agent using the tunnel established between them. The home agent forwards the packets after decapsulating them. In the reverse direction, the communicating node sends a packet of which the destination address is the address of the MNN, which is fixed. The packet is routed to the home network based on the Internet routing infrastructure. If the mobile router is attached to the home network, the packets from the communicating node will be forwarded to the mobile network by the mobile router. When the mobile router is away from home, the home agent becomes the next hop router to the mobile network as a result of the registration message exchange procedure. The home agent forwards the packet to the mobile network using the tunnel between the home agent and the mobile node.

The most straightforward usage scenario of NEMO BS is placing a mobile router to the transportation system, such as a train, bus, or airplane. The passengers bring their own equipment such as a PDA or laptop PC. Because all the mobility-related management is done by the mobile

FIGURE 1-4

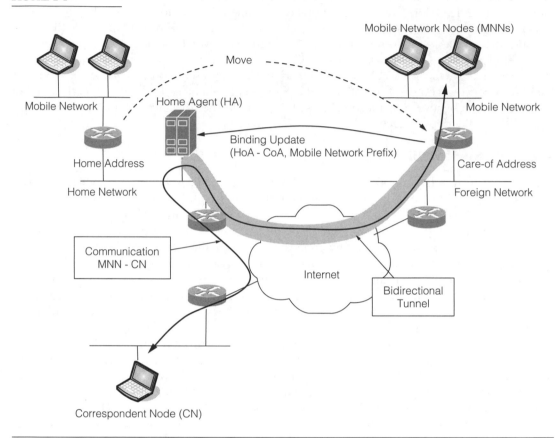

router, passengers do not need to modify their devices to support mobility functions. They may not even be aware of the fact that they are in the moving network because all they see is a logically fixed network provided in the car or cabin. The limitation is that passengers cannot move outside of the network and keep the fixed address achieved from the mobile network. If they want to keep mobility even when they are not in the mobile network, they have to use the basic Mobile IPv6.

The other example of NEMO BS usage is using it as a simple fault-tolerant system for small networks. The more the Internet becomes an important infrastructure of business tasks, the more the reliability of the Internet is required. The Internet is expected to be a dependable infrastructure. Trouble with Internet connectivity sometimes causes serious damage to businesses. It has become common for many business organizations that having multiple Internet connectivity subscriptions to prepare for any connectivity problems. In such cases, when one of the Internet Service Providers (ISPs) has trouble, the organization changes its connection to the alternative ISP it is using.

Changing ISPs requires a major investment, especially when used with IPv6. In IPv6, every organization has its own IPv6 address block. There are two ways to get an IPv6 address block for organizations. One is to obtain it from one of the ISPs by subscribing to their connectivity service. In this case, the address block will be a part of the larger address block that the ISP is using for their customers. If the address block is a part of the address block of the upstream organization, it will be difficult to change the upstream ISP. This is because the routing information is usually aggregated by the upstream organization based on the address block hierarchy to reduce the number of routing information entries advertised to the Internet core network.

The other way is to obtain an address block directly from the Internet Registry and subscribe to a traffic exchange service with ISPs. This makes it easier to change the upstream ISPs; however, acquiring an address block from the Internet Registry is only possible for large organizations.

NEMO BS may be a solution to this problem. Figure 1-5 shows how this works.

The mobile service provider is an operator of the home agent that manages the mobile network of small offices. The office networks are logically connected to the Internet as a part of the mobile service provider network. The actual location of the office network can be anywhere. The office network manager can subscribe as many ISPs as he or she wants. These ISPs are used to get a care-of address for the mobile router of the office network. Of course, the mobile router never actually moves; however, when one of the subscribed ISPs fails, the router can virtually move to another working ISP. Since the internal office network is logically fixed, the users never notice that the external connectivity is changed. This is not a perfect solution for the real multihoming requirement because the traffic of the office network is tunneled to the home agent of the mobile service provider and the home agent may be a single point of failure. However, it may open the way to the redundant connectivity service with less cost.

FIGURE 1-5

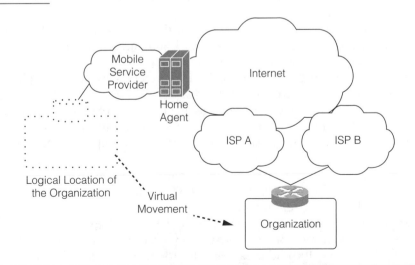

Simple multihoming using NEMO BS.

1.3.3 Proxy Mobile IPv6

As previously noted, one of the problems when deploying Mobile IPv6 is that the user terminals have to be upgraded to add mobility function. The NEMO BS approach reduces the upgrade cost dramatically; however, it has the limitation that all the nodes must be in the same mobile network. Proxy Mobile IPv6 is one of the most recent proposals in the Mobile IPv6 area. It allows unmodified IPv6 terminals to access different subnetworks while keeping their address fixed. All the mobility management procedures are done by the network side on behalf of the user terminals. It seems that Proxy Mobile IPv6 is the best solution for Mobile IPv6 deployment; however, it has a different drawback. Since the mechanism requires the upgrade of the network infrastructure instead of user terminals, the cost of the network operator is high. In addition, because the mobility support is provided by the network, the user terminal cannot move from one network service operator to another network operator while keeping the mobility function. The seamless mobility connectivity is only provided as long as the user terminal is connected to the network attachment points of the same network operator. Regardless of these drawbacks, the Proxy Mobile IPv6 technology attracts many network service providers, especially the telephone carrier-based operators and WiMAX operators. In fact, the next-generation all-IP telecommunication service models, such as the 3GPP and LTE, assume that the terminals are simple IPv6 nodes and the network provides the proxy service. The approach reduces the burden of the terminal manufactures and users, and it is a good starting point as the initial mobility deployment. However, when a user wants to access the other operators, he and his terminal eventually need other mobility technologies, such as Mobile IPv6.

As the name of the protocol indicates, Proxy Mobile IPv6 is similar to the Mobile IPv6 protocol. The difference is that the proxy node manages mobility signal processing instead of user terminals. Proxy Mobile IPv6 introduces two new entities: the Local Mobility Anchor (LMA) and the Mobility Access Gateway (MAG). The role of LMA is similar to that of the home agent. MAG is the proxy node that handles mobility signaling procedures. Since the user terminals do not have any mobility functions, all the access networks must provide MAGs to accept the terminals.

When a node attaches to an access network, it first tries to get an approval to use the network by some means. During this time, the local authentication entity (it can be a MAG) authenticates the node and allocates a dedicated network prefix for the node. Once the network prefix allocated to the node is decided, the MAG of the access network sends a binding update message on behalf of the attached node to the LMA. This binding update message is called as a proxy binding update message. The message includes the network prefix allocated to the attached node. When LMA receives a proxy binding update message, it acts similar to the home agent in the NEMO BS case. The LMA starts advertising the routing information of the network prefix allocated to the node, creates an IPv6-over-IPv6 tunnel between the LMA and the MAG that sent the proxy binding update message, and replies to the MAG with an acknowledgment message. The MAG also establishes an IPv6-over-IPv6 tunnel between the MAG and the LMA after receiving the acknowledgment message.

Once the tunnel establishment has been completed, the MAG advertises the network prefix to the node as a router advertisement message. Since the node is acting as a normal IPv6 node, it configures its own IPv6 address from the router advertisement message and starts communicating with other nodes. All the traffic sent from the node is received by the MAG. The MAG encapsulates the traffic and transfers to the LMA using the tunnel. The LMA decapsulates the packets and forwards them to the final destination. The node communicating with the node moving with the support of Proxy Mobile IPv6 can just send their packets to the fixed address of the node. Because the routing information of the network prefix allocated to the node is advertised from the LMA, all the packets sent to the node go to the LMA. The LMA interrupts the packets and transfers them to the MAG using the tunnel established as the result of the proxy binding message exchange. The MAG decapsulates the packets and forwards them to the access network where the node is attached. The node does not see any difference between the Proxy Mobile IPv6 network and the normal IPv6 network. In this explanation, the MAG and the access network use the stateless address autoconfiguration mechanism; however, any address assignment mechanism for IPv6 can be used, such as DHCPv6. Figure 1-6 depicts the operation.

If the node moves from one access network to another access network, the node starts the access network authentication procedure again. The MAG attached to the new access network detects the node attachment and tries to allocate the network prefix for the node. Since the binding between the node identifier and the network prefix allocated to the node is fixed, the MAG gets the same network prefix information that was being used in the previous access network. The same procedure of exchanging a proxy binding update message and acknowledgment message is performed between the new MAG and the LMA, and a new tunnel is established between the new MAG and the LMA. Because the new MAG advertises the same network prefix to the attached node, the attached node does not notice that the network

FIGURE 1-6

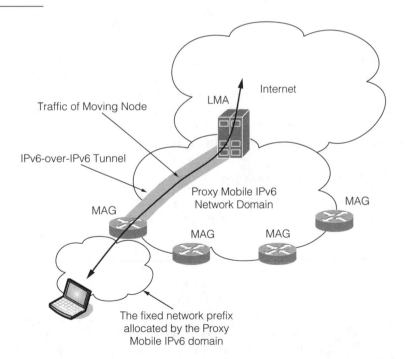

Proxy Mobile IPv6 operation.

FIGURE 1-7

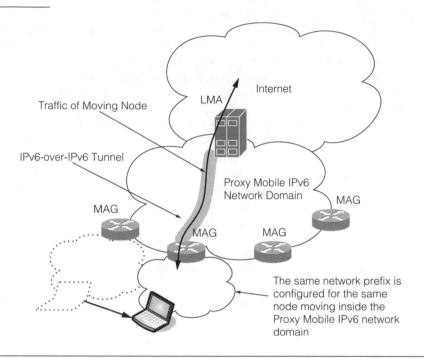

Moving operation in the Proxy Mobile IPv6 domain.

currently attached is a different network from the previous access network. The node can use the same address configured at the previous access network. Figure 1-7 shows the network status when the node moves from one MAG to another MAG. As you can imagine, it is difficult to extend the mechanism to multiple network operators. The addresses used for the moving node are tightly bound to the network operator and strong trust relationships between the LMA and the MAGs are required to secure the mobility signaling messages.

1.3.4 Global Operation Technology

It is frequently pointed out that the Mobile IPv6-based mobility operation has two disadvantages. One is that the home agent can be a single point of failure. The other is that the communication path between a mobile node and its communicating node becomes redundant when the home agent of the mobile node is far away. When operating Mobile IPv6 on a small scale, these problems may not be serious; however, when considering using it on a global scale, these problems cannot be ignored. Currently, there is no proposed solution to these problems from IETF. The following technique is one solution for the global operation.

The concept introduces multiple home sites of which the geographical location and topological location are different. Each home site has the same network prefix that is used as a home network of mobile nodes. Each home site advertises the same routing information of the home network independently. In this configuration, the communication using the addresses allocated

$64.95

from the home network prefix cannot be made. Because the same routing information is advertised from multiple points, the packet addressed to one of the addresses from the home network prefix is forwarded to the nearest home site based on the Internet routing information.

The proposed idea introduces tunnel links between the home sites. Each home agent involved in this mechanism establishes bidirectional tunnels with other home agents located in the different home sites. Every home agent has a routing information entry of other home agents. If one of them receives a packet destined to another home agent, it forwards the packet using the bidirectional tunnel established between it and the destination home agent.

The mobile node operation is performed as follows. It first finds one of its home agent addresses by some means, such as the DNS or AAA mechanism as described in the normal Mobile IPv6 operation procedure. The mobile node sends a binding update message to the achieved home agent address. Because the home network routing information is advertised from multiple points, the message is transferred to the nearest home site. If the home site where the message arrived has the home agent that is specified in the binding update message, then the normal Mobile IPv6 procedure goes on. If the site does not have the home agent—for example, the destination address of the binding update message is Home Agent A but the site only has Home Agent B—then the packet is forwarded to Home Agent A using the bidirectional tunnel established between Home Agents A and B. When Home Agent A receives the message, it notices that the mobile node is attached to the network that is nearer to Home Agent B than Home Agent A. The decision is simple because if Home Agent A is nearer than Home Agent B, then the binding update message should come directly to Home Agent A rather than tunneling from Home Agent B. Home Agent A sends a notification message to the mobile node to change its home agent from A to B. Once the mobile node receives the notification message, the mobile node sends another binding update message of which the destination address is set to Home Agent B. Finally, the mobile node and Home Agent B establish a bidirectional tunnel between them and start the normal Mobile IPv6 operation.

Figure 1-8 shows the message flow of the procedure to find the nearest home agent at the initial registration. A similar procedure is performed when a mobile node that has already registered to one of the home agents moves to another network where there is another home agent nearer than the home agent currently used.

If a home site has trouble and is disconnected from the Internet, the bidirectional tunnels connected to the home site are shut down. The routing information advertisement of the home network from the site in trouble is also stopped. For example, if the site where Home Agent B belongs faces trouble, then the tunnels between Home Agent B and Home Agents A and C are shut down. The routing information of the home network is only advertised from the sites to which Home Agents A and C belong. Even in this case, the mobile node does not need to know the home network status. It simply keeps registering its current location to the home agent. The re-registration message is sent to Home Agent B; however, the message is forwarded to either Home Agent A or C because the routing information for the home network is no longer advertised from the site at which Home Agent B was attached. Assuming that the message is delivered to the site of Home Agent C, there is no node that can receive the binding message. Home Agent C may interrupt the message and suggest the mobile node change the home agent to Home Agent C, or Home Agent C can just ignore the message. In the latter case, the mobile node will try to register Home Agent B and finally detect that it is not available and find another available home agent.

FIGURE 1-8

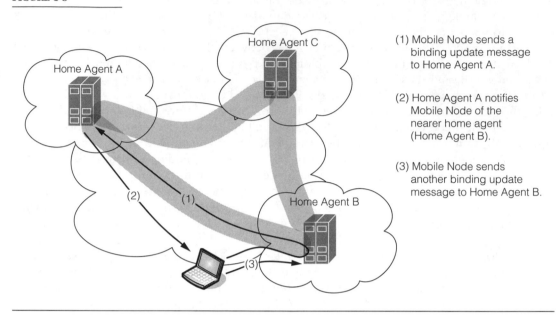

(1) Mobile Node sends a
    binding update message
    to Home Agent A.

(2) Home Agent A notifies
    Mobile Node of the
    nearer home agent
    (Home Agent B).

(3) Mobile Node sends
    another binding update
    message to Home Agent B.

Registration to the nearest home agent.

1.4 Coverage of this Book

This book provides a detailed protocol description of Mobile IPv6 and a complete source code of the Mobile IPv6 protocol stack and its line-by-line explanation. The source code is taken from the KAME project Mobile IPv6 reference code, which has been tested with various other independent implementations through several interoperability test events and confirmed that it is fully interoperable with other standard-based Mobile IPv6 implementations. The code covers all types of nodes defined in the Mobile IPv6 specification: The mobile node, the home agent, and the route optimization operation of the correspondent node are all discussed. We hope that this book will support your understanding of the basic Mobile IPv6 protocol specification and how the specification is translated to the actual source code.

Mobile IPv6 Overview

Mobile Internet Protocol version 6 (IPv6) adds the mobility function to IPv6. Mobile IPv6 is specified in [RFC3775] and [RFC3776]. An IPv6 host that supports the Mobile IPv6 function can move around the IPv6 Internet(*). The host that supports Mobile IPv6 can change its point of attachment to the IPv6 Internet whenever it wants. If a host does not support Mobile IPv6, all the existing connections on the host are terminated when it changes its point of attachment. A connection between two nodes is maintained by the pairing of the source address and the destination address. Since the IPv6 address of an IPv6 node is assigned based on the prefix of the network, the assigned address on a given network becomes invalid when the host leaves that network and attaches itself to another network. The reason for this problem came from the nature of IP addresses. An IP address has two meanings, one is the identifier of the node and the other is the location information of the node. It would not be a big problem as long as IP nodes do not move around the Internet frequently because in that case, the location information would not change frequently and we could use location information as the identifier of a node. However, recent progress of communication technologies and small computers made it possible for IP nodes to move around. It is getting harder and harder to treat location information as an identifier because the location information frequently changes.

(*) There is ongoing work to extend the Mobile IPv6 specification to support the IPv4 Internet [MIP6-NEMO-V4TRAVERSAL]. With this extension, a Mobile IPv6 mobile node can attach to the IPv4 Internet keeping the existing connections with its IPv6 peer nodes. In addition, the mobile node can use a fixed IPv4 address to communicate with other IPv4 nodes regardless of the IP version of the network to which the node is attached.

As such the basic idea of Mobile IPv6 is to provide a second IPv6 address to an IPv6 host as an identifier in addition to the address that is usually assigned to the node from the attached network as a locator. The second address is fixed to the home position of the host and never changes even if the host moves. The fixed address is called a "home address." As long as the host uses its home address as its connection information, the connection between the host and other nodes will not be terminated when the mobile host moves.

The concept of a home address provides another useful feature to a host that supports Mobile IPv6. Any IPv6 nodes on the Internet can access a host that supports Mobile IPv6 by specifying its home address, regardless of the location of the host. Such a feature will make it possible to create a roaming server. Since the home address of the roaming server never changes, we can constantly reach the server at the home address. For example, anyone could run a web server application on a notebook computer, which supports Mobile IPv6, and everyone could access it without any knowledge of where the computer is located.

2.1   Types of Nodes

The Mobile IPv6 specification defines three types of nodes. The first type is the *mobile node*, which has the capability of moving around IPv6 networks without breaking existing connections while moving. A mobile node is assigned a permanent IPv6 address called a *home address*. A home address is an address assigned to the mobile node when it is attached to the *home network* and through which the mobile node is always reachable, regardless of its location on an IPv6 network. Because the mobile node is always assigned the home address, it is always logically connected to the home link. When a mobile node leaves its home network and attaches to another network, the node will get another address called a *care-of address*, which is assigned from the newly attached network. This network, which is not a home network, is called a *foreign network* or a *visited network*. A mobile node does not use a care-of address as an endpoint address when communicating with other nodes since the address may change when the mobile node changes its point of attachment.

A second Mobile IPv6 node type is the *home agent*, which acts as a support node on the home network for Mobile IPv6 mobile nodes. A home agent is a router that has a proxy function for mobile nodes while they are away from home. The destination address of packets sent to mobile nodes are set to the home addresses of the mobile nodes. A home agent intercepts all packets that are addressed to the mobile node's home address, and thus delivered to the home network, on behalf of the mobile nodes.

This forwarding mechanism is the core feature provided by the Mobile IPv6 protocol. All IPv6 nodes that want to communicate with a mobile node can use the home address of the mobile node as a destination address, regardless of the current location of the mobile node. Those packets sent from an IPv6 node to the home address of a mobile node are delivered to the home network by the Internet routing mechanism where the home agent of the mobile node receives the packets and forwards the packets appropriately. For the reverse direction, a mobile node uses its home address as a source address when sending packets. However, a mobile cannot directly send packet nodes of which the source address is a home address from its current location if it is away from home since source addresses are not topologically correct. Sending a packet of which the source address is out of the range of the network address of the sender node is a common technique when an attacker tries to hide its location when he is attacking a specific node. Such a packet may be considered as an attack. Because of this reason, the first hop router may drop such topologically incorrect packets to avoid the risk of the source spoofing attack. To solve this problem, a mobile node uses the IPv6 in IPv6 encapsulation technology. All packets sent from a mobile node while away from home are sent to its home agent using the encapsulation mechanism. The home agent decapsulates the packets and forwards them as if the packets were sent from the home network.

A third type of Mobile IPv6 node is called the *correspondent node*. A correspondent node is an IPv6 node that communicates with a mobile node. A correspondent node does not have to be Mobile IPv6 capable, other than supporting the IPv6 protocol; any IPv6 node can be a correspondent node. Since the Mobile IPv6 specification provides a backward compatibility to all IPv6 nodes that do not support Mobile IPv6, all IPv6 nodes can communicate with mobile nodes without any modification. However, as we have described in the previous paragraph, all packets between a mobile node and a correspondent node must be forwarded basically by the home agent of the mobile node. This process is sometimes redundant, especially when a correspondent node and a mobile node are located on topologically near networks. To solve this redundancy, Mobile IPv6 provides an optimization mechanism called the *route optimization* mechanism, which a correspondent node may support. A mobile node can send packets directly to a correspondent node using the care-of address of the mobile node as a source address. The information of the home address of a mobile node is carried by the newly defined option for the Destination Options Header. Also, a correspondent node can send packets directly to the care-of address of a mobile node. In this case, the information of the home address is carried by the Routing Header. For a general discussion about IPv6 Extension Headers, refer to Chapter 3 of *IPv6 Core Protocol Implementation*, "Internet Protocol Version 6."

A correspondent node may itself be a mobile node. In this case, two moving nodes can communicate with each other without terminating their sessions regardless of their points of attachment to the Internet.

2.2 Basic Operation of Mobile IPv6

A mobile node uses a home address when communicating with other nodes. When a mobile node moves from one network to another network, the node sends a message called a *Binding Update* to its home agent. The message includes the care-of address and the home address of the mobile node. Such information is called *binding information* since it binds a care-of address to the home address of a mobile node.

When a home agent receives the message and accepts the contents of the message, the home agent replies with a *Binding Acknowledgment* message to indicate that the Binding Update message is accepted. The home agent creates a bidirectional tunnel connection from its address to the care-of address of the mobile node. A mobile node also creates a bidirectional tunnel connection from its care-of address to the home agent when it receives the acknowledgment message. After the successful tunnel creation, all packets sent to the home address of the mobile node are intercepted by the home agent at the home network and tunneled to the mobile node. Also, all packets originated at the mobile node are tunneled to its

FIGURE 2-1

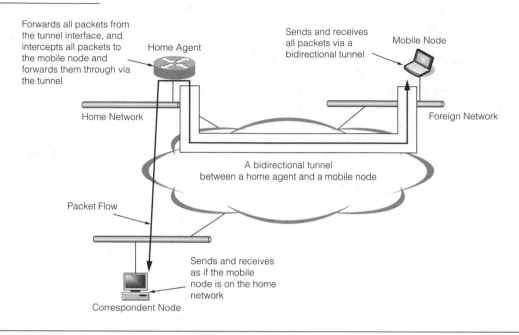

Forwards all packets from
the tunnel interface, and
intercepts all packets to   Home Agent
the mobile node and
forwards them through via
the tunnel

Sends and receives
all packets via a   Mobile Node
bidirectional tunnel

Home Network   Foreign Network

A bidirectional tunnel
between a home agent and a mobile node

Packet Flow

Sends and receives
as if the mobile
node is on the home
network

Correspondent Node

Bidirectional tunneling.

home agent and forwarded from its home network to destination nodes. Figure 2-1 shows
the concept.

The communication path between a mobile node and a peer node described in Figure 2-1
sometimes may not be optimal. Figure 2-2 shows the worst case: a mobile node and a correspondent node are on the same network. The packets exchanged between them are always
sent to the home network of the mobile nodes, even if they are directly accessible to each other
using the local network. For example, when two people whose mobile nodes are originally
located in Japan visit the United States, their traffic always traverses the Pacific Ocean.

If a peer node supports the route optimization mechanism defined in the Mobile IPv6
specification, the mobile node and the peer node can communicate directly without detouring
through the home agent. To optimize the route, the mobile node sends a Binding Update
message to the peer node. After it receives the message, the peer node sends packets directly to
the care-of address of the mobile node. The packets also contain a Routing Header that specifies
their final destination that is set to the home address of the mobile node. The packets are
routed directly to the care-of address of the mobile node. The mobile node receives the packets
and finds that the packets have a Routing Header and performs Routing Header processing,
which involves swapping the destination address in the packets' IPv6 header and the home
address carried in the Routing Header. The mobile node forwards the packets to the final
destination that is the home address at this point, and the packets are delivered to the mobile
node itself. When the mobile node sends packets to the peer node, the mobile node sets its
care-of address as a source address of the packets and inserts its home address into a Destination

FIGURE 2-2

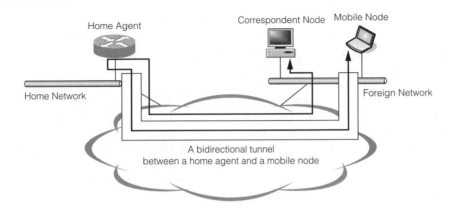

The worst case of bidirectional tunneling.

FIGURE 2-3

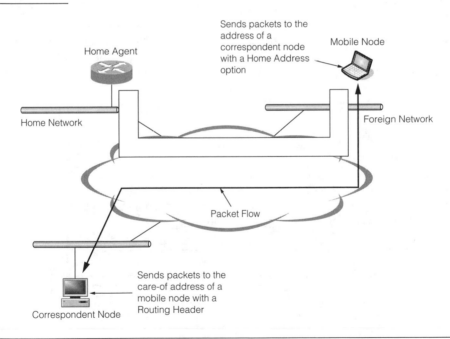

Optimized communication between a mobile node and a correspondent node.

Options Header. The peer node swaps the care-of address and the home address when it receives those packets, and processes the packets as if they were sent from the home address. Figure 2-3 shows the procedure.

As you may notice, a Binding Update message is quite a dangerous message. If a node accepts the message without any verification, an attacker can easily redirect packets sent to the mobile node to the attacker. To prevent this attack, the message is protected in the following two ways:

(1) A Binding Update message to a home agent is protected by the IPsec mechanism.

(2) A Binding Update message to a correspondent node is protected by the return routability procedure described in Section 5.1 of Chapter 5.

The IPsec mechanism is strong enough to prevent this type of attack and we can use the technology between a mobile node and a home agent. However, it is difficult to use the IPsec mechanism between a mobile node and a correspondent node, since the IPsec mechanism requires both nodes to be in the same administrative domain. We can assume that a home agent and a mobile node can share such a secret since they are managed by the same administrative domain in most cases. However, there is usually no such relationship between a mobile node and a correspondent node.

> There is an ongoing action to use the IPsec mechanism between a mobile node and a correspondent node [CN-IPSEC].

The Mobile IPv6 specification defines a new method of creating a shared secret between a mobile node and a correspondent node. The procedure is called the *return routability* procedure. When a mobile node sends a Binding Update message, the most important thing is to

FIGURE 2-4

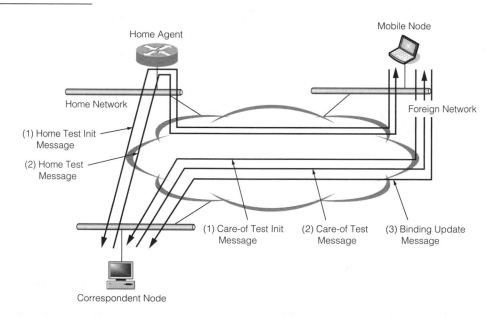

The return routability procedure.

provide a way to prove to the correspondent node that the care-of address and home address are owned by the same mobile node. The return routability procedure provides such an address ownership proof mechanism.

A mobile node sends two messages: one message is sent from its home address and the other message is sent from its care-of address. Respectively, the messages are called a *Home Test Init* message and a *Care-of Test Init* message. A correspondent node replies to both messages with a *Home Test* message to the first and a *Care-of Test* message to the second. These reply messages include values for tokens that are computed from addresses of the mobile node and secret information that is only kept in the correspondent node. A mobile node generates a shared secret from the token values and puts a signature in a Binding Update message using the shared secret. This mechanism ensures that the home address and the care-of address are assigned to the same mobile node. Figure 2-4 shows the procedure. Chapter 4 contains a detailed discussion of Mobile IPv6 operation.

3
Header Extension

[RFC3775] defines new extension headers and several new types and options for existing headers for Mobile Internet Protocol version 6 (IPv6). The specification also defines some header formats of Neighbor Discovery [RFC2461], which are modified for Mobile IPv6. The following is a list of new or modified headers and options. The detailed description of each header and option will be discussed in Sections 3.2–3.7.

Home Address option The Home Address option is a newly defined destination option that carries the home address of a mobile node when packets are sent from a mobile node.

Type 2 Routing Header The Type 2 Routing Header is a newly defined routing header type that carries a home address of a mobile node when packets are sent from a home agent or a correspondent node to a mobile node.

Mobility Header The Mobility Header is a newly defined Extension Header that carries the signaling information of the Mobile IPv6 protocol.

Router Advertisement message The Router Advertisement message is modified to include a flag that indicates whether a router has the home agent function or not.

Prefix Information option The Prefix Information option is one of the Neighbor Discovery options used to distribute prefix information of a network from a router to other nodes connected to the network. In Mobile IPv6, a home agent includes its address in this option as a part of the prefix information. All home agents on the same home network can know all addresses of home agents of the network by listening to this option.

Home Agent Information option The Home Agent Information option is a newly defined Neighbor Discovery option that carries the lifetime and preference information of a home agent.

Advertisement Interval option The Advertisement Interval option is a newly defined Neighbor Discovery option that carries the interval value between unsolicited Router Solicitation messages sent from a router.

Dynamic Home Agent Address Discovery Request/Reply messages The Dynamic Home Agent Address Discovery Request and Reply messages are newly defined Internet Control Message Protocol version 6 (ICMPv6) message types that provide the mechanism to discover the addresses of home agents for a mobile node when the mobile node is away from home.

Mobile Prefix Solicitation/Advertisement messages The Mobile Prefix Solicitation and Advertisement messages are newly defined ICMPv6 message types used to solicit/deliver the prefix information of a home network to a mobile node while the mobile node is away from home.

3.1   Alignment Requirements

Some Extension Headers and Options have alignment requirements when placing these headers in a packet. Basically, the header or option fields are placed at a natural boundary, that is, fields of n bytes in length are placed at multiples of n bytes from the start of the packet. The reason for such a restriction is for performance; accessing the natural boundary is usually faster. For example, the Home Address Option (Figure 3-1) has $8n+6$ alignment requirements, which puts the home address field on an 8-byte boundary.

3.2   Home Address Option

The *Home Address* option is a newly defined Destination option. The alignment requirement of the Home Address option is $8n + 6$. The format of the Home Address option is shown in Figure 3-1. This option is used to specify the home address of a mobile node when the mobile node sends packets while it is away from home. It is used in the following three cases.

(1) When a mobile node sends a Binding Update (BU) message

(2) When a mobile node communicates with peers with route optimization

(3) When a mobile node sends a Mobile Prefix Solicitation message

FIGURE 3-1

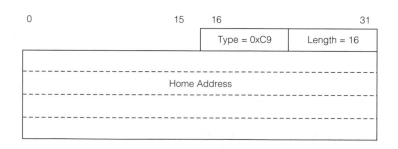

Home Address option.

A mobile node never sends packets with its source address set to its home address directly while it is away from home, since such a source address is topologically incorrect and may be dropped by an intermediate router because of the Ingress Filtering posed on that router. When sending a packet, a mobile node needs to perform one of following procedures.

- Send a packet using a bidirectional tunnel created between a mobile node and its home agent
- Use the Home Address option that includes the home address of the mobile node with the source address of the packet set to the care-of address of the mobile node

The *Type* field is `0xC9`. The first two bits of an option type number determine the action taken on the receiving node when the option is not supported, as discussed in Chapter 3 of *IPv6 Core Protocols Implementation*. In this case, the first two bits are both set. This means that if a node does not recognize the option, the following actions must be taken:

- The packet that includes the option must be dropped
- An ICMPv6 Parameter Problem message must be sent if the destination address of the incoming packet is not a multicast address

This provides a mechanism to detect whether a peer supports the Home Address option. If the peer does not support the option, the mobile node cannot use the route optimization mechanism.

The *Length* field is set to `16`. The *Home Address* field contains a home address of a mobile node.

3.3 Type 2 Routing Header

The *Type 2 Routing Header* is a newly defined routing header type for Mobile IPv6. This Routing Header is used by a home agent or a correspondent node to carry a home address of a mobile node when packets are sent to the mobile node. The format of the Type 2 Routing Header is shown in Figure 3-2.

FIGURE 3-2

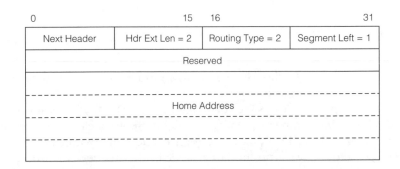

Type 2 Routing Header.

The Type 2 Routing Header is used in the following three cases:

(1) When a node sends a Binding Acknowledgment message

(2) When a home agent or a correspondent is performing route optimization

(3) When a home agent sends a Mobile Prefix Advertisement message

A packet of which the destination address is a home address of a mobile node is never delivered to the mobile node directly when the mobile node is away from home. Such a packet is delivered to the home network of the mobile node.

A node needs to use a Type 2 Routing Header if it wants to send packets directly to a mobile node that is away from home. In this case, the destination address of the packet is set to the care-of address of the mobile node. The home address is carried in the Type 2 Routing Header. A packet is delivered directly to the care-of address of the mobile node, and the mobile node processes the Type 2 Routing Header and delivers the packet to the home address, which is the mobile node itself.

The *Next Header* field is set to the protocol number of the following header. The *Hdr Ext Len* field is fixed at 2, since the length of the Type 2 Routing Header is fixed. The *Routing Type* field is set to 2. The *Segments Left* field is initialized to 1. The *Home Address* field contains one IPv6 address, which is the home address of the mobile node. The usage of this header is very restrictive. We can only specify one intermediate node as a home address. A mobile node that receives this header drops the packet if there is more than one intermediate node specified. Also, the address in the Type 2 Routing Header and the destination address of the IPv6 packet must belong to the same mobile node. That is, the packet can only be forwarded to the mobile node itself.

A new type number is required for Mobile IPv6 to make it easy to support Mobile IPv6 on firewall software. In the early stages of the Mobile IPv6 standardization, a Type 0 Routing Header was used instead of a Type 2 Routing Header. However, many people thought that it would be difficult to distinguish between using a Type 0 Routing Header for carrying a Mobile IPv6 home address or carrying a Routing Header that is being used as a method to perform source routing. We need to pay attention to the usage of source routing since such forwarding is sometimes used as a method for attacking other nodes. Firewall vendors may drop all packets with a Type 0 Routing Header to decrease the risk of such attacks. It is much easier for those vendors to pass only Mobile IPv6 data if we have a new routing header type number for the exclusive use of Mobile IPv6.

3.4   Mobility Header

The *Mobility Header* is a newly introduced extension header to carry Mobile IPv6 signaling messages. The format of the Mobility Header is shown in Figure 3-3. The format of the header is based on the usual extension header format.

The *Payload Proto* field indicates the following header. The field is equivalent to the Next Header field of other extension headers; however, the current specification does not allow the Mobility Header to be followed by other extension headers or by transport headers. That is, the Mobility Header must always be the last header in the header chain of an IPv6 packet. The reason for this restriction is to simplify the interaction between the IPsec mechanism and Mobile IPv6. Some signaling messages used by the Mobile IPv6 protocol must be protected by

FIGURE 3-3

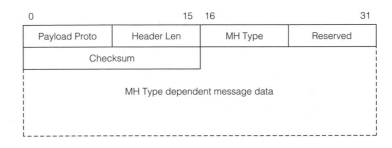

Mobility Header.

TABLE 3-1

Type	Description
0	Binding Refresh Request requests a mobile node to resend a Binding Update message to update binding information
1	Home Test Init starts the return routability procedure for a home address of a mobile node
2	Care-of Test Init starts the return routability procedure for a care-of address of a mobile node
3	Home Test is a response message to the Home Test Init message
4	Care-of Test is a response message to the Care-of Test Init message
5	Binding Update sends a request to create binding information between a home address and a care-of address of a mobile node
6	Binding Acknowledgment is a response message to the Binding Update message
7	Binding Error notifies an error related to the signal processing of the Mobile IPv6 protocol

Mobility Header types.

the IPsec mechanism. It is impossible to protect the Mobility Header if other headers follow it, because with the current IPsec specification, we cannot apply IPsec policies to the intermediate extension headers. Currently, the Payload Proto field is always set to 58 (IPV6-NONXT), which indicates there is no next header. The *Header Len* field indicates the length of a Mobility Header in units of 8 bytes excluding the first 8 bytes. The *MH Type* field indicates the message type of the Mobility Header. Currently, eight kinds of Mobility Header types are defined. Table 3-1 shows all Mobility Header types. The *Reserved* field is reserved for future use. The *Checksum* field stores the checksum value of a Mobility Header message. The algorithm used to compute the checksum value is the same as is used for ICMPv6. The rest of the header is defined depending on the Mobility Header type value. Also, the Mobility Header may have some options called *mobility options*.

3.4.1 Binding Refresh Request Message

The *Binding Refresh Request (BRR)* message is used when a correspondent needs to extend the lifetime of binding information for a mobile node. A mobile node that has received a

FIGURE 3-4

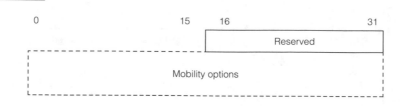

Binding Refresh Request message.

BRR message should send a BU message to the correspondent node to update the binding information held in the correspondent node. The format of the BRR message is shown in Figure 3-4.

The BRR message is sent from a correspondent node to a mobile node. The source address of the IPv6 packet is the address of the correspondent node which is sending the BRR message. The destination address of the IPv6 packet is the home address of a mobile node, which is requested to resend a BU message. The BRR message must have neither a Type 2 Routing Header nor a Home Address option. That is, the message is tunneled by the home agent to the destination mobile node, if the destination mobile node is away from home. Currently, no mobility options are defined for the BRR message.

3.4.2   Home Test Init Message

The *Home Test Init (HoTI)* message is used to initiate the return routability procedure. The format of the HoTI message is shown in Figure 3-5.

The HoTI message is sent from a mobile node to a correspondent node when the mobile node wants to optimize the path between itself and the correspondent node. The source address of the IPv6 packet is the home address of the mobile node and the destination address of the IPv6 packet is the address of the correspondent node.

The *Home Init Cookie* field is filled with a random value generated in the mobile node. The cookie is used to match a HoTI message and a *Home Test* (HoT) message, which is sent from a correspondent node in response to the HoTI message. The HoTI message must have neither

FIGURE 3-5

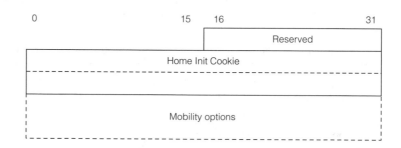

Home Test Init message.

a Type 2 Routing Header nor a Home Address option. The HoTI message is always tunneled from a mobile node to its home agent and forwarded to a correspondent node. Currently, no mobility options are defined for the HoTI message.

3.4.3 Care-of Test Init Message

The *Care-of Test Init (CoTI)* message is used to initiate the return routability procedure. The format of the CoTI message is shown in Figure 3-6.

The CoTI message is sent from a mobile node to a correspondent node when a mobile node wants to optimize the path between itself and the correspondent node. The source address of the IPv6 packet is the care-of address of the mobile node and the destination address of the IPv6 packet is the address of the correspondent node.

The *Care-of Init Cookie* is filled with a random value generated in the mobile node. The cookie is used to match a CoTI message and a *Care-of Test* (CoT) message, which is sent from the correspondent node in response to the CoTI message. A CoTI message must have neither a Type 2 Routing Header nor a Home Address option. A CoTI message is always directly sent from a mobile node to a correspondent node. Currently, no mobility options are defined for the CoTI message.

3.4.4 Home Test Message

The HoT message is used as a reply to a HoTI message sent from a mobile node to a correspondent node. This message includes a token that is used to compute a shared secret to protect the BU message. The format of the HoT message is shown in Figure 3-7.

The HoT message is sent from a correspondent node to a mobile node as a response to a HoTI message that was previously sent from the mobile node. The source address of the IPv6 packet is the address of the correspondent node and the destination address is the home address of the mobile node.

The *Home Nonce Index* indicates an index value of the nonce value in the home nonce array, which is maintained in the correspondent node. The Home Init Cookie is a copy of the value of the Home Init Cookie field of the corresponding HoTI message. A mobile node can match a previously sent HoTI message and the received Home Test message by comparing the cookie values. If there is no corresponding HoTI message, the received HoT message is

FIGURE 3-6

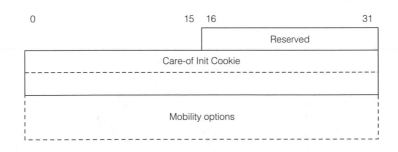

Care-of Test Init message.

FIGURE 3-7

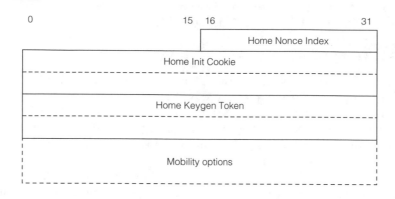

Home Test message.

dropped. The *Home Keygen Token* is a token value that is used to compute a shared secret to secure the BU message. The algorithm used is described in Section 5.1 of Chapter 5. Currently, no mobility options are defined for the HoT message.

3.4.5   Care-of Test Message

The CoT message is used as a reply to a CoTI message sent from a mobile node to a correspondent node. This message includes a token value that is used to compute a shared secret to protect the BU message. The format of the CoT message is shown in Figure 3-8.

The CoT message is sent from a correspondent node to a mobile node as a response to the CoTI message that was previously sent from the mobile node. The source address of the IPv6 packet is the address of the correspondent node and the destination address is the care-of address of the mobile node.

FIGURE 3-8

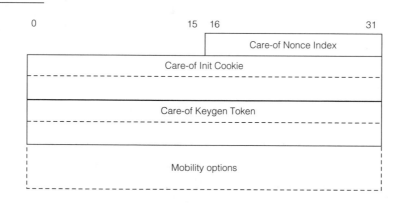

Care-of Test message.

The *Care-of Nonce Index* indicates the index value of the nonce value in the care-of nonce array which is maintained in the correspondent node. The Care-of Init Cookie is a copy of the value of the Care-of Init Cookie field of the corresponding CoTI message. A mobile node can match a previously sent CoTI message and the received CoT message by comparing the cookie values. The *Care-of Keygen Token* is a token value that is used to compute a shared secret to secure the BU message later. The algorithm used is described in Section 5.1 of Chapter 5. Currently, no mobility options are defined for the CoT message.

3.4.6 Binding Update Message

The BU message is used by a mobile node to notify a correspondent node or a home agent of the binding information of a care-of address and a home address of the mobile node. A mobile node sends the BU message with its care-of address and its home address whenever it changes its point of attachment to the Internet and changes its care-of address. The node that receives the message will create an entry to keep the binding information. Figure 3-9 shows the BU message.

The BU message is sent from a mobile node to a home agent or a correspondent node. The source address of the IPv6 packet is the care-of address of the mobile node and the destination address is the address of the home agent or the correspondent node. To include the information of the home address of the mobile node, the BU message contains a Destination Options Header that has a Home Address option as described in Section 3.2.

The *Sequence Number* field contains a sequence number for a BU message to avoid a replay attack. The *flag* fields of the BU message may contain the flags described in Table 3-2.

The *Lifetime* field specifies the proposed lifetime of the binding information. When a BU message is used for home registration, the value must not be greater than the remaining lifetime of either the home address or the care-of address of the mobile node, which is sending the BU message. The value is in units of 4 s.

The BU message may have the following mobility options.

- The Binding Authorization Data option
- The Nonce Indices option
- The Alternate Care-of Address option

Each option is described in Section 3.5.

FIGURE 3-9

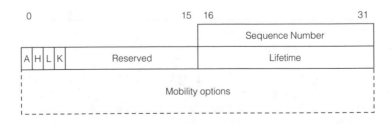

Binding Update message.

TABLE 3-2

Flag	Description
A	*Acknowledge* requires a Binding Acknowledgment message as a response to a Binding Update message. When the H flag is set, the A flag must be set. Note that a Binding Acknowledgment message may be sent to indicate an error even if the A flag is not set.
H	*Home Registration* means that this Binding Update message is a message for home registration.
L	*Link-local Address Compatibility* means that the link-local address of a mobile node has the same interface ID with its home address.
K	*Key Management Mobility Capability* means the IKE SA information survives on movements.

The flags of the Binding Update message.

3.4.7   Binding Acknowledgment Message

The *Binding Acknowledgment (BA)* message is sent as a response to a BU message sent from a mobile node. The format of the BA message is shown in Figure 3-10.

A BA message is sent from a home agent or a correspondent node to a mobile node. The source address of the BA message is the address of the home agent or the correspondent node and the destination address is the care-of address of the mobile node. To deliver a BA message to the home address of a mobile node that is away from home, a Type 2 Routing Header, which contains the home address of the mobile node, is necessary.

The *Status* field specifies the result of the processing of the received BU message. Table 3-3 is a list of currently specified status codes. The field immediately after the Status field is the flag field. Currently, only the K flag is defined. Table 3-4 describes the K flag. The *Sequence Number* field indicates the copy of the last valid sequence number that was contained in the last BU message. The field is also used as an indicator of the latest sequence number when a mobile node sends a BU message with a smaller sequence number value. This situation may occur when a mobile node reboots and loses the sequence number information of recent binding information. The Lifetime field indicates the approved lifetime for the binding information. Even if a mobile node requests a large lifetime value in the Lifetime field in the BU message, the requested lifetime is not always approved by the receiving node. The actual lifetime can be determined by the node that receives the BU message.

The BA message may have the following mobility options:

- The Binding Authorization Data option
- The Binding Refresh Advice option

Each option is described in Section 3.5.

3.4.8   Binding Error Message

The *Binding Error* (BE) message is used to indicate an error that occurs during the mobility signaling processing. The format of a BE message is shown in Figure 3-11.

The BE message is sent from a node that supports Mobile IPv6. The source address of the IPv6 packet is the address of the node that sends the BE message. The BE message must have neither a Type 2 Routing Header nor a Home Address option.

FIGURE 3-10

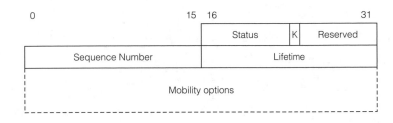

Binding Acknowledgment message.

TABLE 3-3

Code	Description
0	Binding Update accepted
1	Accepted but prefix discovery necessary
128	Reason unspecified
129	Administratively prohibited
130	Insufficient resources
131	Home registration not supported
132	Not home subnet
133	Not home agent for this mobile node
134	Duplicate Address Detection failed
135	Sequence number out of window
136	Expired home nonce index
137	Expired care-of nonce index
138	Expired nonces
139	Registration type change disallowed

The status codes of the Binding Acknowledgment message.

TABLE 3-4

Flag	Description
K	Key Management Mobility Capability means the IKE SA information cannot survive on movements

The flag of the Binding Acknowledgment message.

The Status field indicates the kind of error as described in Table 3-5. The *Home Address* field contains the home address of a mobile node if the packet that causes the error is sent from a mobile node. Otherwise, the field contains an unspecified address. Currently, no mobility options are defined for the BE message.

FIGURE 3-11

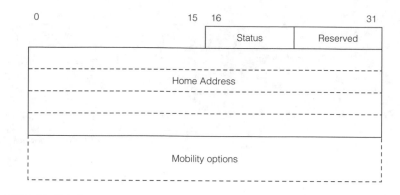

Binding Error message.

TABLE 3-5

Status	Description
1	A Home Address option is received without existing binding information
2	Unrecognized Mobility Header type value is received

Status value of a Binding Error message.

3.5   Mobility Options

The Mobility Options are the options used with the Mobility Header to provide supplemental information. Figure 3-12 shows the format of the Mobility Option.

The Mobility Option format is the same format used by the Hop-by-Hop options and Destination options. The first byte indicates the type of the option. The second byte indicates the length of the following data. Currently, six options are defined as described in Table 3-6.

3.5.1   Pad1 Option

The *Pad1* option is used when 1 byte of padding is needed to meet the alignment requirements of other Mobility Options. This option does not have any effect and must be ignored on the receiver side. The format of the Pad1 option is a special format that does not meet the standard format described in Figure 3-12. Figure 3-13 shows the format of the Pad1 option.

3.5.2   PadN Option

The *PadN* option is used when two or more bytes of padding are needed to meet the alignment requirements of other Mobility Options. This option does not have any effect and must be ignored on the receiver side. The format of the PadN option is described in Figure 3-14.

The *Option Length* field is set to the size of the required padding length minus 2. The *Option Data* field consists of a zero-cleared byte stream of which the length is the required padding size minus 2. A receiver must ignore the contents of the Option Data field when processing this option.

FIGURE 3-12

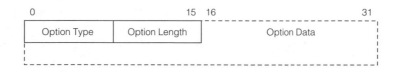

Mobility Option.

TABLE 3-6

Type	Description
0	Pad1
1	PadN
2	Binding Refresh Advice
3	Alternate Care-of address
4	Nonce indices
5	Binding Authorization data

Mobility Options.

FIGURE 3-13

Pad1 option.

FIGURE 3-14

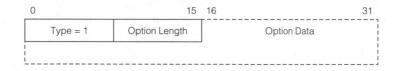

PadN option.

3.5.3   Binding Refresh Advice Option

The *Binding Refresh Advice* option is used to specify the recommended interval between BU messages for updating the binding information. The option is used with the BA message that is sent from a home agent to a mobile node that the home agent serves. The format of the Binding Refresh Advice option is shown in Figure 3-15. The alignment requirement of the Binding Refresh Advice option is $2n$.

FIGURE 3-15

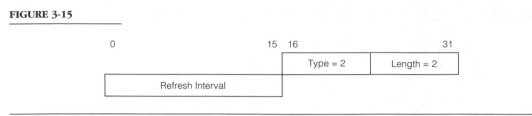

Binding Refresh Advice option.

The *Length* field is set to 2. The *Refresh Interval* field indicates the interval value. The value is specified in units of 4 s.

3.5.4   Alternate Care-of Address Option

The *Alternate Care-of Address* option is used in two cases with the BU message. The first case is when a mobile node wants to bind its home address to an address other than the source address of the BU message. Usually, the source address of the IPv6 packet is used as a care-of address if the Alternate Care-of Address option does not exist. The second case is to protect the care-of address information from on-path attackers. The BU message for home registration must be protected by an IPsec ESP or AH. However, the ESP does not protect the IPv6 header itself. That is, the source address, which is used as a care-of address, is not protected by the ESP. Adding this option to a Binding Update message will protect the care-of address information since this option is included in a Mobility Header and the Mobility Header is covered by the ESP. If we use the AH, the option can be omitted. The format of the Alternate Care-of Address option is shown in Figure 3-16. The alignment requirement of the Alternate Care-of Address option is $8n + 6$.

The Length field is set to 16, which is the length of an IPv6 address. The Alternate Care-of Address field contains the address that should be used as a care-of address instead of the source address of the BU message.

3.5.5   Nonce Indices Option

The *Nonce Indices* option is used to specify nonce values that are used to compute the Authenticator value specified by the Binding Authorization Data option. This option is used with the Binding Authorization Data option. The alignment requirement of the Nonce Indices option is $2n$. The format of the Nonce Indices option is shown in Figure 3-17.

The Length field is set to 4. The value of the Home Nonce Index and Care-of Nonce Index fields are copied from the Home Nonce Index field of the HoT message and Care-of Nonce Index field of the CoT message which a mobile node has previously received.

3.5.6   Binding Authorization Data Option

The *Binding Authorization Data* option stores a hash value computed over the BU or the BA message. The option does not have any alignment requirement; however, because it has to be placed at the end of the message, it eventually has an $8n + 2$ requirement. The format of the Binding Authorization Data option is shown in Figure 3-18.

FIGURE 3-16

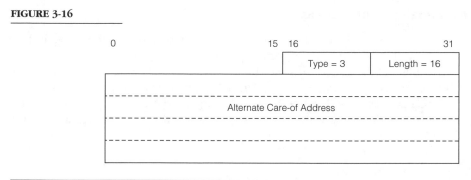

Alternate Care-of Address option.

FIGURE 3-17

Nonce Indices option.

FIGURE 3-18

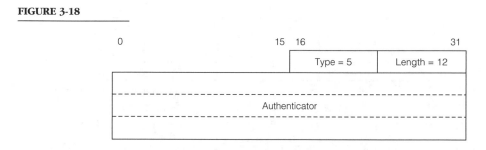

Binding Authorization Data option.

The Length field depends on the length of the *Authenticator* field. At this moment, the length is 12 because the procedure to compute the authenticator produces a 96-bit Authenticator value. The algorithm used for this computation is discussed in Section 5.1 of Chapter 5.

3.6 Neighbor Discovery Messages

The Mobile IPv6 specification modifies the Router Advertisement message and the Prefix Information option so that we can distribute information about a home agent. Two new Neighbor Discovery options are introduced.

3.6.1   Router Advertisement Message

The *Router Advertisement* message is modified to include the newly defined Home Agent flag. Figure 3-19 shows the modified Router Advertisement message.

The H flag in the flags field is added. A router that is acting as a Mobile IPv6 home agent must specify the H flag so that other home agents can detect there is another home agent on the same network. This information is used on each home agent when creating a list of home agent addresses. A-mobile node may use this option to create the list of home agents when it is at home.

3.6.2   Prefix Information Option

The *Prefix Information* option is an option defined in [RFC2461]. The option is used with the Router Advertisement message to distribute the prefix information to the nodes on the attached network. Figure 3-20 shows the format of this option.

FIGURE 3-19

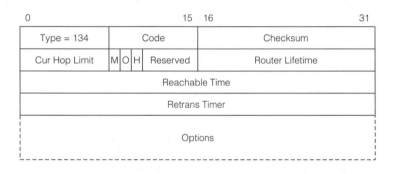

Modified Router Advertisement message.

FIGURE 3-20

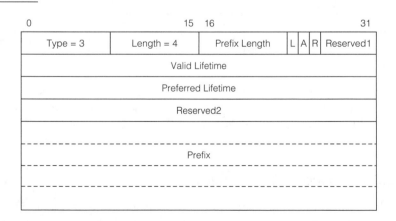

Prefix Information option.

In [RFC2461], this option only carries the information of the prefix part. In the Mobile IPv6 specification, the option is modified to include the address of the home agent including the interface identifier part. The R flag is added in the flags field for that purpose. If the R flag is set, the Prefix field includes a full IPv6 address of the home agent, not only the prefix part. A node that receives this option with the R flag can discover the address of a home agent on the network. This information is used when each home agent creates a list of home agent addresses. The mechanism is described in Chapter 6.

3.6.3   Advertisement Interval Option

The *Advertisement Interval* option is used to supply the interval at which Router Advertisement messages are sent from a home agent. The Router Advertisement message is used as a hint of the reachability of the router. A mobile node assumes it has not moved to other networks as long as the same router is reachable on the attached network. A mobile node can detect the unreachability of a router by listening for the Router Advertisement message since a router periodically sends these messages. However, such detection is usually difficult since the interval between Router Advertisement messages varies on each network. This option explicitly supplies the interval between Router Advertisement messages. The interval is set to a lower value than the usual IPv6 Router Advertisement messages. A mobile node can determine a router is unreachable if the router does not send a Router Advertisement message for the period specified in this option. The format of the Advertisement Interval option is shown in Figure 3-21.

The *Type* field is set to 7. The *Length* field is fixed at 1. The *Reserved* field must be cleared by the sender and must be ignored by the receiver. The Advertisement Interval field is a 32-bit unsigned integer that specifies the interval value between Router Advertisement messages in units of 1 s.

3.6.4   Home Agent Information Option

The *Home Agent Information* option is a newly defined Neighbor Discovery option to distribute the information of a home agent. This option is used with the Router Advertisement message sent from a home agent. The format of the Home Agent Information option is shown in Figure 3-22.

The Type field is set to 8. The Length field is fixed at 1. The *Reserved* field must be cleared by the sender and must be ignored by the receiver. The *Home Agent Preference* field specifies the preference value of a home agent that sends this option. The value is a 16-bit unsigned integer. Higher values mean the home agent is more preferable. This value is used to order the addresses of the home agent list that is maintained on each home agent on the home network.

FIGURE 3-21

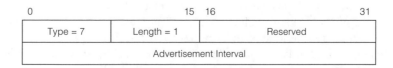

Advertisement Interval option.

FIGURE 3-22

0	15	16	31
Type = 8	Length = 1	Reserved	
Home Agent Preference		Home Agent Lifetime	

Home Agent Information option.

The home agent list is sent to a mobile node when the mobile node requests the latest list of home agents. The *Home Agent Lifetime* field contains the lifetime of the home agent. The value is a 16-bit unsigned integer and stored in units of 1 s. This value specifies how long the router can provide the home agent service. If there is no Home Agent Information option sent by a home agent, the preference value is considered 0 and the lifetime is considered the same value as the router lifetime.

3.7   ICMPv6 Messages

The Mobile IPv6 specification defines four new types of the ICMPv6 message.

3.7.1   Dynamic Home Agent Address Discovery Request

A mobile node sometimes requests the latest list of home agents on its home network. When requesting the list, a mobile node sends the *Dynamic Home Agent Address Discovery Request* message, which is a newly defined ICMPv6 message. The format of the Dynamic Home Agent Address Discovery Request message is shown in Figure 3-23.

The source address of the IPv6 packet is the care-of address of a mobile node. The destination address is the *home agent anycast address*. The algorithm to construct the home agent anycast address is shown in Figure 3-24. There are two patterns to compute the anycast address: one is for the prefix of which the prefix length is 64 and the other is for the prefix of which the prefix length is not 64. The home agent anycast address is a combination of a prefix and the anycast identifier `ffff:ffff:ffff:ffff:ffff:ffff:ffff:fffe`, which is reserved for the home agent anycast address. The important point when generating the anycast address is if the prefix length is 64, the interface identifier part of the generated anycast address must satisfy the EUI-64 requirements. That is, the universal/local bit must be cleared since the anycast address may be assigned to multiple home agents. In this case, we must use `fdff:ffff:ffff:fffe` as an anycast identifier. The interface identifier of the home agent anycast address is defined in [RFC2526].

The Type field is set to 144. The *Code* field is set to 0. No other code value is defined. The *Checksum* field is a checksum value computed as specified in the ICMPv6 specification [RFC2463]. The *Identifier* field contains an identifier to match the request message and the reply message. The Reserved field must be cleared by the sender and must be ignored by the receiver. The procedure of Dynamic Home Agent Address Discovery is discussed in Chapter 6.

FIGURE 3-23

0		15 16	31
Type = 144	Code = 0	Checksum	
Identifier		Reserved	

Dynamic Home Agent Address Discovery Request message.

FIGURE 3-24

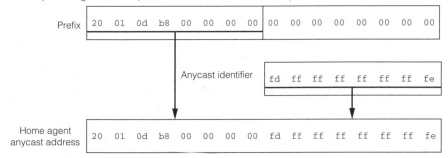

For prefixes that prefix length is 64 bits (ex. `2001:0db8:0000:0000::/64`)

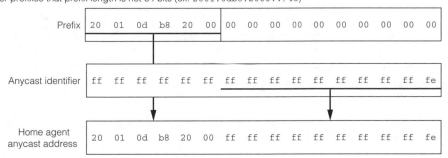

For other prefixes that prefix length is not 64 bits (ex. `2001:0db8:2000::/48`)

Computation of the home agent anycast address.

3.7.2 Dynamic Home Agent Address Discovery Reply

The *Dynamic Home Agent Address Discovery Reply* message is used as a response message to the Dynamic Home Agent Address Discovery Request message. Each home agent maintains the list of home agents on its home network by listening to Router Advertisement messages sent by other home agents and updating the list as necessary. When a home agent receives a Dynamic Home Agent Address Discovery Request message, the node will reply to the mobile node that has sent the request message with a Dynamic Home Agent Address Discovery Reply

FIGURE 3-25

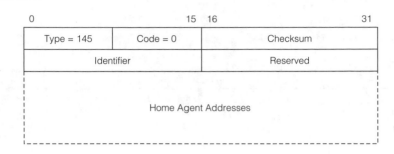

Dynamic Home Agent Address Discovery Reply message.

message including the latest list of home agents. The format of the Dynamic Home Agent Address Discovery Reply message is shown in Figure 3-25.

The source address of the IPv6 packet is set to one of the addresses of the home agent that replies to this message. The source address must be an address recognized as the home agent's address because a mobile node may use the source address as the home agent's address in the following Mobile IPv6 signaling process. The destination address is copied from the source address field of a Dynamic Home Agent Address Discovery Request message.

The Type field is set to 145. The Code field is set to 0. No other code values are defined. The Checksum field is a checksum value computed as specified in the ICMPv6 specification [RFC2463]. The value of the Identifier field is copied from the Identifier field of the corresponding Dynamic Home Agent Address Discovery Request message. The Reserved field must be cleared by the sender and must be ignored by the receiver. The *Home Agent Addresses* field contains the list of addresses of home agents on the home network. The order of the list is decided based on the preference value of each home agent. To avoid fragmentation of the message, the maximum number of addresses in the list is restricted to not exceed the path MTU value from a home agent to a mobile node. The procedure of Dynamic Home Agent Address Discovery is discussed in Chapter 6.

3.7.3　Mobile Prefix Solicitation

The *Mobile Prefix Solicitation* message is a newly defined ICMPv6 message that is sent when a mobile node wants to know the latest prefix information on its home network. This message is typically sent to extend the lifetime of the home address before it expires. The format of the Mobile Prefix Solicitation message is shown in Figure 3-26.

FIGURE 3-26

0	15	16	31
Type = 146	Code = 0	Checksum	
Identifier		Reserved	

Mobile Prefix Solicitation message.

The source address of the IPv6 packet is set to the current care-of address of the mobile node. The destination address is set to the address of the home agent with which the mobile node is currently registered. This message must contain the Home Address option to carry the home address of the mobile node. This message should be protected by the IPsec ESP header to prevent the information from being modified by attackers.

The Type field is set to 146. The Code field is set to 0. No other code values are defined. The Checksum field is a checksum value computed as specified in the ICMPv6 specification [RFC2463]. The Identifier field contains a random value that is used to match the solicitation message and the advertisement message. The Reserved field is cleared by the sender and must be ignored by the receiver.

3.7.4 Mobile Prefix Advertisement

The *Mobile Prefix Advertisement* message is a newly defined ICMPv6 message that is used to supply the prefix information of a home network to mobile nodes. This message is used as a response message to a Mobile Prefix Solicitation message sent from a mobile node. Also, this message may be sent from a home agent to each mobile node that has registered with the home agent to notify the mobile node of updates to the prefix information of the home network, even if the mobile nodes do not request the information explicitly. The format of the Mobile Prefix Advertisement message is shown in Figure 3-27.

The source address of the IPv6 packet is one of the addresses of the home agent. The destination address is copied from the source address field of the Mobile Prefix Solicitation message if the message is in response to a solicitation message. Otherwise, the destination address is the registered care-of address of a mobile node. A Type 2 Routing Header must be included in this message to contain the home address of a mobile node. This message should be protected by the IPsec ESP header to prevent being modified by attackers.

The Type field is set to 147. The Code field is set to 0. No other code values are defined. The Checksum field is a checksum value computed as specified in the ICMPv6 specification [RFC2463]. If this message is in response to a solicitation message, the value of the Identifier field is copied from the Identifier field of the Mobile Solicitation message. If the message is not a response message, this field can be set to any value. A mobile node that receives a Mobile Prefix Advertisement, which has an unmatched identifier, should send the Mobile Prefix Solicitation message to confirm the prefix information. The M and O flags are copied from

FIGURE 3-27

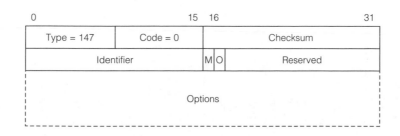

Mobile Prefix Advertisement message.

the configuration of a home network. That is, if the home network is being operated with a managed address configuration mechanism (e.g., Dynamic Host Configuration Protocol for IPv6 [DHCPv6]), the M flag is set. Also if the home network provides stateful configuration parameters (e.g., domain name server [DNS] server addresses via DHCPv6), the O flag is set. Currently, the exact processing procedure of these flags is not defined in the Mobile IPv6 specification. A future document will define the exact processing mechanism. The Reserved field must be cleared by the sender and must be ignored by the receiver. This message will have the modified Prefix Information option described in Section 3.6.

Procedure of Mobile IPv6

In this chapter, we discuss the detailed procedure of the Mobile Internet Protocol version 6 (IPv6) protocol operation.

4.1 Protocol Constants and Variables

Table 4-1 shows a list of the variables used in the Mobile IPv6 protocol. Some of these variables are constant, while others may have their values modified.

4.2 Home Registration

When a mobile node is at home, the node acts as a fixed IPv6 node. Figure 4-1 shows the situation.

A mobile node gets its IPv6 addresses from its home network. The addresses assigned on the home network are called home addresses. When a mobile node sends a packet, the source address of the packet is set to one of the home addresses of the mobile node. The destination address of the packet is the address of the peer node. When the peer node sends a packet to the mobile node, the source and the destination address are set to the peer address and the home address, respectively.

When a mobile node moves to a foreign network, the mobile node will get address(es) from the foreign network. These addresses are called care-of addresses. If the mobile node detects that it is on a foreign network, the node creates an entry that keeps the state of the mobile node and maintains it. The entry is called a *binding update list* entry. It contains the information of

TABLE 4-1

Name	Description
INITIAL_DHAAD_TIMEOUT	The initial timeout value when retransmitting a Dynamic Home Agent Address Discovery Request message (constant: 3 s)
DHAAD_RETRIES	The maximum number of retries for a Dynamic Home Agent Address Discovery Request message (constant: 4 times)
InitialBindackTimeoutFirstReg	The initial timeout value when retransmitting a Binding Update message when a mobile node moves from a home network to a foreign network for the first time (configurable: default to 1.5 s)
INITIAL_BINDACK_TIMEOUT	The initial timeout value when retransmitting a Binding Update message when updating the existing binding information of a peer node (constant: 1 s)
MAX_BINDACK_TIMEOUT	The maximum timeout value for retransmitting a Binding Update message (constant: 32 s)
MAX_UPDATE_RATE	The maximum number of Binding Update messages, which a mobile node can send in 1 s (constant: 3 times)
MAX_NONCE_LIFETIME	The maximum lifetime of nonce values (constant: 240 s)
MAX_TOKEN_LIFETIME	The maximum lifetime of Keygen Token values (constant: 210 s)
MAX_RR_BINDING_LIFETIME	The maximum lifetime for binding information created by the Return Routability procedure (constant: 420 s)
MaxMobPfxAdvInterval	The maximum interval value between Mobile Prefix Advertisement messages (modifiable: default to 86 400 s)
MinMobPfxAdvInterval	The minimum interval value between Mobile Prefix Advertisement messages (modifiable: default to 600 s)
PREFIX_ADV_TIMEOUT	The timeout value when retransmitting a Mobile Prefix Advertisement message (constant: 3 s)
PREFIX_ADV_RETRIES	The maximum number of retransmissions of Mobile Prefix Advertisement messages (constant: 3 times)
MinDelayBetweenRAs	The minimum interval value between Router Advertisement messages (modifiable: default to 3 s, minimum 0.03 s)

Protocol constants and variables.

the home address and one of the care-of addresses of the node, the lifetime of the entry, and so on. The detailed contents of this entry are discussed in Section 10.29 of Chapter 10.

The mobile node sends a Binding Update message to its home agent to notify the home agent of its current location. The source address of the message is set to the care-

FIGURE 4-1

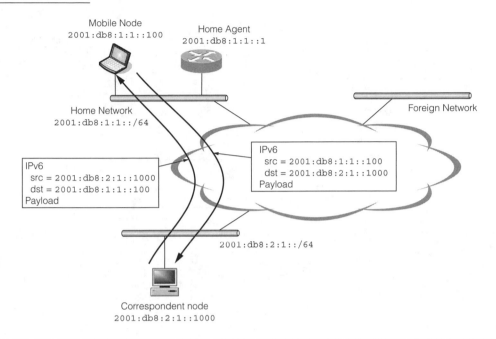

Packet exchange while a mobile node is home.

of address picked from the list of available care-of addresses. The destination address is the address of the home agent. The message also includes a Home Address option that contains the home address of the mobile node. This message must be protected by the IPsec ESP mechanism.

When a home agent receives a Binding Update message, it adds the information to its internal database. The information kept in a home agent is called a *binding cache* (the detailed structure of this information is discussed in Section 10.28 of Chapter 10). The home agent replies with a Binding Acknowledgment message in response to the Binding Update message. If the mobile node does not receive the acknowledgment message, it resends a Binding Update message until it gets an acknowledgment message. This procedure is called *Home registration*. Figure 4-2 shows the procedure.

A Binding Update message includes a sequence number. If a home agent already has a corresponding binding cache entry and the sequence number of the received Binding Update message is smaller than the sequence number kept in the cache entry, the home agent returns a Binding Acknowledgment message with an error status of 135 and the latest sequence number. The mobile node resends a Binding Update message with a correct sequence number to complete home registration. The comparison of sequence numbers is based on modulo 2^{16} since the sequence number is represented as a 16-bit variable. For example, if the current sequence number is 10015, then the numbers 0 through 10014 and 42783 through 65535 are considered less than 10015 (Figure 4-3).

FIGURE 4-2

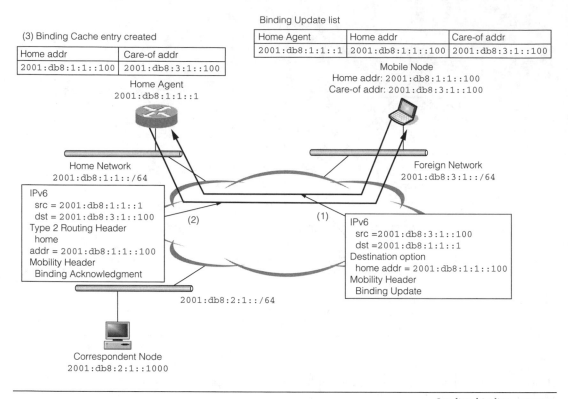

(3) Binding Cache entry created

Home addr	Care-of addr
2001:db8:1:1::100	2001:db8:3:1::100

Binding Update list

Home Agent	Home addr	Care-of addr
2001:db8:1:1::1	2001:db8:1:1::100	2001:db8:3:1::100

Mobile Node
Home addr: 2001:db8:1:1::100
Care-of addr: 2001:db8:3:1::100

Home Agent
2001:db8:1:1::1

Home Network
2001:db8:1:1::/64

Foreign Network
2001:db8:3:1::/64

IPv6
  src = 2001:db8:1:1::1
  dst = 2001:db8:3:1::100
Type 2 Routing Header
  home
addr = 2001:db8:1:1::100
Mobility Header
  Binding Acknowledgment

(2)

(1)

IPv6
  src =2001:db8:3:1::100
  dst =2001:db8:1:1::1
Destination option
  home addr = 2001:db8:1:1::100
Mobility Header
  Binding Update

2001:db8:2:1::/64

Correspondent Node
2001:db8:2:1::1000

Sending binding messages.

FIGURE 4-3

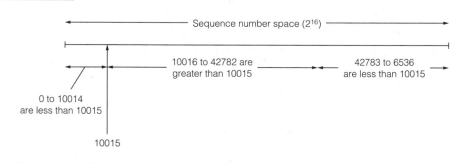

Sequence number space (2^{16})

10016 to 42782 are
greater than 10015

42783 to 6536
are less than 10015

0 to 10014
are less than 10015

10015

Sequence number comparison.

A mobile node must set the H and A flags to indicate that it is requesting home registration when it registers its current location with its home agent. In addition to the flags, a mobile node must set the L flag if the home address of the mobile node has the same interface identifier as is used in its link-local address. Setting the L flag will create a binding cache entry for the link-local address of the mobile node and protect that address from being used by other nodes on its home network.

When a mobile node sets the A flag, the node resends a Binding Update message until it receives a Binding Acknowledgment message. The initial retransmission timeout value is determined based on whether this registration is the first home registration or if it is updating the home registration entry. If the message is for the first home registration, the initial retransmission timeout is set to `InitialBindackTimeoutFirstReg` seconds. Otherwise, the initial retransmission timeout is `INITIAL_BINDACK_TIMEOUT` seconds. The difference is due to running the Duplicate Address Detection (DAD) procedure at the home agent. The first time a mobile node registers its location, the home agent must make sure that the home address (and the link-local address, if the L flag is set) is not used on the home network by some other node by performing the DAD procedure. Usually, the DAD procedure takes 1 s. This is why the initial timeout must be greater than 1 s. The timeout value is increased exponentially on every retransmission with the maximum retransmission timeout being `MAX_BINDACK_TIMEOUT` seconds. If a mobile node does not receive a Binding Acknowledgment after the last retransmission, the mobile node may perform a Dynamic Home Agent Address Discovery to find another home agent on the home network.

A Binding Update message includes an Alternate Care-of Address option to protect the care-of address information. The Binding Update message is protected by an ESP IPsec header, but the ESP header does not cover the source address field of an IPv6 header that contains the care-of address of a mobile node. A mobile node needs to put its care-of address in the Alternate Care-of Address option as a part of the Binding Update message in order for it to be covered by the ESP header.

The lifetime field of a Binding Update message is set to the smaller lifetime of either the care-of address or the home address of a mobile node. If a home agent accepts the requested lifetime, the acknowledgment message includes the same value. A home agent can reduce the lifetime based on the local policy of the home agent. A Binding Acknowledgment message may include a Binding Refresh Advice option.

A mobile node maintains its binding update list entry for home registration by sending a Binding Update message periodically.

4.3 Bidirectional Tunneling

When a mobile node and a home agent complete the exchange of the binding information, these nodes create a tunnel connection between them. The endpoint addresses of the tunnel connection are the address of the home agent and the care-of address of the mobile node. This tunnel connection is used to hide the location of the mobile node from correspondent nodes. The peer node does not notice whether the mobile node is at home or in any foreign networks. Note that the packets sent to the link-local address of the mobile node are not forwarded to the mobile node even if the L flag is set in the Binding Update message from the mobile node. The flag is used to protect the link-local address to be used with other nodes on the home link but not to be used to forward the link-local packets to other links.

A mobile node usually uses its home address as a logical endpoint address when sending packets. This ensures that the communication between a mobile node and other nodes survives when the mobile node moves from one network to another network since a home address

FIGURE 4-4

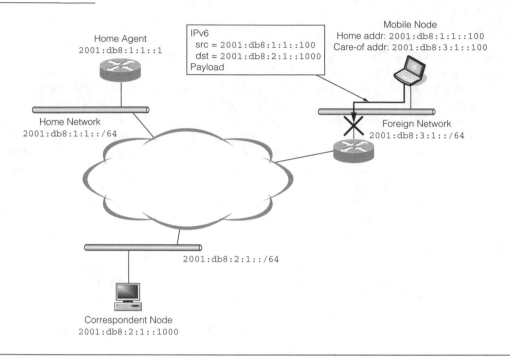

Topologically incorrect packets may be dropped.

never changes. However, a mobile node cannot simply send a packet with its source address set to the home address of the node. Such a packet is topologically incorrect and the router that serves the foreign network may discard the packet based on its local security policy. Figure 4-4 shows the procedure.

To avoid this problem, a mobile node sends packets of which the source address is the home address of the node by using the tunnel connection created between the mobile node and its home agent. Figure 4-5 shows the procedure.

A packet is encapsulated within another IPv6 header whose source and destination addresses are the care-of address of the mobile node and the address of mobile node's home agent, respectively. The packet is decapsulated at the home agent, and the home agent forwards the packet to the final destination. The packet looks as if it is being sent from a node that is attached to the home network.

When a correspondent node sends packets to the mobile node, the tunnel connection is also used in reverse direction. All packets of which the destination address is the home address of the mobile node are delivered to the home network of the mobile node. These packets are intercepted by the home agent of the mobile node if the home agent has a valid binding cache entry for the mobile node and sent to the mobile node using IPv6 in IPv6 tunneling. The source and destination addresses of the outer IPv6 header are the address of the home agent and the care-of address of the mobile node, respectively. Figure 4-6 shows the flow.

FIGURE 4-5

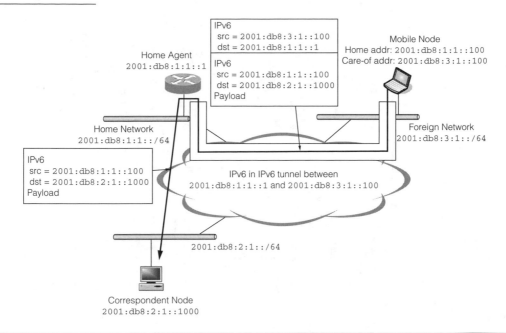

Sending packets by a tunnel connection from a mobile node to a home agent.

FIGURE 4-6

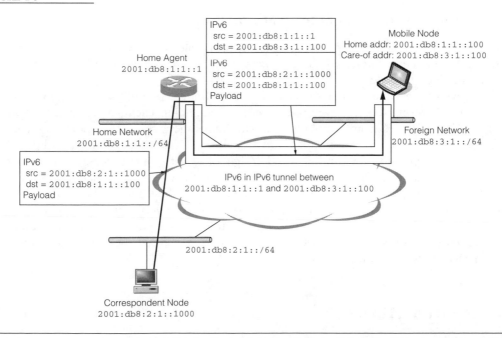

Sending packets by a tunnel connection from a home agent to a mobile node.

FIGURE 4-7

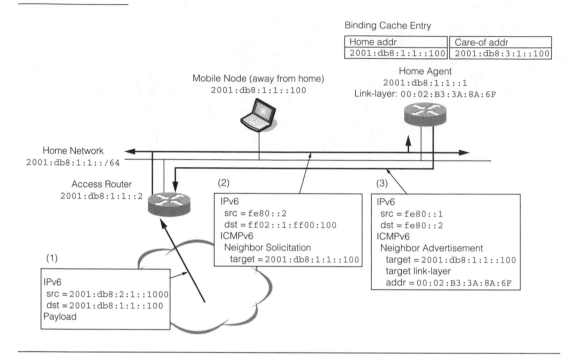

Intercepting packets.

4.4   Intercepting Packets for a Mobile Node

A home agent needs to intercept packets sent to a mobile node, which the home agent is serving, and then needs to forward these packets using a tunnel connection between the home agent and the mobile node.

To receive packets that are sent to a mobile node, a home agent utilizes the proxy Neighbor Discovery mechanism. When a home agent creates a binding cache entry after receiving a Binding Update message from a mobile node, the home agent starts responding to Neighbor Solicitation messages sent to the home address or the solicited node multicast address of the home address. The home agent replies with a Neighbor Advertisement message in response to these solicitation messages. In the advertisement message, the home agent includes its own link-layer address as a target link-layer address. As a result, all packets sent to the home address of the mobile node are sent to the link-layer address of the home agent. The home agent forwards the received packets to the tunnel connection constructed between the home agent and the mobile node as described in the previous section. Figure 4-7 shows the behavior of the proxy Neighbor Discovery mechanism.

4.5   Returning Home

When a mobile node returns home, it must clear any of its binding information registered on a home agent and correspondent nodes. The procedure to deregister binding information is almost the same as that of registering the information. The message used to deregister the binding is a Binding Update message.

First, a mobile node must send a Binding Update message to its home agent. The source address of the message must be a care-of address of a mobile node however, in this case, the source address is set to the home address of a mobile node since the care-of address and the home address are the same when a mobile node is home. The message also contains a Home Address option that contains the home address. The lifetime field is set to 0 to indicate deregistration. Also, the message contains an Alternate Care-of Address option to hold a care-of address (which is a home address in this case). The message must be protected by the IPsec ESP mechanism.

In some situations, a mobile node may not know the link-layer address of its home agent, which is necessary when sending a packet to the home agent. In this case, a mobile node must perform the Neighbor Discovery procedure but we need to take care of one thing. If a home agent has a valid binding cache entry for the mobile node's link-local address, the mobile node cannot use its link-local address during the Neighbor Discovery procedure because the home agent is acting as a proxy server of the address. Such usage may be considered address duplication. When a mobile node needs to resolve the link-layer address of its home agent when returning home, it sends a Neighbor Solicitation message from an unspecified address. When the home agent receives such a solicitation message, it replies with a Neighbor Advertisement message to an all-node multicast address as described in the Neighbor Discovery specification [RFC2461]. A mobile node can learn the link-layer address of the home agent by listening to the advertisement message.

If a home agent accepts the Binding Update message, it replies with a Binding Acknowledgment message. A home agent also stops its proxy function for the mobile node and shuts down the tunnel connection between the home agent and the mobile node. Finally, it removes the binding cache entry for the mobile node.

A mobile node also shuts down the tunnel connection between itself and its home agent after receiving a Binding Acknowledgment message from its home agent. This procedure is called *home deregistration*.

There is a possibility that the signaling messages may be dropped because of communication errors. If a Binding Update message sent from a mobile node for deregistration is lost, the mobile node will resend another Binding Update message until it receives a Binding Acknowledgment message. If a Binding Acknowledgment message is lost, the situation is slightly complicated because the binding cache entry for the mobile node that sent a deregistration message has already been removed from the home agent when the Binding Acknowledgment message was sent. The mobile node will resend a Binding Update message because it has not received a corresponding Binding Acknowledgment message. When a home agent receives a Binding Update message for deregistration from a mobile node but it does not have a corresponding binding cache entry, it will reply to the mobile node with a Binding Acknowledgment message with status code 133. When a mobile node that has returned home receives a Binding Acknowledgment message with status code 133, the mobile node should consider that the acknowledgment message has been lost and complete the deregistration procedure.

A mobile node may deregister its address from its home agent even when it does not return to home (for example, when the mobile node stops its mobility function on a foreign network). In this case, a similar procedure is used to deregister the address. The Binding Update message sent from the mobile node will have a different home address and care-of address but the lifetime field will be set to 0. The home agent will remove its binding cache entry and stop intercepting packets for the mobile node.

5

Route Optimization

When a mobile node communicates with other nodes, all packets are forwarded by a home agent if the mobile node is away from home. This causes a communication delay, especially if the mobile node and its peer node are located on networks that are topologically close and the home agent is far away. The worst case is when both nodes are on the same network.

The Mobile Internet Protocol version 6 (IPv6) specification provides a solution to this problem. If the peer node supports the Mobile IPv6 correspondent node function, the path between a mobile node and the peer node can be optimized. To optimize the path, a mobile node sends a Binding Update message to the correspondent node. The message must not have the H and L flags set because the message is not requesting home registration. The A flag may be set; however, it is not mandatory. If the A flag is set, a correspondent node replies with a Binding Acknowledgment message in response to the Binding Update message. Note that even if the A flag is not set, a correspondent node must reply to the mobile node with a Binding Acknowledgment message when an error occurs during the message processing except in the authentication error case.

A Binding Update message must be protected by the return routability procedure, discussed in the next section. The message must contain a Binding Authorization Data option. The option contains a hash value of the Binding Update message, which is computed with the shared secret generated as a result of the return routability procedure. If the hash value is incorrect, the message is dropped. Similarly, a Binding Acknowledgment message sent from a correspondent node must include a Binding Authorization Data option to protect the contents.

Once the exchange of a Binding Update message (and a Binding Acknowledgment message, if the A flag is set) has completed, a mobile node starts exchanging route-optimized packets with a correspondent node. The source address field of the packets is set to the care-of address of the mobile node. The mobile node cannot set the source address to its home address directly, since intermediate routers may drop a packet of which the source address is not topologically correct to prevent source spoofing attacks. The home address information is kept in a Home Address option of a Destination Options header of the packet.

When a correspondent node receives a packet that has a Home Address option, it checks to see if it has a binding cache entry related to the home address. If there is no such entry, the correspondent node responds with a Binding Error message with status code 0. A mobile node needs to resend a Binding Update message to create a binding cache entry in the correspondent node if it receives a Binding Error message. This validation procedure prevents any malicious nodes from using forged care-of addresses on behalf of the legitimate mobile node.

If the Home Address option is valid, the correspondent node accepts the incoming packet and swaps the home address in the option and the source address of the packet. As a result, the packets passed to the upper layer protocols have the home address as the source address. The upper layer protocols and applications need not care about any address changes for the mobile node since this address swap is done in the Internet Protocol version 6 (IPv6) layer.

When a correspondent node sends a packet to a mobile node, it uses the Type 2 Routing Header. A home address of a mobile node is put in the Routing Header and the destination address of the IPv6 packet is set to the care-of address of the mobile node. The packet does not go to the home network. Instead, the packet is routed to the foreign network where the mobile node is currently attached since the destination address is set to the care-of address. The processing of a Type 2 Routing Header is similar to the processing of a Type 0 Routing Header except for some validation checks. A mobile node checks that the Routing Header contains only one address in the intermediate nodes field and ensures that the address is assigned to the mobile node itself. If the address specified in the Routing Header is not an address of the mobile node, the mobile node discards the packet, as the packet may be an attempt to force the mobile node to forward the packet. A mobile node drops any packets that contain an invalid Type 2 Routing Header.

5.1   Return Routability

A mobile node and a correspondent node need to share secret information before exchanging binding information. When a mobile node sends a Binding Update message, it computes a hash value of the message, using the shared information, and puts the value in the message. A correspondent node verifies the hash value by recomputing it and drops the packet if the value computed on the correspondent node and the value specified in the message are different. In the same manner, a Binding Acknowledgment message sent from a correspondent node to a mobile node is protected by the hash mechanism. The shared information is created by the return routability procedure. In this chapter, we discuss the detailed procedure of the return routability mechanism.

5.2   Sending Initial Messages

Only a mobile node can initiate the return routability procedure. When a mobile node wants to start route-optimized communication, it sends two initial messages. One is a Home Test Init message and the other is a Care-of Test Init message. There is no strict specification as to when a mobile node should send these messages. A mobile node can initiate the procedure whenever it needs to optimize the route. In the KAME implementation, for example, a mobile node sends

FIGURE 5-1

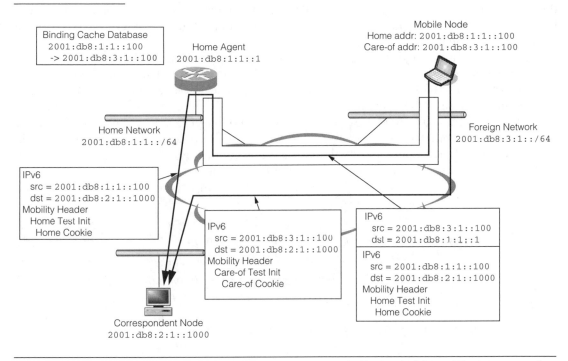

Home Test Init and Care-of Test Init message flow.

these messages when the mobile node receives a packet from a correspondent node via a bidirectional tunnel between the mobile node and its home agent.

A Home Test Init message is sent from the home address of a mobile node. A packet of which the source address is a home address cannot be sent directly from a foreign network. A Home Test Init message is sent through a tunnel connection between a mobile node and its home agent. A correspondent node will receive the message as if it were sent from the home network of the mobile node.

A Care-of Test Init message is sent from the care-of address of a mobile node. This message can be sent directly from a foreign network.

Both messages contain a random value called a cookie. The cookie in a Home Test Init message is called the Home Init Cookie and the cookie in a Care-of Test Init message is called the Care-of Test Init Cookie. These cookie values are used to match messages that a mobile node receives in response to the Home Test Init/Care-of Test Init messages from the correspondent node.

Figure 5-1 shows the packet flow of the Home Test Init and Care-of Test Init messages.

5.3 Responding to Initial Messages

When a correspondent node that supports the return routability procedure receives a Home Test Init or a Care-of Test Init message from a mobile node, the correspondent node replies to the mobile node with a Home Test message and a Care-of Test message.

A Home Test message is sent to the home address of a mobile node. The message is delivered to the home network of the mobile node and intercepted by the home agent of the mobile node. The mobile node receives the message from a tunnel connection between the node and its home agent.

A Care-of Test message is sent to the care-of address of a mobile node directly.

Both messages contain a copy of the cookie value, which is contained in the Home Test Init/Care-of Test Init message, so that a mobile node can check to see if the received messages are sent in response to the initial messages.

A Home Test and a Care-of Test message have two other pieces of information: the nonce index and the Keygen Token. A correspondent node keeps an array of nonce values and node keys. The nonce index values specify the nonce values in the array. The nonce values and the node key values are never exposed outside of a correspondent node. This information must be kept in the correspondent node. The Keygen Token is computed from a nonce value and a node key using the following algorithms.

$$Home\ Keygen\ Token = First(64, HMAC\_SHA1(K_{cn},$$
$$(the\ home\ address\ of\ a\ mobile\ node$$
$$|the\ nonce\ specified\ by\ the\ home\ nonce\ index$$
$$|0)))$$

$$Care\text{-}of\ Keygen\ Token = First(64, HMAC\_SHA1(K_{cn},$$
$$(the\ care\text{-}of\ address\ of\ a\ mobile\ node$$
$$|the\ nonce\ specified\ by\ the\ care\text{-}of\ nonce\ index$$
$$|1))),$$

where

"|" denotes concatenation;

$First\ (x,\ y)$ function returns the first x bits from y;

$HMAC\_SHA1(key,\ data)$ function returns

an HMAC SHA-1 hash value against "*data*" using "*key*" as a key;

K_{cn} is a node key of a correspondent node.

These tokens are used to generate a shared secret that is used to compute the hash values of a Binding Update message on a mobile node and a Binding Acknowledgment message on a correspondent node. To prevent a replay attack, a correspondent node must generate a new nonce value and node key and revoke the old nonce value and node key periodically. The maximum lifetime of all nonce values is restricted to MAX_NONCE_LIFETIME seconds. The lifetime of generated tokens is also restricted to MAX_TOKEN_LIFETIME seconds.

The array that keeps the nonce values and node keys are shared between mobile nodes with which the correspondent node is communicating. In theory, it is possible to use different values per mobile node, however, it introduces a vulnerability in management of the values. That is, a malicious node can easily consume the memory of the correspondent node

FIGURE 5-2

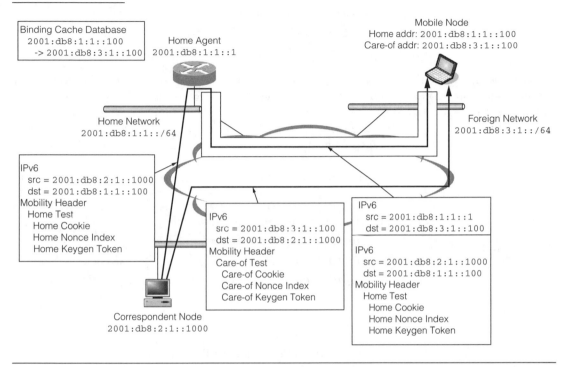

Home Test and Care-of Test message flow.

sending bogus Home Test Init or Care-of Test Init messages with a lot of fake mobile node's addresses.

When a mobile node sends a Binding Update message, it includes nonce index values. A correspondent node must keep the history of these values and must be able to regenerate Keygen Tokens from the index values.

Figure 5-2 shows the packet flow of the Home Test and the Care-of Test messages.

5.4 Computing a Shared Secret

A shared secret is computed as follows:

$$K_{bm} = SHA1(home\ keygen\ token\ |\ care\text{-}of\ keygen\ token)$$

...(if a mobile node is at a foreign network)

or

$$K_{bm} = SHA1(home\ keygen\ token)$$

,..(if a mobile node is at home),

where

"|" denotes concatenation of data;

K_{bm} is a shared secret computed from token values;

*SHA*1(*data*) computes a SHA-1 hash value against "*data*."

Depending on the location of a mobile node, the shared secret is computed differently. If a mobile node is in a foreign network, the secret is computed from both a Home Keygen Token and a Care-of Keygen Token. If a mobile node is at home, only a Home Keygen Token is used, because the home address and the care-of address of the mobile node are the same. In this case, we need to check only one of them. The procedure when returning to home is discussed in Section 5.7.

A mobile node computes a hash value using the secret information computed above. The algorithm is as follows:

$$Mobility\ Data = the\ care\text{-}of\ address\ of\ a\ mobile\ node$$
$$|\ the\ address\ of\ a\ correspondent\ node$$
$$|\ the\ Mobility\ Header\ message$$
$$Authenticator = First(96, HMAC\_SHA1(K_{bm}, Mobility\ Data)),$$

where

"|" denotes concatenation of data;

"*the Mobility Header message*" is either a Binding Update

or a Binding Acknowledgment message;

First(*x, y*) function returns the first *x* bits from *y*;

*HMAC_SHA*1(*key, data*) computes a HMAC SHA-1 hash value

against "*data*" using "*key*" as a key.

The hash value is called an Authenticator. The original data of the hash value consists of a care-of address, a home address, and a Mobility Header message. When sending a Binding Update message, the Mobility Header message is the contents of the Binding Update message. When computing the hash value, all mobility options are included as a part of the Mobility Header, except the Authenticator field of the Binding Authorization Data option. The checksum field of a Mobility Header message is considered zero and it must be cleared before computing the hash value.

5.5 Verifying Message

A mobile node sends a Binding Update message with a Binding Authorization Data option, which includes the Authenticator value computed by the procedure described in the previous paragraph, and a Nonce Index option that contains the home nonce index and the care-of nonce index, which have been used when generating a shared secret to compute the Authenticator. When creating a Binding Update message as a result of the return routability procedure, the lifetime of the binding information is limited to `MAX_RR_BINDING_LIFETIME` seconds.

When a correspondent node receives a Binding Update message, it first checks the existence of a Binding Authorization Data option and a Nonce Index option. If these options do not exist, the message is dropped.

The correspondent node generates a Home Keygen Token and a Care-of Keygen Token from the nonce index values included in the Nonce Index option of the incoming Binding Update message. From the tokens, the correspondent node can generate the shared secret, which was used by the mobile node when it created the Binding Update message. A correspondent node verifies the message by computing a hash value of the message using the same algorithm described previously. If the result is different from the Authenticator value of the Binding Authorization Data option, which was computed in the mobile node, the incoming message is dropped.

In some cases, a mobile node may use older nonce index values, which a correspondent node has not kept any more. In this case, the correspondent node replies with a Binding Acknowledgment message with a status code 136 to 138 (see Table 3-3 in Chapter 3), which indicates the specified nonce index is not valid. The mobile node which receives such an error status performs the return routability procedure to get the latest nonce values.

If the incoming Binding Update message is valid, the correspondent node creates a binding cache entry for the mobile node and, if the A flag is set in the Binding Update message, replies with a Binding Acknowledgment message. The Binding Acknowledgment message also includes a Binding Authorization Data option and a Nonce Index option to protect the message. Figure 5-3 describes the packet flow of the Binding Update and the Binding Acknowledgment messages between a mobile node and a correspondent node.

FIGURE 5-3

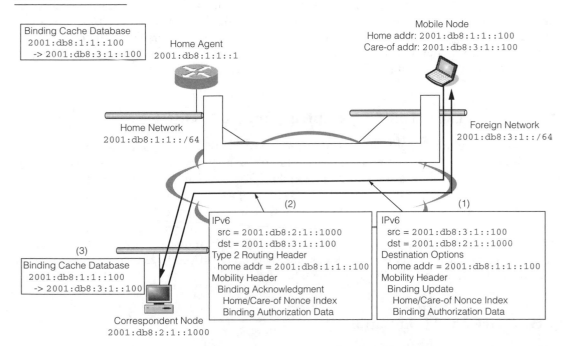

Exchanging binding information between a mobile node and a correspondent node.

5.6   Security Considerations

The return routability procedure provides an authorization mechanism for mobile nodes to inject binding cache entries to correspondent nodes. A correspondent node can ensure that the home address and the care-of address provided by a Binding Update message are bound to a single mobile node. But it cannot determine what the mobile node is.

For the purpose of route optimization, the provided feature is sufficient. The problem when creating a binding cache entry is that if an attacker can create a binding cache entry with the home address of a victim mobile node and the care-of address of the attacker, all traffic to the victim node is routed to the attacker. The return routability procedure at least prevents this problem.

The messages exchanged between a mobile node and a correspondent node are protected by a hash function. The tokens used to generate a shared secret are exchanged by the Home Test and Care-of Test messages. That means anyone can generate the shared secret once he acquires these tokens. The Mobile IPv6 specification stipulates that the tunnel connection between a mobile node and a home agent used to send or receive the Home Test Init and the Home Test messages must be protected by the IPsec ESP mechanism. This is done by using the IPsec tunnel mode communication between them. As a result, an attacker cannot eavesdrop on the contents of the Home Test message that includes a Home Keygen Token value; however, the path between the home network of the mobile node and the correspondent node is not protected. If the attacker is on this path, the Home Keygen Token value can be examined.

To generate a shared secret, an attacker must get both a Home Keygen Token and a Care-of Keygen Token. One possible way to get both the tokens is to attach to the network between the home agent and the correspondent node of the victim mobile node. In this case, the attacker can eavesdrop on the Home Keygen token sent to the victim and can request a Care-of Keygen token by sending a fake Care-of Test Init message from the attacker's address. However, even if the attacker can get access to such a network, the situation is no worse than the normal IPv6 (not Mobile IPv6) communication. If the attacker can get access between two nodes, it can do more than just examine traffic, as with a man-in-the-middle attack.

5.7   Deregister Binding for Correspondent Nodes

After successful home deregistration as discussed in Section 4.5 of Chapter 4, a mobile node may perform the return routability procedure for all correspondent nodes for which it has binding update list entries. The return routability procedure from a home network is slightly different from the procedure done in a foreign network since the care-of address and the home address of a mobile node are the same. In this case, a mobile node and correspondent nodes only exchange a Home Test Init and a Home Test message, and a shared secret is generated only from a Home Keygen Token as described in Section 5.1. These messages are not tunneled to the home agent because the tunnel link has already been destroyed by the home deregistration procedure performed before this return routability procedure.

5.8   Backward Compatibility

When we consider deploying a new technology, we need to take care of the backward compatibility with legacy nodes. Mobile IPv6 will not be deployed if it cannot communicate with many old IPv6 nodes that do not understand it.

To ensure backward compatibility, the Mobile IPv6 specification defines a tunnel mechanism. A mobile node can send and receive packets using a tunnel between a mobile node and its home agent, as if the mobile node were at home. As long as a mobile node uses the tunnel, no backward compatibility issues occur.

However, a mobile node may initiate the return routability procedure to optimize the route between itself and a correspondent node. A mobile node cannot know beforehand if the peer node, with which the mobile node is currently communicating, supports Mobile IPv6. So, a mobile node may send a Home Test Init or a Care-of Test Init message, even if the peer node does not support Mobile IPv6. These messages use the Mobility Header, which is a new extension header introduced by the Mobile IPv6 specification. The old IPv6 nodes do not know of the extension header and cannot recognize the protocol number (in this case, 135). When a node receives an unrecognized protocol number, the node will generate an ICMPv6 Parameter Problem message with code 2 indicating that the incoming packet has an unrecognized next header value. The ICMPv6 message also indicates the position where an error occurred. In this case, the error messages point to the next header field of the header located before the Mobility Header. The generation of an ICMPv6 message for an unrecognized header is defined in the IPv6 base specification. We can assume all IPv6 nodes have this functionality.

If a mobile receives an ICMPv6 Parameter Problem message with code 2, and the error position indicates the protocol number of a Mobility Header, the mobile node stops performing the return routability procedure and uses only tunnel communication. Figure 5-4 shows the packet exchange.

FIGURE 5-4

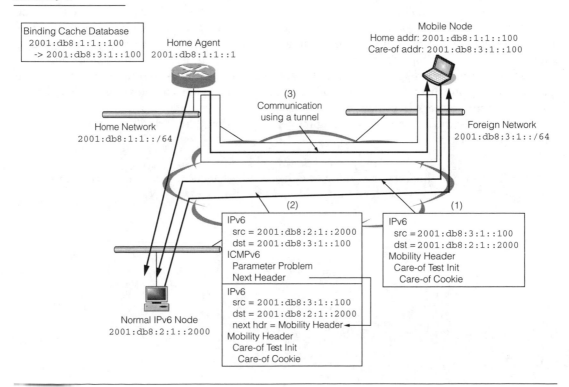

5.9 Movement Detection

When a mobile node attaches to a network, it must detect whether or not it has moved. There are several pieces of information that can be used to detect the movement of a node. The Mobile IPv6 specification talks about a basic movement detection method that uses Neighbor Unreachability Detection of a default router of a mobile node. As described in the Neighbor Discovery specification, an IPv6 node keeps a list of default routers on the attached network. If the routers become unreachable, it can be assumed that the node is attached to a different network.

When performing Neighbor Unreachability Detection for default routers, we need to take care of one thing. The Neighbor Unreachability Detection is done by sending a Neighbor Solicitation message to the target router. Usually, the address of the target router is a link-local address, since a normal Router Advertisement does not contain the global address of the router. A node usually does not know the global address of routers. However, a link-local address is unique only on a single link. This means that even if a mobile node moves from one network to another network, the mobile node may not be able to detect the unreachability of the default router if routers on the different links use the same link-local address. A mobile node needs to use other information as much as possible.

One of the other pieces of information that can be used for the unreachability detection is a global address from a Prefix Information option, which is extended by the Mobile IPv6 specification. If a Router Advertisement message contains the extension, a mobile node should perform Neighbor Unreachability Detection against the global address. Of course, this can be used only with routers that support Mobile IPv6 extension.

Another method is collecting all prefix information on a network. The prefix value is unique to each network. In this method, a mobile node keeps collecting prefix information. If prefix information which was advertised before can no longer be seen, the node may have moved to another network. The important thing is that the mobile node must not decide its movement by receiving only one advertisement message because there may be several routers which advertise different prefix information on the network. In that case, a single router advertisement does not show the entire network information.

There is no standard way of detecting movement of a mobile node. It is highly implementation dependent.

> The IETF DNA working group is trying to enhance the detection mechanism so that mobile nodes can detect their location or movement faster and more precisely.

Dynamic Home Agent Address Discovery

A mobile node may not know the address of its home agent when it wants to send a Binding Update message for home registration. For example, if a mobile node reboots on a foreign network, there is no information about the home agent unless such information is preconfigured.

The Dynamic Home Agent Address Discovery mechanism is used to get the address information of home agents when a mobile node is in a foreign network. A mobile node sends a Dynamic Home Agent Address Discovery request message when it needs to know the address of its home agent. The source address of the message is a care-of address of a mobile node and the destination address of the message is a home agent anycast address, which can be computed from the home prefix. This message does not contain a Home Address option, since this message may be sent before the first home registration is completed. A mobile node cannot use its home address before home registration is completed.

On the home network, home agents maintain the list of global addresses of all home agents on the home network by listening to each other's Router Advertisement messages. As described in Section 3.6 of Chapter 3, a home agent advertises its global address with a modified Prefix Information option. Figure 6-1 shows the concept.

Every home agent has a special anycast address, called a home agent anycast address, which is computed as described previously in Figure 3-24 of Chapter 3. A Dynamic Home Agent Address Discovery request message is delivered to one of the home agents in a home network thanks to the anycast address mechanism. The home agent which receives the message will reply to the mobile node with a Dynamic Home Agent Address Discovery reply message containing all of the home agent addresses which the home agent currently knows. The address list is ordered by the preference value of each home agent. If there are multiple home agents with the same preference value, the addresses should be ordered randomly every time for load

68

FIGURE 6-1

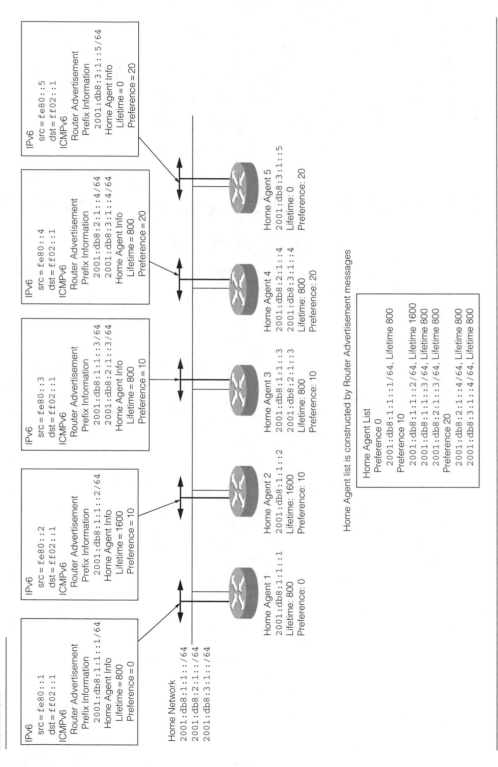

Home agent list generated in the home network.

FIGURE 6-2

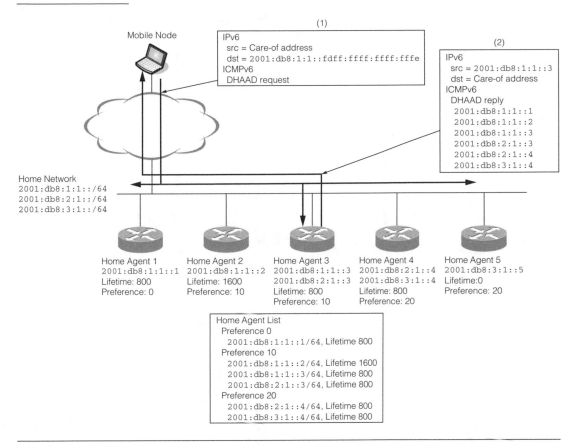

Replying to a Dynamic Home Agent Address Discovery Reply message.

balancing. To avoid packet fragmentation, the total length of the message must be smaller than the path MTU to the mobile node. If the list is too long to include in one packet, the home agents that have low preference values are excluded from the reply message. Figure 6-2 shows the procedure.

If a mobile node does not receive a reply message, the node will resend a request message. The initial timeout value for the retransmission is INITIAL_DHAAD_TIMEOUT seconds. The timeout value is increased exponentially at every retransmission. The maximum number of retransmissions is restricted to DHAAD_RETRIES times.

In theory, the Home Agent Address Discovery mechanism can be used as a mechanism to notify mobile nodes of available home agents on its home network. However, as we discuss in Chapter 8, adding/removing the home agent causes IPsec configuration problems. In the recent discussion at the IETF, the dynamic home agent assignment and security setup are considered part of other infrastructure-based mechanisms [RFC4640].

7

Mobile Prefix
Solicitation/Advertisement

An Internet Protocol version 6 (IPv6) address has a lifetime value. The lifetime is derived from the lifetime of the prefix. If the home address of a mobile node is going to expire, the mobile node sends a Mobile Prefix Solicitation message to get the latest information about home prefixes. The source address of the message is set to the care-of address of the mobile node. The destination of the message is the address of the home agent with which the mobile node is currently registered. The message must include a Home Address option, which contains the home address of the mobile node (i.e., a Mobile Prefix Solicitation message can be sent only after successful home registration.) Since the home registration procedure requires the information of a home network, this prefix discovery mechanism cannot be used to find the home prefixes when a mobile node is booting up on a foreign network, but can be only used to know new home prefixes or deprecated home prefixes.

When a home agent receives a Mobile Prefix Solicitation message from a mobile node, the node must reply to the mobile node with a Mobile Prefix Advertisement message. The source address of the message must be the destination address of the corresponding solicitation message. The destination address of the message must be the source address of the corresponding solicitation message, that is, the care-of address of a mobile node. A Type 2 Routing Header that contains the home address of a mobile node must exist. The list of modified Prefix Information options follows the advertisement message header.

Unlike the Router Advertisement messages, the list of Prefix Information options sent from the home agents on the same home network must be consistent. To make sure of the consistency, every home agent must be configured to have the same prefix information of its home network, or must listen to Router Advertisement messages from other home agents and construct a merged list of prefix information. A mobile node sends a solicitation message to the home agent with which the mobile node is currently registered. If the prefix information returned in response to the solicitation message differs for each home agent, the mobile node may incorrectly consider that some prefix information has disappeared.

A home agent may send a Mobile Prefix Advertisement message even if a mobile node does not request the prefix information in the following cases:

- The state of the flags of the home prefix that a mobile node is using changes.
- The valid or preferred lifetime of a home prefix is reconfigured.
- A new home prefix is added.
- The state of the flags or lifetime values of a home prefix that is not used by any mobile node changes.

When either of the first two conditions occur, a home agent must send an unsolicited Mobile Prefix Advertisement. When the third condition occurs, a home agent should send an unsolicited Mobile Prefix Advertisement message. When the last condition occurs, a home agent may send the message. A mobile node updates its prefix information and home addresses derived from updated prefixes when it receives this unsolicited Mobile Prefix Advertisement.

When sending an advertisement message, a home agent must follow the following scheduling algorithm to avoid network congestion:

- If a mobile node sends a solicitation message, a home agent sends an advertisement message immediately.
- Otherwise, a home agent schedules the next transmission time as follows:

$$MaxScheduleDelay = MIN(MaxMobPfxAdvInterval, preferred\ lifetime)$$
$$RandomDelay = MinMobPfxAdvInterval$$
$$+ (RANDOM()\ \%$$
$$ABS(MaxScheduleDelay - MinMobPfxAdvInterval)),$$

where

$MIN(a, b)$ returns the smaller of a or b;

$RANDOM()$ generates a random value from 0 to the maximum possible integer value;

$ABS(a)$ returns an absolute value of a.

The next advertisement will be sent after *RandomDelay* seconds.

When a mobile node receives an unsolicited Mobile Prefix Advertisement message, it must send a Mobile Prefix Solicitation message as an acknowledgment of that message. Otherwise, a home agent will resend the unsolicited advertisement message every `PREFIX_ADV_TIMEOUT` seconds. The maximum number of retransmissions is restricted to `PREFIX_ADV_RETIRES` times. The Mobile Prefix Solicitation and Advertisement message should be protected by the IPsec mechanism.

In theory, the Mobile Prefix Solicitation/Advertisement mechanism can be used as a mechanism to renumber the home network of mobile nodes, however, as discussed in Chapter 8, renumbering the home addresses has IPsec configuration problems. A mobile node and its home agent must negotiate which home address should be used, and the IPsec policy database on both nodes need to be updated because the database has home address information. The Mobile IPv6 specification does not specify any address transition procedure in its base specification.

Relationship with IPsec

Mobile Internet Protocol version 6 (IPv6) uses the IPsec mechanism to protect the Mobile IPv6 signaling messages. The specifications on how to protect messages are defined in [RFC3776].

The messages directly exchanged between a mobile node and a home agent are protected by the IPsec transport mode mechanism. The Binding Update and Binding Acknowledgment messages must be protected by the IPsec ESP or AH. The Mobile Prefix Solicitation and Advertisement messages should be protected by the IPsec mechanism.

The messages exchanged between a mobile node and a correspondent node, and relayed by the home agent, are protected by the IPsec tunnel mode mechanism. The Home Test Init and Home Test messages must be protected by the IPsec ESP header with the IPsec tunnel mode. As we show in this chapter, the tunnel mode policy entries must be able to support the Mobility Header type specific policy rule. More precisely, it must be able to send and receive the Home Test Init and Home Test messages only via the IPsec tunnel. This is necessary when two mobile nodes communicate with route optimization. If a mobile node cannot specify the Home Test Init/Home Test messages as policy specification, a Binding Update message to the other mobile node (this node is actually treated as a correspondent node) is incorrectly tunneled to the home agent of the mobile node that is sending the Binding Update message.

Note that the Dynamic Home Agent Address Discovery Request and Reply messages cannot be protected because the mobile node does not know the home agent address before exchanging these messages. The address information is required to set up the IPsec security policy database to protect messages.

Tables 8-1 and 8-2 summarize the policy entries required for a mobile node and a home agent, respectively.

The Security Associations for each policy can be configured by a manual operation. The IKE mechanism can be used to create these Security Associations dynamically, however, it requires a modification to the IKE program. Usually, the addresses of a Security Association IKE

TABLE 8-1

Mode	IPsec protocol	Target source	Target destination	Target protocol	Tunnel source	Tunnel destination
Transport	ESP (or AH)	Home address	Home agent	MH (Binding Update)	–	–
Transport	ESP (or AH)	Home agent	Home address	MH (Binding Acknowledgment)	–	–
Transport	ESP (or AH)	Home address	Home agent	ICMPv6 (Mobile Prefix Solicitation)	–	–
Transport	ESP (or AH)	Home agent	Home address	ICMPv6 (Mobile Prefix Advertisement)	–	–
Tunnel	ESP	Home address	Any	MH (Home Test Init)	Care-of address	Home agent
Tunnel	ESP	Any	Home address	MH (Home Test)	Home agent	Care-of address

Security policy entries required for a mobile node.

TABLE 8-2

Mode	IPsec protocol	Target source	Target destination	Target protocol	Tunnel source	Tunnel destination
Transport	ESP (or AH)	Home agent	Home address	MH (Binding Update)	–	–
Transport	ESP (or AH)	Home address	Home agent	MH (Binding Acknowledgment)	–	–
Transport	ESP (or AH)	Home agent	Home address	ICMPv6 (Mobile Prefix Solicitation)	–	–
Transport	ESP (or AH)	Home address	Home agent	ICMPv6 (Mobile Prefix Advertisement)	–	–
Tunnel	ESP	Any	Home address	MH (Home Test Init)	Home agent	Care-of address
Tunnel	ESP	Home address	Any	MH (Home Test)	Care-of address	Home agent

Security policy entries required for a home agent.

configures are derived from the addresses that are used to perform the IKE negotiation. In the Mobile IPv6 case, when a mobile node moves from its home network to a foreign network, the home address cannot be used until the home registration procedure has been completed. But, we need a Security Association between the home address and the home agent address to complete the home registration procedure. The IKE program must use a care-of address for IKE negotiation and create a Security Association for addresses that are not used in the IKE negotiation. Currently, few IKE implementations support this function.

There are other problems that are caused by the design of the IPsec policy configuration mechanism. The IPsec policy configuration is usually static; however, in the Mobile IPv6 operation we need to change policies in the following situations:

- When a new home agent is installed, a mobile node needs to install new transport and tunnel mode policy entries for the new home agent.

- When renumbering occurs, a mobile node and a home agent need to update their home prefix information in the policy database.

The use of IPsec with Mobile IPv6 has many unresolved issues. More research is required to achieve flexible operation of the combination of these technologies.

9

Code Introduction

In this chapter, we describe the detailed Mobile Internet Protocol version 6 (IPv6) code implemented as a part of the KAME distribution. The code used in this book is based on the snapshot code generated on July 12, 2004. The code discussed in this book refers to the IPv6 source and destination address fields to access address information, while the information is stored in the `ip6_aux{}` structure separately in older codes. We believe such inconsistency will not confuse readers in understanding the implementation of Mobile IPv6, since the framework of the KAME code itself has not been modified drastically.

9.1 Statistics

Statistics are stored in the `mip6stat{}` structure. Table 9-1 describes the statistics variables.

TABLE 9-1

`mip6sta{}` member	Description
`mip6s_mobility`	# of Mobility Header packets received
`mip6s_omobility`	# of Mobility Header packets sent
`mip6s_hoti`	# of Home Test Init packets received
`mip6s_ohoti`	# of Home Test Init packets sent
`mip6s_coti`	# of Care-of Test Init packets received

Continued

TABLE 9-1 (*Cont.*)

mip6sta{} member	*Description*
mip6s_ocoti	# of Care-of Test Init packets sent
mip6s_hot	# of Home Test packets received
mip6s_ohot	# of Home Test packets sent
mip6s_cot	# of Care-of Test packets received
mip6s_ocot	# of Care-of Test packets sent
mip6s_bu	# of Binding Update received
mip6s_obu	# of Binding Update sent
mip6s_ba	# of Binding Acknowledgment received
mip6s_ba_hist[0...255]	Histogram based on the status code of Binding Acknowledgment received
mip6s_oba	# of Binding Acknowledgment sent
mip6s_oba_hist[0...255]	Histogram based on the status code of Binding Acknowledgment sent
mip6s_br	# of Binding Refresh Request received
mip6s_obr	# of Binding Refresh Request sent
mip6s_be	# of Binding Error received
mip6s_be_hist [0...255]	Histogram based on the status code of Binding Error received
mip6s_obe	# of Binding Error sent
mip6s_obe_hist[0...255]	Histogram based on the status code of Binding Error sent
mip6s_hao	# of Home Address option received
mip6s_unverifiedhao	# of received Home Address options that do not have corresponding binding cache information
mip6s_ohao	# of Home Address option sent
mip6s_rthdr2	# of Type 2 Routing Header received
mip6s_orthdr2	# of Type 2 Routing Header sent
mip6s_revtunnel	# of packets that came from bidirectional tunnel
mip6s_orevtunnel	# of packets that are sent to bidirectional tunnel
mip6s_checksum	# of Mobility Header packets in which checksum value was incorrect
mip6s_payloadproto	# of Mobility Header packets in which payload protocol number is other than IPV6-NONXT
mip6s_unknowntype	# of Mobility Header packets in which type value is unknown
mip6s_nohif	# of packets in which destination address is not my home address
mip6s_nobue	# of packets that have no corresponding binding update information
mip6s_hinitcookie	# of Home Test packets in which cookie does not match the stored cookie

Continued

TABLE 9-1 (*Cont.*)

mip6sta{} member	*Description*
mip6s_cinitcookie	# of Care-of Test packets in which cookie does not match the stored cookie
mip6s_unprotected	# of Binding Update/Binding Acknowledgment packets that are not protected by IPsec
mip6s_haopolicy	# of Binding Update/Binding Acknowledgment packets in which HAO is not protected by IPsec or Authentication Data suboption
mip6s_rrauthfail	# of failure of the Return Routability procedure
mip6s_seqno	# of failure of sequence number mismatch
mip6s_paramprobhao	# of ICMPv6 Parameter Problem packets against HAO option
mip6s_paramprobmh	# of ICMPv6 Parameter Problem packets against Mobility Header packets
mip6s_invalidcoa	# of packets which care of address was not acceptable
mip6s_invalidopt	# of packets that contained invalid mobility options
mip6s_circularrefered	# of Binding Update packets that request binding a care-of address of one node with a home address of another node

Mobile IPv6 statistics.

Mobile IPv6-related Structures

In this chapter, we introduce all structures used by the Mobile Internet Protocol version 6 (IPv6) stack.

As discussed in Section 3.4, a new extension header, Mobility Header, is introduced in [RFC3775]. Mobility Header has a type field to specify different message types based on the function of each message. In [RFC3775], eight type values are defined. Each message format is described in Sections 10.2–10.10. Mobility Header may have option data as discussed in Section 3.5. The related structures are described in Sections 10.11–10.16.

The message format of the Home Address option and the Type 2 Routing Header (Sections 3.2 and 3.3), which are used for route optimized communication, are described in Sections 10.18 and 10.19.

[RFC3775] extends some Neighbor Discovery messages as discussed in Section 3.6. The extended Neighbor Discovery structures are described in Sections 10.20–10.23.

The Dynamic Home Agent Address Discovery messages discussed in Section 3.7 are used to discover the addresses of a mobile node's home agent. The message formats are described in Sections 10.24 and 10.25.

Finally, the Mobile Prefix Solicitation/Advertisement message formats used to distribute home prefix information from a home agent to a mobile node are described in Sections 10.26 and 10.27.

This chapter also describes some internal structures that are not related to any messages defined in [RFC3775]. These structures are used for the kernel internal use.

10.1 Files

Table 10-1 shows the files that define Mobile IPv6-related structures.

TABLE 10-1

File	Description
`${KAME}/kame/sys/net/if_hif.h`	Home virtual interface structures
`${KAME}/kame/sys/netinet/icmp6.h`	Dynamic Home Agent Address Discovery and Mobile Prefix Solicitation/Advertisement structures
`${KAME}/kame/sys/netinet/ip6.h`	Home Address option structure
`${KAME}/kame/sys/netinet/ip6mh.h`	Mobility Header structures
`${KAME}/kame/sys/netinet6/mip6_var.h`	All structures that are used in the Mobile IPv6 stack
`${KAME}/kame/kame/had/halist.h`	Home agent information structure used by the home agent side

Files that define Mobile IPv6-related structures.

10.2 Mobility Header Message: `ip6_mh{}` Structure

The `ip6_mh{}` structure defined in `ip6mh.h` represents the Mobility Header (Section 3.4) described in Listing 10-1. The structure definitions of protocol headers and options are documented in [RFC4584].

Listing 10-1

```
                                                                        ip6mh.h
36      struct ip6_mh {
37              u_int8_t  ip6mh_proto;    /* following payload protocol (for PG) */
38              u_int8_t  ip6mh_len;      /* length in units of 8 octets */
39              u_int8_t  ip6mh_type;     /* message type */
40              u_int8_t  ip6mh_reserved;
41              u_int16_t ip6mh_cksum;    /* sum of IPv6 pseudo-header and MH */
42              /* followed by type specific data */
43      } __attribute__((__packed__));
                                                                        ip6mh.h
```

36–43 The `ip6_mh{}` structure is the base structure for all the Mobility Headers. `ip6mh_proto` is a protocol number of an upper layer protocol that follows the Mobility Header. At this moment, the RFC specifies that there should not be any upper layer protocol headers after Mobility Headers. This field should be always set to `IPV6-NONXT` (decimal 58), which means there are no following headers. This field can be considered a reserved field for future use in piggybacking the mobility signals on upper layer packets. `ip6mh_len` is the length of a Mobility Header in units of 8 bytes, not including the first 8 bytes. `ip6mh_type` indicates the message type. Currently, eight message types are defined. Table 10-2 shows the current message types. `ip6mh_reserved` is a reserved field for future use. This field should be zero-cleared when sending and must be ignored when receiving. `ip6mh_cksum` keeps the checksum value of an MH message. The computation procedure is the same as the one for the ICMPv6 message.

TABLE 10-2

Name	Value	Description
IP6_MH_TYPE_BRR	0	Binding Refresh Request message: sent from a correspondent node to a mobile node when it wants to extend its binding lifetime
IP6_MH_TYPE_HOTI	1	Home Test Init message: sent from a mobile node to a correspondent node when it initiates the return routability procedure to confirm the home address ownership and reachability
IP6_MH_TYPE_COTI	2	Care-of Test Init message: sent from a mobile node to a correspondent node when it initiates the return routability procedure to confirm the care-of address ownership and reachability
IP6_MH_TYPE_HOT	3	Home Test message: sent from a correspondent node to a mobile node in response to a Home Test Init message
IP6_MH_TYPE_COT	4	Care-of Test message: sent from a correspondent node to a mobile node in response to a Care-of Test Init message
IP6_MH_TYPE_BU	5	Binding Update message: sent from a mobile node to a home agent or a correspondent node to inform binding information between the home and care-of address of the mobile node
IP6_MH_TYPE_BACK	6	Binding Acknowledgment message: sent from a home agent or a correspondent node in response to a Binding Update message. Note that the KAME Mobile IPv6 correspondent node docs not send this message except in error cases
IP6_MH_TYPE_BERROR	7	Binding Error message: sent when an error occurs while a node is processing Mobile IPv6 messages
IP6_MH_TYPE_MAX	7	(maximum type value)

MH-type numbers.

10.3 Binding Refresh Request Message: `ip6_mh_binding_request{}` Structure

The `ip6_mh_binding_request{}` structure represents the Binding Refresh Request message described in Figure 3-4 of Section 3.4. Listing 10-2 shows the definition of the `ip6_mh_binding_request{}` structure.

Listing 10-2

─── ip6mh.h

```
61      struct ip6_mh_binding_request {
62              struct ip6_mh ip6mhbr_hdr;
63              u_int16_t     ip6mhbr_reserved;
64              /* followed by mobility options */
65      } __attribute__((__packed__));
66      #ifdef _KERNEL
67      #define ip6mhbr_proto ip6mhbr_hdr.ip6mh_proto
68      #define ip6mhbr_len ip6mhbr_hdr.ip6mh_len
69      #define ip6mhbr_type ip6mhbr_hdr.ip6mh_type
```

```
70        #define ip6mhbr_reserved0 ip6mhbr_hdr.ip6mh_reserved
71        #define ip6mhbr_cksum ip6mhbr_hdr.ip6mh_cksum
72        #endif /* _KERNEL */
```
── ip6mh.h

61–72 The ip6mhbr_hdr field is common for all Mobility Header messages. To make it easy to access the member fields in ip6mhbr_hdr, some macros are defined in lines 67–71. Note that these shortcuts, bracketed by the _KERNEL macro, can be used only from the inside kernel. Any application program that uses the header fields must use the ip6mhbr_hdr member field for compatibility. In the Binding Refresh Request message, the type number, ip6mhbr_type, is set to IP6_MH_TYPE_BRR. The ip6mhbr_reserved field is reserved for future use. A sender must clear this field when sending a message and a receiver must ignore this field.

10.4   Home Test Init Message: `ip6_mh_home_test_init{}` Structure

The ip6_mh_home_test_init{} structure represents the Home Test Init message described in Figure 3-5 of Section 3.4. Listing 10-3 shows the definition of the ip6_mh_home_test_init{} structure.

Listing 10-3
── ip6mh.h

```
75        struct ip6_mh_home_test_init {
76                struct ip6_mh ip6mhhti_hdr;
77                u_int16_t      ip6mhhti_reserved;
78                union {
79                        u_int8_t  __cookie8[8];
80                        u_int32_t __cookie32[2];
81                } __ip6mhhti_cookie;
82                /* followed by mobility options */
83        } __attribute__((__packed__));
84        #ifdef _KERNEL
85        #define ip6mhhti_proto ip6mhhti_hdr.ip6mh_proto
86        #define ip6mhhti_len ip6mhhti_hdr.ip6mh_len
87        #define ip6mhhti_type ip6mhhti_hdr.ip6mh_type
88        #define ip6mhhti_reserved0 ip6mhhti_hdr.ip6mh_reserved
89        #define ip6mhhti_cksum ip6mhhti_hdr.ip6mh_cksum
90        #define ip6mhhti_cookie8 __ip6mhhti_cookie.__cookie8
91        #endif /* _KERNEL */
92        #define ip6mhhti_cookie __ip6mhhti_cookie.__cookie32
```
── ip6mh.h

75–92 The ip6mhhti_hdr field is common to all Mobility Header messages. There are five macro definitions for shortcut access to the common header part in lines 85–89. Also, there are two other macros to access the cookie value. The application programming interface (API) specification only defines ip6mhhti_cookie, which is used to access the values in a unit of 32 bits. There are some cases where it is more convenient if we can access the value in an 8-bit unit. The ip6mhhti_cookie8 macro is provided for this purpose. ip6mhhti_type is set to IP6_MH_TYPE_HOTI. ip6mhhti_cookie is an 8-byte cookie value that is generated in a mobile node to bind a Home Test Init message and a Home Test message. Two reserved fields (ip6mhhti_reserved0 and ip6mhti_reserved) must be cleared by the sender and must be ignored by the receiver.

10.5 Care-of Test Init Message: `ip6_mh_careof_test_init{}` Structure

The `ip6_mh_careof_test_init{}` structure represents the Care-of Test Init message described in Figure 3-6 of Section 3.4. Listing 10-4 shows the definition of the `ip6_mh_careof_test_init{}` structure.

Listing 10-4

—— ip6mh.h
```
95      struct ip6_mh_careof_test_init {
96              struct ip6_mh ip6mhcti_hdr;
97              u_int16_t     ip6mhcti_reserved;
98              union {
99                      u_int8_t  __cookie8[8];
100                     u_int32_t __cookie32[2];
101             } __ip6mhcti_cookie;
102             /* followed by mobility options */
103     } __attribute__((__packed__));
104     #ifdef _KERNEL
105     #define ip6mhcti_proto ip6mhcti_hdr.ip6mh_proto
106     #define ip6mhcti_len ip6mhcti_hdr.ip6mh_len
107     #define ip6mhcti_type ip6mhcti_hdr.ip6mh_type
108     #define ip6mhcti_reserved0 ip6mhcti_hdr.ip6mh_reserved
109     #define ip6mhcti_cksum ip6mhcti_hdr.ip6mh_cksum
110     #define ip6mhcti_cookie8 __ip6mhcti_cookie.__cookie8
111     #endif /* _KERNEL */
112     #define ip6mhcti_cookie __ip6mhcti_cookie.__cookie32
```
—— ip6mh.h

95–112 The `ip6mhcti_hdr` field is common to all Mobility Header messages. The structure is almost the same as the `ip6_mh_home_test_init{}` structure. The only difference is the name of its member fields. In the Home Test Init message, `ip6mhhti_` is used as a prefix for each member field while `ip6mhcti_` is used in the Care-of Test Init message.

10.6 Home Test Message: `ip6_mh_home_test{}` Structure

The `ip6_mh_home_test{}` structure represents the Home Test message described in Figure 3-7 of Section 3.4. Listing 10-5 shows the definition of the `ip6_mh_home_test{}` structure.

Listing 10-5

—— ip6mh.h
```
115     struct ip6_mh_home_test {
116             struct ip6_mh ip6mhht_hdr;
117             u_int16_t     ip6mhht_nonce_index; /* idx of the CN nonce list array */
118             union {
119                     u_int8_t  __cookie8[8];
120                     u_int32_t __cookie32[2];
121             } __ip6mhht_cookie;
122             union {
123                     u_int8_t  __keygen8[8];
124                     u_int32_t __keygen32[2];
125             } __ip6mhht_keygen;
126             /* followed by mobility options */
127     } __attribute__((__packed__));
128     #ifdef _KERNEL
129     #define ip6mhht_proto ip6mhht_hdr.ip6mh_proto
```

```
130       #define ip6mhht_len ip6mhht_hdr.ip6mh_len
131       #define ip6mhht_type ip6mhht_hdr.ip6mh_type
132       #define ip6mhht_reserved0 ip6mhht_hdr.ip6mh_reserved
133       #define ip6mhht_cksum ip6mhht_hdr.ip6mh_cksum
134       #define ip6mhht_cookie8 __ip6mhht_cookie.__cookie8
135       #define ip6mhht_keygen8 __ip6mhht_keygen.__keygen8
136       #endif /* _KERNEL */
137       #define ip6mhht_cookie __ip6mhht_cookie.__cookie32
138       #define ip6mhht_keygen __ip6mhht_keygen.__keygen32
```
—— ip6mh.h

115–138 The `ip6mhht_hdr` field is common for all Mobility Header messages. The `ip6_mh_home_test{}` structure also has several shortcuts, which can be used only in the kernel, to access the member fields in the common part of the MH message. In the Home Test message, the type number (`ip6mhht_type`) is set to `IP6_MH_TYPE_HOT`, the `ip6mhht_cookie` field is used to store the cookie value that is sent from a mobile node in the Home Test Init message. The cookie value is copied from the Home Test Init message to the Home Test message. A mobile node uses the cookie value to determine if the received Home Test message was sent in response to a Home Test Init message, which the mobile node sent previously. The `ip6mhht_keygen` field contains the keygen token value that is computed inside a correspondent node using secret information from the correspondent node. The `ip6mhht_cookie8` and `ip6mhht_keygen8` fields are provided as methods to access the cookie and token values as byte arrays.

10.7   Care-of Test Message: `ip6_mh_careof_test{}` Structure

The `ip6_mh_careof_test{}` structure represents the Care-of Test message described in Figure 3-8 of Section 3.4. Listing 10-6 shows the definition of the `ip6_careof_test{}` structure.

Listing 10-6
—— ip6mh.h

```
141       struct ip6_mh_careof_test {
142               struct ip6_mh ip6mhct_hdr;
143               u_int16_t    ip6mhct_nonce_index; /* idx of the CN nonce list array */
144               union {
145                       u_int8_t   __cookie8[8];
146                       u_int32_t  __cookie32[2];
147               } __ip6mhct_cookie;
148               union {
149                       u_int8_t   __keygen8[8];
150                       u_int32_t  __keygen32[2];
151               } __ip6mhct_keygen;
152               /* followed by mobility options */
153       } __attribute__((__packed__));
154       #ifdef _KERNEL
155       #define ip6mhct_proto ip6mhct_hdr.ip6mh_proto
156       #define ip6mhct_len ip6mhct_hdr.ip6mh_len
157       #define ip6mhct_type ip6mhct_hdr.ip6mh_type
158       #define ip6mhct_reserved0 ip6mhct_hdr.ip6mh_reserved
159       #define ip6mhct_cksum ip6mhct_hdr.ip6mh_cksum
160       #define ip6mhct_cookie8 __ip6mhct_cookie.__cookie8
161       #define ip6mhct_keygen8 __ip6mhct_keygen.__keygen8
162       #endif /* _KERNEL */
163       #define ip6mhct_cookie __ip6mhct_cookie.__cookie32
164       #define ip6mhct_keygen __ip6mhct_keygen.__keygen32
```
—— ip6mh.h

141–164 The `ip6mhct_hdr` field is common for all Mobility Header messages. This structure is almost identical to the `ip6_mh_home_test{}` structure. The difference is the name of each member field. In the `ip6_mh_careof_test{}` structure, each member field has `ip6mhct_` as a prefix while the `ip6_mh_home_test{}` structure uses `ip6mhht_`. The `ip6mhct_cookie` and `ip6mhct_keygen` fields are equivalent to the `ip6mhht_cookie` and `ip6mht_keygen` fields. The `ip6mhct_cookie` field stores the cookie value sent from a mobile node in the Care-of Test Init message. The `ip6mhct_keygen` field stores the token value that is used to compute the shared secret between the correspondent node and the mobile node. The `ip6mhct_cookie8` and `ip6mhct_keygen8` fields point to the same contents as do the `ip6mhct_cookie` and `ip6mhct_keygen` fields, but they allow access to the fields as byte streams.

10.8 Binding Update Message: `ip6_mh_binding_update{}` Structure

The `ip6_mh_binding_update{}` structure represents the Binding Update message described in Figure 3-9 of Section 3.4. Listing 10-7 shows the definition of the `ip6_mh_binding_update{}` structure.

Listing 10-7

--- ip6mh.h
```
167     struct ip6_mh_binding_update {
168             struct ip6_mh ip6mhbu_hdr;
169             u_int16_t      ip6mhbu_seqno;    /* sequence number */
170             u_int16_t      ip6mhbu_flags;    /* IP6MU_* flags */
171             u_int16_t      ip6mhbu_lifetime; /* in units of 4 seconds */
172             /* followed by mobility options */
173     } __attribute__((__packed__));
174     #ifdef _KERNEL
175     #define ip6mhbu_proto ip6mhbu_hdr.ip6mh_proto
176     #define ip6mhbu_len ip6mhbu_hdr.ip6mh_len
177     #define ip6mhbu_type ip6mhbu_hdr.ip6mh_type
178     #define ip6mhbu_reserved0 ip6mhbu_hdr.ip6mh_reserved
179     #define ip6mhbu_cksum ip6mhbu_hdr.ip6mh_cksum
180     #endif /* _KERNEL */
```
--- ip6mh.h

167–180 The `ip6mhbu_hdr` field is common for all Mobility Header messages. As with the other messages, there are macro definitions to access inside members of the `ip6mhbu_hdr` field. The `ip6mhbu_type` field is set to `IP6_MH_TYPE_BU`. The `ip6mhbu_seqno` field keeps a sequence number of the binding information, which is used to prevent a replay attack from malicious nodes. A mobile node must increase the sequence number when sending a Binding Update message. The home agent of the mobile node will discard any messages that have an old sequence number. (See Section 4.2 for the procedure and Listing 14.8 for its implementation.)

The `ip6mhbu_flags` field keeps the flag values of the message. Currently, there are four flags in the specification. Table 3-2 (page 34) describes the flags and their meanings. Table 10-3 shows the list of macros for the flags. `IP6MU_CLONED` macro is a special macro that is used to represent the internal state of binding information. `IP6MU_CLONED` flag is set when a home agent receives a Binding Update message with the L flag set. The home agent will create two binding entries when it accepts the message. One is to bind the home address and the care-of address of the mobile node. The other is to bind the

TABLE 10-3

Name	Description
IP6MU_ACK	Acknowledgment (A flag)
IP6MU_HOME	Home Registration (H flag)
IP6MU_LINK	Link-Local Address Compatibility (L flag)
IP6MU_KEY	Key Management Mobility Capability (K flag)
IP6MU_CLONED	(internal use only) means the binding information is generated by the L flag processing

Macro definitions for flags of Binding Update message.

link-local address (which is generated automatically using the interface identifier part of the home address) and the care-of address of the mobile node to protect the link-local address of the mobile node. The latter binding cache entry will have IP6MU_CLONED flag set to mark that the entry is created as a side effect of the L flag. This flag is an implementation-dependent flag and should never be sent to the wire with any Mobility messages.

The ip6mhbu_lifetime field indicates the proposed lifetime of the binding information. The value is units of 4 s to allow for a mobile node to specify a longer time with smaller data. A mobile node sets the value based on the remaining lifetimes of its home and care-of addresses. Note that the value is just a proposal from the mobile node to the node receiving the Binding Update message. The actual approved lifetime may be reduced from the proposed lifetime based on the local policy of the recipient node.

10.9   Binding Acknowledgment Message: ip6_mh_binding_ack{} Structure

The ip6_mh_binding_ack{} structure represents the Binding Acknowledgment message described in Figure 3-10 of Section 3.4. Listing 10-8 shows the definition of the ip6_mh_binding_ack{} structure.

Listing 10-8

—— ip6mh.h

```
199     struct ip6_mh_binding_ack {
200             struct ip6_mh ip6mhba_hdr;
201             u_int8_t        ip6mhba_status;   /* status code */
202             u_int8_t        ip6mhba_flags;
203             u_int16_t       ip6mhba_seqno;    /* sequence number */
204             u_int16_t       ip6mhba_lifetime; /* in units of 4 seconds */
205             /* followed by mobility options */
206     } __attribute__((__packed__));
207     #ifdef _KERNEL
208     #define ip6mhba_proto ip6mhba_hdr.ip6mh_proto
209     #define ip6mhba_len ip6mhba_hdr.ip6mh_len
210     #define ip6mhba_type ip6mhba_hdr.ip6mh_type
211     #define ip6mhba_reserved0 ip6mhba_hdr.ip6mh_reserved
212     #define ip6mhba_cksum ip6mhba_hdr.ip6mh_cksum
213     #endif /* _KERNEL */
```

—— ip6mh.h

199–213 The `ip6mhba_hdr` is common for all Mobility Header messages. Similar to other messages, some macros are defined to make it easy to access the members of the `ip6mhba_hdr` field. The `ip6mhba_type` field is set to `IP6_MH_TYPE_BACK`, the `ip6mhba_status` field indicates the result of processing a Binding Update message. Table 10-4 shows all the status codes defined in [RFC3775] which can be contained in the `ip6mhba_status` field. The `ip6mhba_flags` field is the flags. Currently, only the `IP6_MH_BA_KEYM` flag is defined as described in Table 10-5. `ip6mhba_seqno` is the latest sequence number for binding information stored in a correspondent node or a home agent. `ip6mhba_lifetime` indicates the approved lifetime for the binding information. A mobile node requests the lifetime of binding information by specifying field

TABLE 10-4

Name	Description
IP6_MH_BAS_ACCEPTED	Binding Update is accepted
IP6_MH_BAS_PRFX_DISCOV	Binding Update is accepted, but need to perform the prefix discovery procedure
IP6_MH_BAS_ERRORBASE	(internal use only) The base value that indicated error status
IP6_MH_BAS_UNSPECIFIED	Binding Update is rejected
IP6_MH_BAS_PROHIBIT	Administratively prohibited
IP6_MH_BAS_INSUFFICIENT	Insufficient resources
IP6_MH_BAS_HA_NOT_SUPPORTED	Home registration function is not provided
IP6_MH_BAS_NOT_HOME_SUBNET	Binding Update is received on another network interface, which is not a home subnet
IP6_MH_BAS_NOT_HA	Binding Update was sent to a wrong home agent
IP6_MH_BAS_DAD_FAILED	Duplicate Address Detection for home address of a mobile node failed
IP6_MH_BAS_SEQNO_BAD	The sequence number specified in a Binding Update message is smaller than the number stored in the binding information on a correspondent node or a home agent
IP6_MH_BAS_HOME_NI_EXPIRED	Home Nonce Index is already expired
IP6_MH_BAS_COA_NI_EXPIRED	Care-of Nonce Index is already expired
IP6_MH_BAS_NI_EXPIRED	Both Home/Care-of Nonce Index are expired
IP6_MH_BAS_REG_NOT_ALLOWED	A mobile node tried to change its registration type (home registration/not home registration)

Macro definitions for the Binding Acknowledgment status field.

TABLE 10-5

Name	Description
IP6_MH_BA_KEYM	Key Management Mobility Capability (K flag)

Macro definitions for the Binding Acknowledgment flag field.

`ip6mhbu_lifetime` in a Binding Update message; however, the requested value is not always appropriate. A node that receives a Binding Update decides the proper lifetime and sets the value in the `ip6mhba_lifetime` field. The value is units of 4 s (that is, the value 100 means 400 s).

10.10   Binding Error Message: `ip6_mh_binding_error{}` Structure

The `ip6_mh_binding_error{}` structure represents a Binding Error message described in Figure 3-11 of Section 3.4. Listing 10-9 shows the definition of the `ip6_mh_binding_error{}` structure.

Listing 10-9

─── ip6mh.h
```
236     struct ip6_mh_binding_error {
237             struct ip6_mh   ip6mhbe_hdr;
238             u_int8_t        ip6mhbe_status;            /* status code */
239             u_int8_t        ip6mhbe_reserved;
240             struct in6_addr ip6mhbe_homeaddr;
241             /* followed by mobility options */
242     } __attribute__((__packed__));
243     #ifdef _KERNEL
244     #define ip6mhbe_proto ip6mhbe_hdr.ip6mh_proto
245     #define ip6mhbe_len ip6mhbe_hdr.ip6mh_len
246     #define ip6mhbe_type ip6mhbe_hdr.ip6mh_type
247     #define ip6mhbe_reserved0 ip6mhbe_hdr.ip6mh_reserved
248     #define ip6mhbe_cksum ip6mhbe_hdr.ip6mh_cksum
249     #endif /* _KERNEL */
```
─── ip6mh.h

236–249 The `ip6mhbe_hdr` field is common for all MH messages. As with the other messages, some macros are defined to make it easy to access the members of field `ip6mhbe_hdr`. The `ip6mhbe_type` field is set to `IP6_MH_TYPE_BERROR`, the `ip6mhbe_status` field indicates the reason for the error. Table 10-6 shows the list of macro names for the status codes currently defined. The `ip6mhbe_homeaddr` field contains the home address of the mobile node that caused this error, if the home address is known, otherwise this field is set to the unspecified address.

10.11   Mobility Option Message Structures

Mobility Options carry additional information in addition to the base Mobility Header messages. Currently, six options are defined. Table 10-7 shows the macro names for these option types. Each option has already been explained in Section 3.5.

TABLE 10-6

Name	Description
IP6_MH_BES_UNKNOWN_HAO	The home address that was included in the received packet was not valid
IP6_MH_BES_UNKNOWN_MH	The type number of the received MH message was unknown

Macro definitions for the Binding Error status field.

TABLE 10-7

Name	Description
IP6_MHOPT_PAD1	Pad1, the padding option to fill one byte space
IP6_MHOPT_PADN	PadN, the padding option to fill from 2 bytes to 253 bytes space
IP6_MHOPT_BREFRESH	Binding Refresh Advice, the option contains the suggested interval to resend a Binding Update message to update the binding
IP6_MHOPT_ALTCOA	Alternate Care-of Address, the option contains the care-of address that should be used as the care-of address instead of the address specified in the source address field of the Binding Update message
IP6_MHOPT_NONCEID	Nonce Indices, the option contains indices of Home Nonce and Care-of Nonce to specify nonce values that are used to authenticate Binding Update and Binding Acknowledgment messages
IP6_MHOPT_BAUTH	Binding Authorization Data, the option contains the computed hash value of Binding Update and Binding Acknowledgment message

Mobility options.

10.12 Mobility Option Message: `ip6_mh_opt{}` Structure

The `ip6_mh_opt{}` structure is a generic structure for all mobility options and is shown in Listing 10-10.

Listing 10-10

—— ip6mh.h

```
256     struct ip6_mh_opt {
257             u_int8_t ip6mhopt_type;
258             u_int8_t ip6mhopt_len;
259             /* followed by option data */
260     } __attribute__((__packed__));
```

—— ip6mh.h

256–260 The `ip6mhopt_type` field specifies the type number of the option. The `ip6mhopt_len` field indicates the length of the option excluding the first two bytes (`ip6mhopt_type` and `ip6mhopt_len`). This structure is also used for the PadN mobility option. When used as a PadN option, the `ip6mhopt_type` field is set to `IP6_MHOPT_PADN` and the `ip6mhopt_len` field is set to the length of the required padding minus 2 bytes.

10.13 Binding Refresh Advice Option: `ip6_mh_opt_refresh_advice{}` Structure

The `ip6_mh_opt_refresh_advice{}` structure represents the Binding Refresh Advice option and is shown in Listing 10-11.

Listing 10-11

—— ip6mh.h

```
271     struct ip6_mh_opt_refresh_advice {
272             u_int8_t ip6mora_type;
```

```
273            u_int8_t ip6mora_len;
274            u_int8_t ip6mora_interval[2];    /* Refresh Interval (units of 4 sec)
275      } __attribute__((__packed__));
```
—— ip6mh.h

271–275 The ip6mora_type field is set to IP6_MHOPT_BREFRESH. The ip6mora_len field is set to 2. The ip6mora_interval field indicates the suggested interval value to use when resending a Binding Update message to update the binding information. The interval value is specified in units of 4 s.

> In the recent specification, the ip6mora_interval field is defined as u_int16_t.

10.14 Alternate Care-of Address Option: ip6_mh_opt_altcoa{} Structure

The ip6_mh_opt_altcoa{} structure represents the Alternate Care-of Address option and is shown in Listing 10-12.

Listing 10-12
—— ip6mh.h
```
278      struct ip6_mh_opt_altcoa {
279            u_int8_t ip6moa_type;
280            u_int8_t ip6moa_len;
281            u_int8_t ip6moa_addr[16];          /* Alternate Care-of Address */
282      } __attribute__((__packed__));
```
—— ip6mh.h

278–282 The ip6moa_type field is set to IP6_MHOPT_ALTCOA. The ip6moa_len field is set to 16. The ip6moa_addr field contains an IPv6 address, which should be used as a care-of address. Usually, the address specified in the IPv6 source address field of a Binding Update message is used as a care-of address in the mobility processing. This option specifies an alternate care-of address to be used as a care-of address. For example, when a mobile node has two network interfaces, interface A and interface B, and the node needs to register the address of interface B as a care-of address, using interface A to send a Binding Update message, the mobile node sets interface A's address to the IPv6 source address field and specifies interface B's address as an Alternate Care-of Address option.

10.15 Nonce Index Option: ip6_mh_opt_nonce_index{} Structure

The ip6_mh_opt_nonce_index{} structure represents the Nonce Indices option and is shown in Listing 10-13.

Listing 10-13
—— ip6mh.h
```
285      struct ip6_mh_opt_nonce_index {
286            u_int8_t ip6moni_type;
287            u_int8_t ip6moni_len;
288            union {
289                    u_int8_t __nonce8[2];
290                    u_int16_t __nonce16;
291            } __ip6moni_home_nonce;
292            union {
```

```
293                      u_int8_t __nonce8[2];
294                      u_int16_t __nonce16;
295                } __ip6moni_coa_nonce;
296          } __attribute__((__packed__));
297    #ifdef _KERNEL
298    #define ip6moni_home_nonce8 __ip6moni_home_nonce.__nonce8
299    #define ip6moni_coa_nonce8 __ip6moni_coa_nonce.__nonce8
300    #endif /* _KERNEL */
301    #define ip6moni_home_nonce __ip6moni_home_nonce.__nonce16
302    #define ip6moni_coa_nonce __ip6moni_coa_nonce.__nonce16
```
—— ip6mh.h

285–302 This option is used with the Binding Authorization Data option. The `ip6moni_type` field is set to `IP6_MH_OPT_NONCE_ID` and the `ip6moni_len` field is set to 4. The `ip6moni_home_nonce` field contains the index of the home nonce value of the nonce array maintained by the correspondent node to which a mobile node is sending this option. `ip6moni_coa_nonce` contains the index to the care-of nonce value of the nonce array that is kept on a correspondent node. The index values for each nonce value have been passed to a mobile node by the Home Test and Care-of Test messages. The nonce values are random numbers periodically generated on a correspondent node to associate the Home Test/Care-of Test messages and the Binding Update message generated based on those messages. The `ip6moni_home_nonce8` and `ip6moni_coa_nonce8` fields provide a way to access those values as a byte array. These shortcuts can be used only inside the kernel.

10.16 Authentication Data Option: `ip6_mh_opt_auth_data{}` Structure

The `ip6_mh_opt_auth_data{}` structure represents a Binding Authorization Data option that is used with the Nonce Indices option. Listing 10-14 shows the definition of the `ip6_mh_opt_auth_data{}` structure.

Listing 10-14
—— ip6mh.h

```
304    /* Binding Authorization Data */
305    struct ip6_mh_opt_auth_data {
306          u_int8_t ip6moad_type;
307          u_int8_t ip6moad_len;
308          /* followed by authenticator data */
309    } __attribute__((__packed__));
```
—— ip6mh.h

304–309 The `ip6moad_type` field is set to `IP6_MHOPT_BAUTH`, and the `ip6moad_len` field is set to 12 because the size of authenticator data computed by the return routability procedure (Section 14.6) is 12 bytes.

10.17 The Internal Mobility Option: `mip6_mobility_options{}` Structure

As described in Section 3.5, four mobility options, except padding options, are defined in the Mobile IPv6 specification. The `mip6_mobility_options{}` structure is an internal structure that is used when parsing the mobility options included in a Mobility Header message and can be seen in Listing 10-15.

Listing 10-15

———————————————————————————————————mip6_var.h
```
310     struct mip6_mobility_options {
311             u_int16_t valid_options;           /* shows valid options in this
   structure */
312             struct in6_addr mopt_altcoa;              /* Alternate CoA */
313             u_int16_t       mopt_ho_nonce_idx;        /* Home Nonce Index */
314             u_int16_t       mopt_co_nonce_idx;        /* Care-of Nonce Index */
315             caddr_t mopt_auth;                        /* Authenticator */
316             u_int16_t       mopt_refresh;             /*  Refresh Interval */
317     };
318
319     #define MOPT_ALTCOA     0x0001
320     #define MOPT_NONCE_IDX  0x0002
321     #define MOPT_AUTHDATA   0x0004
322     #define MOPT_REFRESH    0x0008
```
———————————————————————————————————mip6_var.h

Note: Line 311 is broken here for layout reasons. However, it is a single line of code.

311–316 The valid_options field is a bit field, which indicates what kind of options are
contained in the structure. The macros defined in lines 319–322 are used to specify which
option value is included. If multiple options exist in one Mobility Header, the logical *OR*
of each value is stored in the valid_options field. The mopt_altcoa field contains
the address value from the Alternate Care-of Address option. The mopt_ho_nonce_idx
and mopt_co_nonce_idx fields store the home and care-of nonce index, which are
contained in the Nonce Index option. The mopt_auth field points to the address of the
Binding Authorization Data option if the option exists. The mopt_refresh field contains
the value specified in the Binding Refresh Advice option.

10.18   Home Address Option: ip6_opt_home_address{} Structure

The ip6_opt_home_address{} structure represents the Home Address option described in
Section 3.2 and is shown in Listing 10-16.

Listing 10-16

———————————————————————————————————ip6.h
```
232     struct ip6_opt_home_address {
233             u_int8_t ip6oh_type;
234             u_int8_t ip6oh_len;
235             u_int8_t ip6oh_addr[16];/* Home Address */
236             /* followed by sub-options */
237     } __attribute__((__packed__));
```
———————————————————————————————————ip6.h

232–237 The ip6oh_type field is an option type, in which 0xC9 is used to indicate the Home
Address option. The ip6oh_len field is the length of the value of this option and its
value is set to 16. The ip6oh_addr field contains the home address of a mobile node.
At this moment, the Home Address option does not have any options.

10.19   Type 2 Routing Header: ip6_rthdr2{} Structure

The ip6_rthdr2{} structure represents the Type 2 Routing Header described in Section 3.3.
Listing 10-17 shows the definition of the ip6_rthdr2{} structure.

Listing 10-17

———————————————————————————————————— ip6.h
```
259    struct ip6_rthdr2 {
260            u_int8_t  ip6r2_nxt;           /* next header */
261            u_int8_t  ip6r2_len;           /* always 2 */
262            u_int8_t  ip6r2_type;          /* always 2 */
263            u_int8_t  ip6r2_segleft;       /* 0 or 1 */
264            u_int32_t ip6r2_reserved;      /* reserved field */
265            /* followed by one struct in6_addr */
266    } __attribute__((__packed__));
```
———————————————————————————————————— ip6.h

259–266 The `ip6_rthdr2{}` structure is similar to the `ip6_rthdr0{}` structure that represents a Type 0 Routing Header. The `ip6r2_nxt` field contains the protocol number that follows immediately after this header. The `ip6r2_len` contains the length of this header excluding the first 8 bytes. The value is specified in units of 8 bytes. Since the Type 2 Routing Header contains only one address field of an intermediate node, the `ip6r2_len` field is always set to 2. The `ip6r2_type` field contains the type of this routing header and is set to 2. The `ip6r2_segleft` field is the number of unprocessed intermediate nodes. The value is always initialized to 1. The `ip6r2_reserved` field is cleared by the sender and must be ignored by the receiver. An IPv6 address, which is the home address of a mobile node, follows this structure immediately.

10.20 The Modified Router Advertisement Message: `nd_router_advert{}` Structure

As described in Section 3.6, a new flag for the Router Advertisement message is defined to advertise that the router is a home agent. Listing 10-18 shows the definition of the `nd_router_advert{}` structure.

Listing 10-18

———————————————————————————————————— icmp6.h
```
326    struct nd_router_advert {      /* router advertisement */
327            struct icmp6_hdr       nd_ra_hdr;
328            u_int32_t              nd_ra_reachable;    /* reachable time */
329            u_int32_t              nd_ra_retransmit;   /* retransmit timer */
330            /* could be followed by options */
331    } __attribute__((__packed__));
332
333    #define nd_ra_type            nd_ra_hdr.icmp6_type
334    #define nd_ra_code            nd_ra_hdr.icmp6_code
335    #define nd_ra_cksum           nd_ra_hdr.icmp6_cksum
336    #define nd_ra_curhoplimit     nd_ra_hdr.icmp6_data8[0]
337    #define nd_ra_flags_reserved  nd_ra_hdr.icmp6_data8[1]
```
———————————————————————————————————— icmp6.h

326–337 The definition of the `nd_router_advert{}` structure is described in Section 5.6.2 of *IPv6 Core Protocols Implementation*. Mobile IPv6 extends the flag values used by the `nd_ra_flags_reserved` field by defining one new flag.

Table 10-8 shows the flags used in `nd_router_advert` structure. The `ND_RA_FLAG_HOME_AGENT` is the newly defined flag, which indicates that the router is a home agent. By examining this flag, a receiver of a Router Advertisement message can know whether the sender of the message node is a home agent or not.

TABLE 10-8

Name	Description
ND_RA_FLAG_MANAGED	The link provides a managed address configuration mechanism
ND_RA_FLAG_OTHER	The link provides other stateful configuration mechanisms
ND_RA_FLAG_HOME_AGENT	The router is a home agent

Router Advertisement flags.

10.21   The Modified Prefix Information Option: `nd_opt_prefix_info{}` Structure

As described in Section 3.6, Mobile IPv6 modifies the `nd_opt_prefix_info{}` structure to carry the address of a home agent in a Router Advertisement message. Listing 10-19 shows the definition of the `nd_opt_prefix_info{}` structure.

Listing 10-19

```
                                                                    icmp6.h
488     struct nd_opt_prefix_info {      /* prefix information */
489            u_int8_t        nd_opt_pi_type;
490            u_int8_t        nd_opt_pi_len;
491            u_int8_t        nd_opt_pi_prefix_len;
492            u_int8_t        nd_opt_pi_flags_reserved;
493            u_int32_t       nd_opt_pi_valid_time;
494            u_int32_t       nd_opt_pi_preferred_time;
495            u_int32_t       nd_opt_pi_reserved2;
496            struct in6_addr nd_opt_pi_prefix;
497     } __attribute__((__packed__));
                                                                    icmp6.h
```

488–497 The definition of the `nd_opt_prefix_info{}` structure is described in Section 5.7.2 of *IPv6 Core Protocols Implementation*. Mobile IPv6 extends the flag values used by `nd_opt_pi_flags_reserved` by defining one new flag.

Table 10-9 shows the available flags for the `nd_opt_prefix_info{}` structure. The `ND_OPT_PI_FLAG_ROUTER` flag is the newly defined flag to specify the address of a home agent. A receiver node of a Router Advertisement message can obtain the address of a home agent from the Prefix Information Option that has this flag.

TABLE 10-9

Name	Description
ND_OPT_PI_FLAG_ONLINK	The prefix can be considered as an onlink prefix
ND_OPT_PI_FLAG_AUTO	The prefix information can be used for the stateless address autoconfiguration
ND_OPT_PI_FLAG_ROUTER	The `nd_opt_pi_prefix` field specifies the address of a home agent

Prefix Information flags.

10.22 Advertisement Interval Option: `nd_opt_adv_interval{}` Structure

The `nd_opt_adv_interval{}` structure represents the Advertisement Interval option described in Section 3.6. Listing 10-20 shows the definition of the Advertisement Interval option.

Listing 10-20
                                                                                    ___ icmp6.h

```
518     struct nd_opt_adv_interval {     /* Advertisement interval option */
519             u_int8_t           nd_opt_ai_type;
520             u_int8_t           nd_opt_ai_len;
521             u_int16_t          nd_opt_ai_reserved;
522             u_int32_t          nd_opt_ai_interval;
523     } __attribute__((__packed__));
```
                                                                                    ___ icmp6.h

518–523 The `nd_opt_ai_type` field is set to 7, which indicates the option type for the Advertisement Interval. The `nd_opt_ai_len` field is set to 1. The `nd_opt_ai_reserved` field is cleared by the sender and must be ignored by the receiver. The `nd_opt_ai_interval` field specifies the interval between the Router Advertisement message in units of a second.

10.23 Home Agent Information Option: `nd_opt_homeagent_info{}` Structure

The `nd_opt_homeagent_info{}` structure represents the Home Agent Information option described in Section 3.6. Listing 10-21 shows the definition of the `nd_opt_homeagent_info{}` structure.

Listing 10-21
                                                                                    ___ icmp6.h

```
525     struct nd_opt_homeagent_info {  /* Home Agent info */
526             u_int8_t           nd_opt_hai_type;
527             u_int8_t           nd_opt_hai_len;
528             u_int16_t          nd_opt_hai_reserved;
529             u_int16_t          nd_opt_hai_preference;
530             u_int16_t          nd_opt_hai_lifetime;
531     } __attribute__((__packed__));
```
                                                                                    ___ icmp6.h

525–531 `nd_opt_hai_type` is set to 8, which indicates the Home Agent Information option. `nd_opt_hai_len` is fixed to 1. `nd_opt_hai_reserved` is cleared by the sender and must be ignored by the receiver. `nd_opt_hai_preference` indicates the preference value of the home agent that is sending this option. `nd_opt_hai_lifetime` indicates the time that this home agent can provide the home agent functions in units of 1 s.

10.24 Dynamic Home Agent Address Discovery Request Message: `mip6_dhaad_req{}` Structure

The `mip6_dhaad_req{}` structure represents the Dynamic Home Agent Address Discovery Request message described in Section 3.7. Listing 10-22 shows the definition of the `mip6_dhaad_req{}` structure.

Listing 10-22

```
                                                                    icmp6.h
420     struct mip6_dhaad_req {        /* HA Address Discovery Request */
421             struct icmp6_hdr       mip6_dhreq_hdr;
422     } __attribute__((__packed__));

424     #define mip6_dhreq_type        mip6_dhreq_hdr.icmp6_type
425     #define mip6_dhreq_code        mip6_dhreq_hdr.icmp6_code
426     #define mip6_dhreq_cksum       mip6_dhreq_hdr.icmp6_cksum
427     #define mip6_dhreq_id          mip6_dhreq_hdr.icmp6_data16[0]
428     #define mip6_dhreq_reserved    mip6_dhreq_hdr.icmp6_data16[1]
                                                                    icmp6.h
```

420–428 The mip6_dhaad_req{} structure uses the standard ICMPv6 header icmp6_hdr{}. The mip6_dhreq_type field is set to 144, which indicates a Dynamic Home Agent Address Discovery Request. The mip6_dhreq_code field is set to 0. The mip6_dhreq_cksum field contains the checksum value computed in the same way as the ICMPv6 checksum. The mip6_dhreq_id field is a random identifier set by the sender to match request messages with reply messages. The mip6_dhreq_reserved field must be cleared by the sender and must be ignored by the receiver.

10.25   Dynamic Home Agent Address Discovery Reply Message: mip6_dhaad_rep{} Structure

The mip6_dhaad_rep{} structure represents the Dynamic Home Agent Address Discovery Reply message described in Section 3.7. Listing 10-23 shows the definition of the mip6_dhaad_rep{} structure.

Listing 10-23

```
                                                                    icmp6.h
430     struct mip6_dhaad_rep {        /* HA Address Discovery Reply */
431             struct icmp6_hdr       mip6_dhrep_hdr;
432             /* could be followed by home agent addresses */
433     } __attribute__((__packed__));

435     #define mip6_dhrep_type        mip6_dhrep_hdr.icmp6_type
436     #define mip6_dhrep_code        mip6_dhrep_hdr.icmp6_code
437     #define mip6_dhrep_cksum       mip6_dhrep_hdr.icmp6_cksum
438     #define mip6_dhrep_id          mip6_dhrep_hdr.icmp6_data16[0]
439     #define mip6_dhrep_reserved    mip6_dhrep_hdr.icmp6_data16[1]
                                                                    icmp6.h
```

430–439 The mip6_dhaad_rep{} structure uses the standard ICMPv6 header icmp6_hdr{}. The mip6_dhrep_type is set to 145 to indicate that the message is a Dynamic Home Agent Address Discovery Reply message. The mip6_dhrep_code field is set to 0. The mip6_dhrep_cksum is the checksum value computed in the same way as the ICMPv6 checksum. The mip6_dhrep_id field is the identifier copied from the mip6_dhreq_id field in the mip6_dhaad_req{} structure to match this reply message with the request message. The mip6_dhrep_reserved field must be cleared by the sender and must be ignored by the receiver. The address(es) of the home agent(s) of the home network follows just after this structure.

10.26   Mobile Prefix Solicitation Message: `mip6_prefix_solicit{}` Structure

The `mip6_prefix_solicit{}` structure represents the Mobile Prefix Solicitation message described in Section 3.7. Listing 10-24 shows the definition of the `mip6_prefix_solicit{}` structure.

Listing 10-24

icmp6.h

```
441     struct mip6_prefix_solicit {      /* Mobile Prefix Solicitation */
442            struct icmp6_hdr           mip6_ps_hdr;
443     } __attribute__((__packed__));

445     #define mip6_ps_type              mip6_ps_hdr.icmp6_type
446     #define mip6_ps_code              mip6_ps_hdr.icmp6_code
447     #define mip6_ps_cksum             mip6_ps_hdr.icmp6_cksum
448     #define mip6_ps_id                mip6_ps_hdr.icmp6_data16[0]
449     #define mip6_ps_reserved          mip6_ps_hdr.icmp6_data16[1]
```

icmp6.h

441–449 The `mip6_prefix_solicit{}` structure uses the standard ICMPv6 header `icmp6_hdr{}`. The `mip6_ps_type` is set to 146, which indicates the Mobile Prefix Solicitation. The `mip6_ps_code` field is set to 0. The `mip6_ps_cksum` field is the checksum value computed in the same manner as the ICMPv6 checksum. The `mip6_ps_id` field is a random value set by the sender to match request messages and reply messages. The `mip6_ps_reserved` field must be cleared by the sender and must be ignored by the receiver.

10.27   Mobile Prefix Advertisement Message: `mip6_prefix_advert{}` Structure

The `mip6_prefix_advert{}` structure represents the Mobile Prefix Advertisement message described in Section 3.7. Listing 10-25 shows the definition of the `mip6_prefix_advert{}` structure.

Listing 10-25

icmp6.h

```
451     struct mip6_prefix_advert {       /* Mobile Prefix Advertisement */
452            struct icmp6_hdr           mip6_pa_hdr;
453            /* followed by options */
454     } __attribute__((__packed__));

456     #define mip6_pa_type              mip6_pa_hdr.icmp6_type
457     #define mip6_pa_code              mip6_pa_hdr.icmp6_code
458     #define mip6_pa_cksum             mip6_pa_hdr.icmp6_cksum
459     #define mip6_pa_id                mip6_pa_hdr.icmp6_data16[0]
460     #define mip6_pa_flags_reserved    mip6_pa_hdr.icmp6_data16[1]
```

icmp6.h

451–460 The `mip6_prefix_advert{}` structure uses the standard ICMPv6 header structure defined as `icmp6_hdr{}`. The `mip6_pa_type` field is set to 147, which indicates the

TABLE 10-10

Name	Description
MIP6_PA_FLAG_MANAGED	Set if the home network provides the managed address configuration
MIP6_PA_FLAG_OTHER	Set if the home network provides the other stateful autoconfiguration parameters

Flags used in Mobile Prefix Advertisement.

Mobile Prefix Advertisement. The mip6_pa_code field is set to 0. The mip6_pa_cksum field is the checksum value computed in the same manner as the ICMPv6 checksum. The mip6_pa_id is the identifier copied from the mip6_ps_id field of the mip6_prefix_solicit{} structure to match the reply message with the request message. The mip6_pa_flags_reserved field may contain flags as defined in Table 10-10.

10.28   Binding Cache Entry: mip6_bc{} Structure

The mip6_bc{} structure represents the binding information, called *binding cache*, which is stored on a correspondent node or a home agent. Listing 10-26 shows the definition of the mip6_bc{} structure.

Listing 10-26

mip6_var.h

```
91      struct mip6_bc {
92              LIST_ENTRY(mip6_bc)    mbc_entry;
93              struct in6_addr        mbc_phaddr;     /* peer home address */
94              struct in6_addr        mbc_pcoa;       /* peer coa */
95              struct in6_addr        mbc_addr;       /* my addr (needed?) */
96              u_int8_t               mbc_status;     /* BA statue */
97              u_int8_t               mbc_send_ba;    /* nonzero means BA should be
   sent */
98              u_int32_t              mbc_refresh;    /* Using for sending BA */
99              u_int16_t              mbc_flags;      /* recved BU flags */
100             u_int16_t              mbc_seqno;      /* recved BU seqno */
101             u_int32_t              mbc_lifetime;   /* recved BU lifetime */
102             time_t                 mbc_expire;     /* expiration time of this BC. */
103             u_int8_t               mbc_state;      /* BC state */
104             struct ifnet           *mbc_ifp;       /* ifp that the BC belongs to. */
105                                                    /* valid only when BUF_HOME. */
106             const struct encaptab *mbc_encap;      /* encapsulation from MN */
107             void                   *mbc_dad;       /* dad handler */
108             time_t                 mbc_mpa_exp;    /* expiration time for sending
   MPA */
109                                                    /* valid only when BUF_HOME. */
110             struct mip6_bc         *mbc_llmbc;
111             u_int32_t              mbc_refcnt;
112             u_int                  mbc_brr_sent;

114             struct callout         mbc_timer_ch;

118      };
```

mip6_var.h

Note: Lines 97 and 108 are broken here for layout reasons. However, these are each a single line of code.

91–118 The `mip6_bc{}` structure is defined as a list element. The `mbc_entry` field links the list of binding cache entries, which is managed as a hash table. The implementation is discussed in Section 14.3.

The `mbc_phaddr` field is a home address of a mobile node and the `mbc_pcoa` field is a care-of address of the same mobile node. The `mbc_addr` field is the address that received the Binding Update message from the mobile node.

The `mbc_status` field keeps the status code of the Binding Acknowledgment message sent to the mobile node in response to the Binding Update message from the mobile node. The `mbc_send_ba` field is an internal flag that indicates whether the Binding Acknowledgment message should be sent after the processing of the Binding Update message has finished. The `mbc_refresh` field is used to store the refresh advice value, which is used by the Binding Refresh Advice option. The `mbc_flags` field contains a copy of flags, which is set in the Binding Update message. Table 10-3 shows the list of flags. The `IP6MU_CLONED` flag is used only in the `mip6_bc{}` structure. The `mbc_seqno` field is a copy of the valid sequence number that was most recently received. The `mbc_lifetime` field is an approved lifetime for this binding cache entry. `mbc_expire` indicates the absolute time that this binding cache expires.

The `mbc_state` field indicates the internal state of this cache. Table 10-11 shows the list of state values. The `mbc_ifp` field points to the network interface to which this cache belongs. The interface must be the same interface to as the `mbc_addr` is assigned. The `mbc_encap` field is a pointer to the `encaptab{}` structure that represents the bidirectional tunneling between a home agent and a mobile node. The `mbc_dad` field is used when a home agent is performing the DAD process for the home address of a mobile node. The `mbc_mpa_exp` field is intended to be used for unsolicited Mobile Prefix Advertisements, but the member is not currently used. The `mbc_llmbc` field is a pointer to another `mip6_bc{}` structure that keeps the binding information for the link-local address of the mobile node whose care-of address is the same as the address of this binding cache. The `mbc_refcnt` field is a reference counter for this cache. The value is set to 1 when no other entry refers to this entry. The value increases when other binding cache entries point to this entry using the `mbc_llmbc` field. The `mbc_brr_sent` field stores the number of Binding Refresh Request (BRR) messages sent to the mobile node to update the cache information. Once the cache is updated, the field is cleared. The `mbc_timer_ch` field is the handler of the timer entry for this cache.

TABLE 10-11

Name	Description
MIP6_BC_FSM_STATE_BOUND	Binding is valid
MIP6_BC_FSM_STATE_WAITB	Binding is valid, but the node should send BRR messages
MIP6_BC_FSM_STATE_WAITB2	Binding is valid, but this entry will be removed at the next timer call

List of states of a binding cache.

10.29   Binding Update List Entry: `mip6_bu{}` Structure

The `mip6_bu{}` structure represents the binding information kept on a mobile node. Such information is called a *binding update list entry*. Listing 10-27 shows the definition of the `mip6_bu{}` structure.

Listing 10-27

```
                                                                   mip6_var.h
150     struct mip6_bu {
151             LIST_ENTRY(mip6_bu) mbu_entry;
152             struct in6_addr      mbu_paddr;        /* peer addr of this BU */
153             struct in6_addr      mbu_haddr;        /* HoA */
154             struct in6_addr      mbu_coa;          /* CoA */
155             u_int16_t            mbu_lifetime;     /* BU lifetime */
156             u_int16_t            mbu_refresh;      /* refresh frequency */
157             u_int16_t            mbu_seqno;        /* sequence number */
158             u_int16_t            mbu_flags;        /* BU flags */
159             mip6_cookie_t        mbu_mobile_cookie;
160             u_int16_t            mbu_home_nonce_index;
161             mip6_home_token_t    mbu_home_token;  /* home keygen token */
162             u_int16_t            mbu_careof_nonce_index;
163             mip6_careof_token_t mbu_careof_token; /* careof keygen token */
164             u_int8_t             mbu_pri_fsm_state; /* primary fsm state */
165             u_int8_t             mbu_sec_fsm_state; /* secondary fsm state */
166             time_t               mbu_expire;       /* expiration time of this BU */
167             time_t               mbu_retrans;      /* retrans/refresh timo value */
168             u_int8_t             mbu_retrans_count;
169             time_t               mbu_failure;      /* failure timo value */
170             u_int8_t             mbu_state;        /* local status */
171             struct hif_softc     *mbu_hif;         /* back pointer to hif */
172             const struct encaptab *mbu_encap;
173     };
                                                                   mip6_var.h
```

150–173 The `mip6_bu{}` structure is defined as a list element. The `mbu_entry` field links the list of binding update entries. A mobile node may have more than one binding update list. Each binding update list is kept in an hif virtual interface structure, which is described in Section 10.32. The `mbu_paddr` field is an address of a correspondent node or a home agent, for which this binding update list entry is created. The `mbu_haddr` and `mbu_coa` fields are a home address and a care-of address of this mobile node.

The `mbu_lifetime` field is the lifetime and indicates the time for which this binding information is valid. The `mbu_refresh` field is the time at which the mobile node should update its binding information by sending a Binding Update message. These fields are represented in units of 1s. The `mbu_seqno` field is a sequence number for this binding information and is incremented when a mobile node sends a Binding Update message to the correspondent node or the home agent specified by the address `mbu_paddr`. The `mbu_flags` field keeps the flags for this binding information. Table 10-3 shows the list of flag values. The `mbu_mobile_cookie` field is a cookie value that is used in the Home Test Init and the Care-of Test Init messages. The `mbu_home_nonce_index` and `mbu_home_token` fields contain the home nonce index and the home keygen token sent with a Home Test message from a correspondent node. The `mbu_careof_nonce_index` and `mbu_careof_token` fields contain the value of the care-of nonce index and the care-of keygen token sent with a Care-of Test message.

The `mbu_pri_fsm_state` and `mbu_sec_fsm_state` fields indicate the current state of this binding update list entry. The former stores the state of registration of the binding information, the latter stores the state of the return routability procedure for the binding information. The list of states is described in Tables 10-12 and 10-13.

The `mbu_expire` field is the time at which this binding update list entry expires. The `mbu_retrans` field is the time at which the retransmission timer of the binding update list entry expires. The `mbu_retrans_count` is the number of retransmissions used to compute the exponential backoff for retransmissions. The `mbu_failure` field is the time when the binding update list entry is disabled to prevent infinite retries in the case of disaster; however, this timeout mechanism is not utilized currently.

TABLE 10-12

Name	Description
MIP6_BU_PRI_FSM_STATE_IDLE	Initial state
MIP6_BU_PRI_FSM_STATE_RRINIT	Performing the Return Routability (RR) procedure; no registration exists
MIP6_BU_PRI_FSM_STATE_RRREDO	Performing the RR procedure for re-registration; a valid registration exists
MIP6_BU_PRI_FSM_STATE_RRDEL	Performing the RR procedure for deregistration
MIP6_BU_PRI_FSM_STATE_WAITA	Waiting a Binding Acknowledgment message; no registration exists
MIP6_BU_PRI_FSM_STATE_WAITAR	Waiting a Binding Acknowledgment message for reregistration; a valid registration exists
MIP6_BU_PRI_FSM_STATE_WAITD	Waiting a Binding Acknowledgment message for deregistration
MIP6_BU_PRI_FSM_STATE_BOUND	Registration completed

List of states for registration.

TABLE 10-13

Name	Description
MIP6_BU_SEC_FSM_STATE_START	Initial state
MIP6_BU_SEC_FSM_STATE_WAITHC	Waiting Home Test and Care-of Test messages
MIP6_BU_SEC_FSM_STATE_WAITH	Waiting Home Test message
MIP6_BU_SEC_FSM_STATE_WAITC	Waiting Care-of Test message

List of states for the Return Routability procedure.

TABLE 10-14

Name	Description
MIP6_BU_STATE_DISABLE	The peer node does not support Mobile IPv6
MIP6_BU_STATE_FIREWALLED	The peer address is behind a firewall
MIP6_BU_STATE_NEEDTUNNEL	(MIP6_BU_STATE_DISABLE|MIP6_BU_STATE_FIREWALLED)

Internal states of a binding update list entry.

The `mbu_state` field is the internal state of this binding update list entry. Table 10-14 shows the list of states. The `mbu_hif` field is a pointer to the hif virtual interface to which this binding update list entry belongs. The `mbu_encap` field is a pointer to the `encaptab{}` structure that represents the bidirectional tunnel between a mobile node and its home agent.

10.30   Home Agent Entry: `mip6_ha{}` Structure

The `mip6_ha{}` structure represents the information a mobile node knows about home agents. In the KAME implementation, the list of home agents is created either by receiving Dynamic Home Agent Address Discovery reply messages or by listening to Router Advertisement messages. Listing 10-28 shows the `mip6_ha{}` structure.

Listing 10-28

```
                                                                mip6_var.h
256     struct mip6_ha {
257             TAILQ_ENTRY(mip6_ha)  mha_entry;
258             struct in6_addr    mha_addr ;    /* lladdr or global addr */
259             u_int8_t           mha_flags;    /* RA flags */
260             u_int16_t          mha_pref;     /* home agent preference */
261             u_int16_t          mha_lifetime; /* router lifetime */
262             time_t             mha_expire;
263
264             time_t             mha_timeout;  /* next timeout time. */
265             long               mha_ntick;
....
267             struct callout     mha_timer_ch;
....
271     };
272     TAILQ_HEAD(mip6_ha_list, mip6_ha);
                                                                mip6_var.h
```

256–271 The `mip6_ha{}` structure is defined as a list. The list structure is defined as the `mip6_ha_list{}` structure on line 272. The `mha_entry` field links the `mip6_ha{}` structures. The `mha_addr` field is the IPv6 address of a home agent. The `mha_flags` field is a copy of the flag that is advertised in Router Advertisement messages sent by this home agent. The `mha_pref` field is the value of the home agent preference advertised in the Home Agent Information option. If Router Advertisement messages from this home agent do not contain the Home Agent Information option, the default value 0 is used. The `mha_lifetime` field is the lifetime of this home agent advertised by the Home Agent Information option. If Router Advertisement messages from this home agent do not contain the Home Agent Information option, the value of the router lifetime specified in a Router Advertisement message is used. The `mha_expire` field indicates the time at which the lifetime of this home agent expires. The `mha_timeout` and `mha_ntick` fields indicate the time left to the next timeout. The `mha_timer_ch` field is a handle for the kernel timer mechanism.

10.31 Prefix Entry: `mip6_prefix{}` Structure

A mobile node receives prefix information via two methods: one by receiving Router Advertisement messages that contain Prefix Information options and the other by receiving Mobile Prefix Advertisement messages that contain prefixes for the home network. The prefix information is stored as the `mip6_prefix{}` structure. Listing 10-29 shows the `mip6_prefix{}` structure.

Listing 10-29

mip6_var.h

```
279     struct mip6_prefix {
280             LIST_ENTRY(mip6_prefix) mpfx_entry;
281             struct in6_addr         mpfx_prefix;
282             u_int8_t                mpfx_prefixlen;
283             u_int32_t               mpfx_vltime;
284             time_t                  mpfx_vlexpire;
285             u_int32_t               mpfx_pltime;
286             time_t                  mpfx_plexpire;
287             struct in6_addr         mpfx_haddr;
288             LIST_HEAD(mip6_prefix_ha_list, mip6_prefix_ha) mpfx_ha_list;
289             int                     mpfx_refcnt;
290
291             int                     mpfx_state;
292             time_t                  mpfx_timeout;
293             long                    mpfx_ntick;
....
295             struct callout          mpfx_timer_ch;
....
299     };
300     LIST_HEAD(mip6_prefix_list, mip6_prefix);
```

mip6_var.h

279–299 The `mip6_prefix{}` structure is defined as a list on line 300. The `mpfx_entry` field makes a list of the `mip6_prefix{}` instances. The `mpfx_prefix` and `mpfx_prefixlen` fields indicate the values of IPv6 prefix and prefix length, respectively. The `mpfx_vltime` and `mpfx_vlexpire` fields indicate the valid lifetime of the prefix and the time when the valid lifetime expires. The `mpfx_pltime` and `mpfx_plexpire` fields indicate the preferred lifetime of the prefix and the time when the preferred lifetime expires. The `mpfx_ha_list` field is a list of `mip6_prefix_ha{}` instances. The `mip6_prefix_ha` field indicates the home agent that advertises this prefix information. The structure is described later. The `mpfx_refcnt` field is a reference count for this instance. The `mip6_prefix` instance is pointed to by the `mip6_prefix_ha{}` structure. The instance is used to manage the lifetime of this instance. The instance will be removed when the reference count becomes zero. The `mpfx_state` field indicates the current state of this prefix information. Table 10-15 shows the possible states. This state is used to decide when a mobile node should send a Mobile Prefix Solicitation to update the information of this prefix. The `mpfx_timeout` and `mpfx_ntick` fields indicate the

TABLE 10-15

Name	Description
MIP6_PREFIX_STATE_PREFERRED	The lifetime left of this prefix is longer than the preferred lifetime
MIP6_PREFIX_STATE_EXPIRING	The lifetime left of this prefix is longer than the valid lifetime

The list of states of the `mip6_prefix{}` *structure.*

time left to the next timeout. The `mpfx_timer_ch` field is a handle for the kernel timer mechanism.

The `mip6_prefix_ha{}` structure is a supplement structure to bind prefix information to the home agent that advertises the prefix. Listing 10-30 shows the definition.

Listing 10-30

```
                                                                        mip6_var.h
274     struct mip6_prefix_ha {
275             LIST_ENTRY(mip6_prefix_ha) mpfxha_entry;
276             struct mip6_ha              *mpfxha_mha;
277     };
                                                                        mip6_var.h
```

275–277 The `mpfxha_entry` field links the list of `mip6_prefix_ha{}` instances, which are used in the `mip6_prefix{}` structure. The `mpfxha_mha` field points to an instance of a home agent.

Figure 10-1 describes the relationship between the prefix information structures and home agent information structures. In the figure, there are three prefixes (Prefix `2001:db8:100::/64`, `2001:db8:101::/64` and `2001:db8:200::/64`) and three home agents (`2001:db8:100::1`, `2001:db8:101::1` and `2001:db8:200::1`). Prefix `2001:db8:100::/64` is advertised by Home Agents `2001:db8:100::1` and `2001:db8:101::1`. To point to these two home agents from the prefix instance, it has two instances of the `mip6_prefix_ha{}` structure in the `mpfx_ha_list`. Prefix `2001:db8:200::/64` is advertised by Home Agent `2001:db8:200::1`, and its `mpfx_ha_list` has one instance of the `mip6_prefix_ha{}` structure to point to the home agent. Prefix `2001:db8:101::/64` is almost the same as Prefix `2001:db8:100::/64`. All instances of the `mip6_prefix{}` structure are linked together. The head structure of the prefix list is the `mip6_prefix_list{}` structure defined on line 300 in Listing 10-29. Similarly, all instances of the `mip6_ha{}` structure are linked together. The head structure of the home agent list is the `mip6_ha_list{}` structure from line 272 in Listing 10-28.

10.32   Home Virtual Interface: `hif_softc{}` Structure

In the KAME Mobile IPv6 implementation, the home address of a mobile node is assigned to the hif virtual interface while the mobile node is away from home. The hif interface represents the home network of a mobile node and keeps related information, such as home prefixes. Listing 10-31 shows the definition of the `hif_softc{}` structure.

109

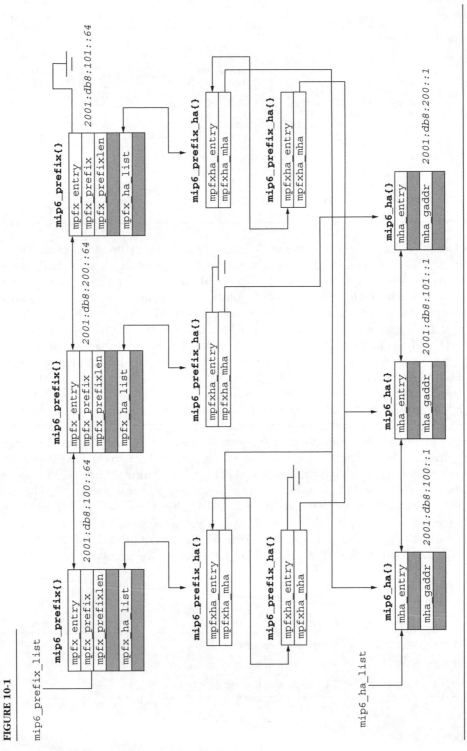

Home agent list generated in the bome network.

Listing 10-31

_____ if_hif.h

```
100     struct hif_softc {
101             struct ifnet hif_if;
102             LIST_ENTRY(hif_softc)   hif_entry;
103             int                     hif_location;         /* cur location */
104             int                     hif_location_prev;    /* XXX */
105             struct in6_ifaddr       *hif_coa_ifa;
106             struct hif_site_prefix_list hif_sp_list;
107             struct mip6_bu_list     hif_bu_list;          /* list of BUs */
108             struct hif_prefix_list  hif_prefix_list_home;
109             struct hif_prefix_list  hif_prefix_list_foreign;
110             u_int16_t               hif_dhaad_id;
111             long                    hif_dhaad_lastsent;
112             u_int8_t                hif_dhaad_count;
113             u_int16_t               hif_mps_id;
114             long                    hif_mps_lastsent;
115             struct in6_addr         hif_ifid;
116     };
```

_____ if_hif.h

100–116 The `hif_if` field is an instance of the `ifnet{}` structure to provide basic ifnet features. The `hif_location` and `hif_location_prev` fields store the current and previous positions of a mobile node. The fields take one of the values described in Table 10-16. The `hif_coa_ifa` field points to the instance of the `in6_ifaddr{}` structure, which contains the current care-of address. The `hif_sp_list` field is a list of the `hif_site_prefix{}` instances, which contains the prefixes that are considered intranet prefixes to provide hints for initiation of the return routability procedure. The `hif_bu_list` field is a list of instances of the `mip6_bu{}` structure, which belong to this interface. The `hif_prefix_list_home` and `hif_prefix_list_foreign` fields are the list of prefixes of a home network and foreign networks, respectively. A mobile node can detect that it is home or foreign by comparing its current care-of address and these prefixes. The `hif_dhaad_id` field stores the last used identifier of the Dynamic Home Agent Address Discovery Request message. The `hif_dhaad_lastsent` field is the time when the last Dynamic Home Agent Address Discovery Request message has been sent. The `hif_dhaad_count` field contains the number of Dynamic Home Agent Address Discovery Request messages. The value is used to perform the exponential backoff calculation done when retransmitting the message. The `hif_mps_id` and `hif_mps_lastsent` fields have the same meaning as the `hif_dhaad_id` and `hif_dhaad_lastsent` fields for the Mobile Prefix Solicitation message. The `hif_ifid` field stores the interface identifier of this

TABLE 10-16

Name	Description
HIF_LOCATION_UNKNOWN	Location is unknown
HIF_LOCATION_HOME	The node is at home network
HIF_LOCATION_FOREIGN	The node is at foreign network

Location.

virtual interface. In the KAME implementation, the home address of a mobile node is generated in the similar manner to that of the stateless address autoconfiguration mechanism. In this mechanism, the mobile node needs to generate a unique identifier value that is used for its lower 64-bit part of the address. The `hif_ifid` field keeps the unique identifier generated from the Media Access Control (MAC) address of the network interface of the node, and it is used as a part of the home address when the node assigns the home address to this virtual interface the first time.

The `hif_prefix{}` structure is used in the `hif_softc{}` structure and is shown in Listing 10-32.

Listing 10-32
——— if_hif.h

```
87      struct hif_prefix {
88              LIST_ENTRY(hif_prefix) hpfx_entry;
89              struct mip6_prefix      *hpfx_mpfx;
90      };
91      LIST_HEAD(hif_prefix_list, hif_prefix);
```
——— if_hif.h

87–90 The `hpfx_entry` field makes a list of `hif_prefix{}` structures represented as the `hif_prefix_list` structure defined on line 91. The `hpfx_mpfx` field is a pointer to the `mip6_prefix{}` structure, which contains the prefix information kept in a mobile node. The `mip6_prefix{}` structure is discussed in Section 10.31.

Figure 10-2 (see page 112) describes the relationship between the `hif_softc{}` structure and other related structures. In this figure, there are two hif virtual interfaces. The mobile node keeps information about three prefix entries as a list of `mip6_prefix_list`. Two of those prefixes are home prefixes of the first hif interface. These two prefixes are considered foreign prefixes for the second interface. Another prefix is a foreign prefix for the first interface and a home prefix for the second interface.

112

FIGURE 10-2

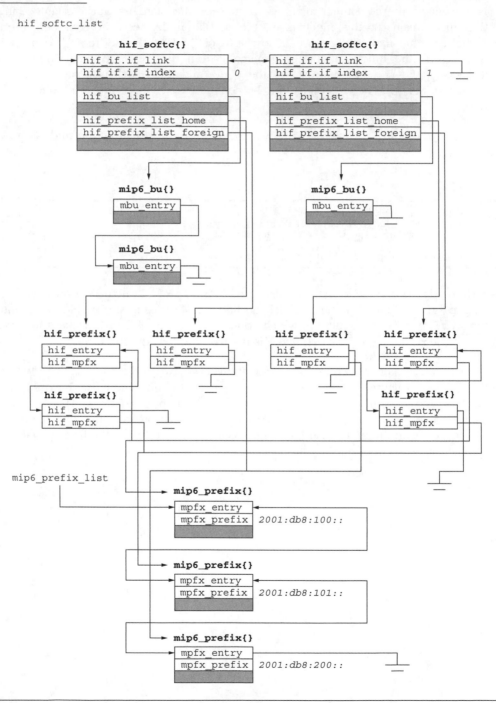

Relationship between `hif_softc{}` *and other structures.*

Macro and Type Definitions

The KAME Mobile Internet Protocol version 6 (IPv6) stack defines several types to increase the readability of the source code. Table 11-1 shows the type definitions for the Mobile IPv6 stack. Table 11-2 shows some constant definitions used by the Mobile IPv6 stack. Table 11-3 shows some utility macros used in the Mobile IPv6 stack.

TABLE 11-1

Type Name	Description
mip6_nonce_t	8-byte unsigned integer array that represents a nonce value
mip6_nodekey_t	20-byte unsigned integer array that represents a nodekey value
mip6_cookie_t	8-byte unsigned integer array that represents a cookie value
mip6_home_token_t	8-byte unsigned integer array that represents a home keygen token value
mip6_careof_token_t	8-byte unsigned integer array that represents a care-of keygen token value

New types introduced in Mobile IPv6.

TABLE 11-2

Name	Value	Description
MIP6_REFRESH_MINLIFETIME	2	Minimum allowed refresh interval
MIP6_REFRESH_LIFETIME_RATE	50	The percentage used when a mobile node calculates refresh interval time from the lifetime of a binding ack, when a home agent does not specify refresh interval
MIP6_MAX_NONCE_LIFE	240	The lifetime of nonces generated by a correspondent node
MIP6_COOKIE_SIZE	8	The size of a cookie value in bytes
MIP6_HOME_TOKEN_SIZE	8	The size of a home keygen token value in bytes
MIP6_CAREOF_TOKEN_SIZE	8	The size of a care-of keygen token value in bytes
MIP6_NONCE_SIZE	8	The size of a nonce value in bytes
MIP6_NODEKEY_SIZE	20	The size of a nodekey in bytes
MIP6_NONCE_HISTORY	10	The number of old nonces kept in a correspondent node; older nonces are removed from a correspondent node and considered as invalid
MIP6_KBM_LEN	20	The size of a key length shared by a mobile node and a correspondent node
MIP6_AUTHENTICATOR_LEN	12	The size of authentication data in Binding Update/Acknowledgment messages
MIP6_MAX_RR_BINDING_LIFE	420	The maximum lifetime of binding information created by the return routability procedure
MIP6_NODETYPE_CORRESPONDENT_NODE	0	Indicates the node is a correspondent node, used by the mip6ctl_nodetype global variable
MIP6_NODETYPE_MOBILE_NODE	1	Indicates the node is a mobile node, used by the mip6ctl_nodetype global variable
MIP6_NODETYPE_HOME_AGENT	2	Indicates the node is a home agent, used by the mip6ctl_nodetype global variable
MIP6_BU_TIMEOUT_INTERVAL	1	The interval between calls to the timer function for binding update list entries

Constants.

TABLE 11-3

Name	Description
GET_NETVAL_S(addr, value)	Copies 2 bytes of data from the memory space specified by addr to value GET_NETVAL_* macros are intended to provide an easy way to write multiple bytes on processor architecture that have alignment restrictions
GET_NETVAL_L(addr, value)	Same as GET_NETVAL_S(), but the length of data is 4 bytes
SET_NETVAL_S(addr, value)	Writes 2 bytes of data to the memory specified by addr from value
SET_NETVAL_L(addr, value)	Same as SET_NETVAL_S(), but the length of data is 4 bytes
MIP6_LEQ(a, b)	Compares a and b using modulo 2^{16}
MIP6_PADLEN(off, x, y)	Computes the length of padding required for Destination/Mobility options; off is the offset from the head of the Destination Options header or Mobility header; x and y represent the alignment requirement (e.g. if the requirement is $8n + 6$, x is 8 and y is 6)
MIP6_IS_BC_DAD_WAIT(bc)	Returns true, if the specified binding cache entry is waiting for DAD completion; otherwise, it returns false
MIP6_IS_BU_BOUND_STATE(bu)	Returns true, if the specified binding update list entry is registered successfully; otherwise, it returns false
MIP6_IS_BU_WAITA_STATE(bu)	Returns true, if the specified binding update list entry is waiting for a Binding Acknowledgment message; otherwise, it returns false
MIP6_IS_BU_RR_STATE(bu)	Returns true, if the specified binding update list entry is performing the return routability procedure
MIP6_BU_DEFAULT_REFRESH_INTERVAL(lifetime)	Computes refresh interval time from the specified lifetime
MIP6_IS_MN	Returns true if a node is a mobile node
MIP6_IS_HA	Returns true if a node is a home agent
MIP6_DEBUG((msg))	Prints a debug message based on when global variable mip6ctl_debug is set to true (see Table 12-1 in Chapter 12); the mip6ctl_debug variable is a configurable variable through the **sysctl** program

Utility macros.

Utility Functions

There are two utility functions that are used from various locations in the Mobile IPv6 code. In this chapter, we discuss these utility functions. Table 12-1 contains the list of functions.

12.1 Global Variables

In the Mobile Internet Protocol version 6 (IPv6) code, the global variables listed in Table 12-2 are used.

12.2 Files

Table 12-3 shows the files that implement utility functions.

12.3 Creation of IPv6 Header

In the Mobile IPv6 code, we need to create IPv6 headers frequently. The `mip6_create_ip6hdr()` function provides a handy way to create an IPv6 header.

TABLE 12-1

Name	Description
mip6_create_ip6hdr()	Create an IPv6 header as a mbuf
mip6_cksum()	Compute a checksum value for MH messages

Utility functions.

TABLE 12-2

Name	Description (related node types)
`struct hif_softc_list hif_softc_list`	The list of home virtual network entries (MN)
`struct mip6_bc_list mip6_bc_list`	The list of binding cache entries (CN, HA)
`struct mip6_prefix_list mip6_prefix_list`	The list of prefix entries (MN)
`struct mip6_ha_list mip6_ha_list`	The list of home agent entries (MN)
`u_int16_t nonce_index`	The current nonce index (CN)
`mip6_nonce_t mip6_nonce[]`	The array that keeps the list of nonce (CN)
`mip6_nonce_t *nonce_head`	A pointer to the current nonce value that is kept in the mip6_nonce[] array (CN)
`mip6_nodekey_t mip6_nodekey[]`	The array that keeps the list of nodekeys (CN)
`u_int16_t mip6_dhaad_id`	The identifier that was used when a Dynamic Home Agent Address Discovery request message was recently sent (MN)
`u_int16_t mip6_mps_id`	The identifier that was used when a Mobile Prefix Solicitation message was recently sent (MN)
`struct mip6_unuse_hoa_list mip6_unuse_hoa`	The list of addresses or port numbers for which we should use a care-of address when we send packets to the destination (MN)
`int mip6ctl_nodetype`	The type of the node (CN, HA, MN)
`int mip6ctl_use_ipsec`	A switch to enable/disable IPsec signal protection (HA, MN)
`int mip6ctl_debug`	A switch to enable/disable printing debug messages (CN, HA, MN)
`u_int32_t mip6ctl_bc_maxlifetime`	The maximum lifetime of binding cache entries for correspondent nodes (CN)
`u_int32_t mip6ctl_hrbc_maxlifetime`	The maximum lifetime of binding cache entries for home registration information (HA)
`u_int32_t mip6ctl_bu_maxlifetime`	The maximum lifetime of binding update entries for correspondent nodes (MN)
`u_int32_t mip6ctl_hrbu_maxlifetime`	The maximum lifetime of binding update entries for home registration information (MN)
`struct mip6_preferred_ifnames mip6_preferred_ifnames`	The list of interface names used when choosing a care-of address. They are ordered by preference (MN).
`struct mip6stat mip6stat`	Mobile IPv6-related statistics (CN, HA, MN)

Global variables.

TABLE 12-3

File	Description
${KAME}/kame/sys/netinet/ip6.h	IPv6 header structure
${KAME}/kame/sys/netinet/ip6mh.h	Mobility Header structures
${KAME}/kame/sys/netinet6/mip6_var.h	All structures that are used in the Mobile IPv6 stack
${KAME}/kame/sys/netinet6/mip6_cncore.c	Implementation of utility functions

Files that implement utility functions.

Listing 12-1

mip6_cncore.c

```
471     struct mbuf *
472     mip6_create_ip6hdr(src, dst, nxt, plen)
473             struct in6_addr *src;
474             struct in6_addr *dst;
475             u_int8_t nxt;
476             u_int32_t plen;
477     {
478             struct ip6_hdr *ip6; /* ipv6 header. */
479             struct mbuf *m; /* a pointer to the mbuf containing ipv6 header. */
480             u_int32_t maxlen;
481
482             maxlen = sizeof(*ip6) + plen;
483             MGETHDR(m, M_DONTWAIT, MT_HEADER);
484             if (m && (max_linkhdr + maxlen >= MHLEN)) {
485                     MCLGET(m, M_DONTWAIT);
486                     if ((m->m_flags & M_EXT) == 0) {
487                             m_free(m);
488                             return (NULL);
489                     }
490             }
491             if (m == NULL)
492                     return (NULL);
493             m->m_pkthdr.rcvif = NULL;
494             m->m_data += max_linkhdr;
495
496             /* set mbuf length. */
497             m->m_pkthdr.len = m->m_len = maxlen;
```

mip6_cncore.c

471–476 The `mip6_create_ip6hdr()` function has four parameters. The `src` and `dst` parameters are used as source and destination addresses for the generated IPv6 header. The `nxt` and `plen` parameters are the next header value and payload length of the generated IPv6 header, respectively.

480–497 An mbuf is allocated by calling the `MGETHDR()` macro. If the allocated mbuf does not have enough length for the requested payload value, then the mbuf is reallocated by requesting a new mbuf as a cluster mbuf via a call to the `MCLGET()` macro. If allocation of a cluster mbuf also fails, a NULL pointer is returned. On line 494, unused space of which the length is `max_linkhdr` is created. This trick will avoid the overhead of another memory allocation when we prepend a link-layer protocol header later. The `max_linkhdr` variable is set to 16 by default. The size of the mbuf (`m_len`) and the header length information, `m_pkthdr.len`, are initialized to the sum of the IPv6 header and payload length.

Listing 12-2

_____mip6_cncore.c

```
499             /* fill an ipv6 header. */
500             ip6 = mtod(m, struct ip6_hdr *);
501             ip6->ip6_flow = 0;
502             ip6->ip6_vfc &= ~IPV6_VERSION_MASK;
503             ip6->ip6_vfc |= IPV6_VERSION;
504             ip6->ip6_plen = htons((u_int16_t)plen);
505             ip6->ip6_nxt = nxt;
506             ip6->ip6_hlim = ip6_defhlim;
507             ip6->ip6_src = *src;
508             ip6->ip6_dst = *dst;
509
510             return (m);
511     }
```

_____mip6_cncore.c

500–510 After the successful allocation of an mbuf, the IPv6 header information is filled. The payload length, `ip6_plen`, is set to `plen` as specified in the parameter list; however, this value will be updated during the output processing of the packet.

12.4   Checksum Computation

The Mobility Header has a checksum field to detect corruption of signaling data. The algorithm to compute the checksum is same as the one used for Transmission Control Protocol (TCP), User Datagram Protocol (UDP), and Internet Control Message Protocol version 6 (ICMPv6) checksum computation.

In TCP, UDP, and ICMPv6, the `in6_cksum()` function provides the computation method. However, the same function cannot be used here to compute the MH checksum, since the `in6_cksum()` function assumes that the packet is passed as an mbuf structure. In the Mobile IPv6 code, a normal memory block is used to prepare MH messages instead of the mbuf structure as in other extension headers. The KAME code processes extension headers with normal memory blocks, not with mbufs, in its output processing. See Chapter 3 of *IPv6 Core Protocols Implementation*, "Internet Protocol version 6." The MH processing code also utilizes normal memory blocks in its output processing since it is treated as one of the extension headers. The `mip6_cksum()` function provides a function to compute the checksum value for MH messages. Most of the `mip6_cksum()` function is copied from the `in6_cksum()` function.

Listing 12-3

_____mip6_cncore.c

```
2677    #define ADDCARRY(x)   (x > 65535 ? x -= 65535 : x)
2678    #define REDUCE do {l_util.l = sum; sum = l_util.s[0] + l_util.s[1];
    ADDCARRY(sum);} while(0);
2679    int
2680    mip6_cksum(src, dst, plen, nh, mobility)
2681            struct in6_addr *src;
2682            struct in6_addr *dst;
2683            u_int32_t plen;
2684            u_int8_t nh;
2685            char *mobility;
2686    {
```

_____mip6_cncore.c

Note: Line 2678 is broken here for layout reasons. However, it is a single line of code.

2679–2685 The `mip6_cksum()` function has five parameters. The `src` and `dst` parameters are the source and destination addresses of the MH message. The `plen` and nh parameters are the payload length and the next header value of the MH message. In the MH case, the nxt parameter is always set to IPPROTO_MH. The mobility parameter is a pointer to the memory that contains the MH message.

Listing 12-4

mip6_cncore.c

```
2687              int sum, i;
2688              u_int16_t *payload;
2689              union {
2690                      u_int16_t uphs[20];
2691                      struct {
2692                              struct in6_addr uph_src;
2693                              struct in6_addr uph_dst;
2694                              u_int32_t uph_plen;
2695                              u_int8_t uph_zero[3];
2696                              u_int8_t uph_nh;
2697                      } uph_un __attribute__((__packed__));
2698              } uph;
2699              union {
2700                      u_int16_t s[2];
2701                      u_int32_t l;
2702              } l_util;
```

mip6_cncore.c

2687–2702 Two structured variables are declared in the declaration section. One is uph which indicates the pseudoheader used for checksum computation, the other is l_util which represents 4 bytes of data as an array of two 16-bit or one 32-bit variable. The l_util variable is used by the REDUCE macro. The REDUCE macro extracts two 16-bit numbers from one 32-bit number and adds those two 16-bit numbers. The result is copied back to the original 32-bit variable again. Figure 12-1 shows the algorithm in the REDUCE macro.

Listing 12-5

mip6_cncore.c

```
2704              bzero(&uph, sizeof(uph));
2705              uph.uph_un.uph_src = *src;
2706              in6_clearscope(&uph.uph_un.uph_src);
2707              uph.uph_un.uph_dst = *dst;
2708              in6_clearscope(&uph.uph_un.uph_dst);
2709              uph.uph_un.uph_plen = htonl(plen);
2710              uph.uph_un.uph_nh = nh;
2711
2712              sum = 0;
2713              for (i = 0; i < 20; i++) {
2714                      REDUCE;
2715                      sum += uph.uphs[i];
2716              }
2717              payload = (u_int16_t *)mobility;
2718              for (i = 0; i < (plen / 2); i++) {
2719                      REDUCE;
2720                      sum += *payload++;
2721              }
```

mip6_cncore.c

2704–2712 The pseudo header information is filled with the parameters passed to function mip6_cksum() and the 1's complement sum is computed over the pseudoheader and the MH message contents.

FIGURE 12-1

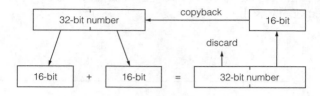

The REDUCE macro.

Listing 12-6

_____ mip6_cncore.c

```
2722              if (plen % 2) {
2723                      union {
2724                              u_int16_t s;
2725                              u_int8_t c[2];
2726                      } last;
2727                      REDUCE;
2728                      last.c[0] = *(char *)payload;
2729                      last.c[1] = 0;
2730                      sum += last.s;
2731              }
2732
2733              REDUCE;
2734              return (~sum & 0xffff);
2735      }
2736  #undef ADDCARRY
2737  #undef REDUCE
```
_____ mip6_cncore.c

2722–2734 If the payload length of the MH message is odd, one byte is appended to the end of the message. The last byte is treated as 0 in the checksum computation. Finally, the 1's complement of the computed result is returned as the checksum value.

13

Common Mobility Header Processing

Because the function of a mobile node and a home agent/correspondent node significantly differs, only a little part of the code that is related to Mobility Header input and error generation processing is shared. In this chapter, we discuss these functions.

13.1 Files

Table 13-1 shows the related files.

13.2 Mobility Header Input

The signaling information used by Mobile IPv6 is carried by the Mobility Header, which is a newly introduced extension header. The Mobility Header was initially designed to be followed

TABLE 13-1

File	Description
${KAME}/kame/sys/netinet/ip6.h	IPv6 header structure
${KAME}/kame/sys/netinet6/in6_proto.c	Mobility Header protocol switch structure
${KAME}/kame/sys/netinet/ip6mh.h	Mobility Header structures
${KAME}/kame/sys/netinet6/mobility6.c	Entry point of Mobility Header input processing
${KAME}/kame/sys/netinet6/mip6_var.h	All structures that are used in the Mobile IPv6 stack

Files related to Mobility Header input processing.

by upper layer protocols, such as Transmission control Protocol (TCP) or User Datagram Protocol (UDP); however, the final specification defines it as a final header with no following headers as explained in Section 13.4.

To accept this new extension header, the protocol switch array for Internet Protocol version 6 IPv6, `inet6sw[]`, is extended as shown in Listing 13-1.

Listing 13-1

in6_proto.c

```
409     #ifdef MIP6
410     { SOCK_RAW,      &inet6domain,     IPPROTO_MH,PR_ATOMIC|PR_ADDR,
411       mobility6_input,       0,              0,            0,
412       0,
413       0,              0,              0,            0,
....
417       &nousrreqs,
....
419     },
420     #endif /* MIP6 */
```

in6_proto.c

410–419 The Mobility Header is defined as a raw socket protocol in the `inet6` domain. The protocol number is `IPPROTO_MH`, decimal 135. All received Mobility Header packets are processed via the `mobility6_input()` function.

> The Mobility Header should be accessible from user space programs via a raw socket to allow application level header management. The KAME Mobile IPv6 code does not provide that function with this version. It can receive but cannot send messages through the raw socket. The latest KAME Mobile IPv6 code, which has been totally rewritten, processes all Mobility Header messages in user space programs using a raw socket.

Listing 13-2

mobility6.c

```
108     int
109     mobility6_input(mp, offp, proto)
110             struct mbuf **mp;
111             int *offp, proto;
112     {
113             struct mbuf *m = *mp;
114             struct m_tag *n; /* for ip6aux */
115             struct ip6_hdr *ip6;
116             struct ip6_mh *mh;
117             int off = *offp, mhlen;
118             int sum;
....
122             ip6 = mtod(m, struct ip6_hdr *);
123
124             /* validation of the length of the header */
....
126             IP6_EXTHDR_CHECK(m, off, sizeof(*mh), IPPROTO_DONE);
127             mh = (struct ip6_mh *)(mtod(m, caddr_t) + off);
```

mobility6.c

108–127 The `mobility6_input()` function has three parameters.

- The `mp` parameter is a double pointer to an mbuf structure that contains the input packet.

- The `offp` parameter is the offset from the head of the IPv6 packet to the head of Mobility Header.

- The `proto` parameter is the protocol number to be processed. (in this case, the number is 135, `IPPROTO_MH`).

The `IP6_EXTHDR_CHECK()` macro makes sure that the content of the Mobility Header is located in contiguous memory. If the `IP6_EXTHDR_CHECK()` macro fails, `IPPROTO_DONE` is returned to terminate the input processing. The mh variable, which is a pointer to an `ip6_mh{}` structure, is initialized to the address offset by `off` bytes. We can access the internal data of the input Mobility Header using the mh variable.

Listing 13-3

_____mobility6.c
```
133               mhlen = (mh->ip6mh_len + 1) << 3;
134               if (mhlen < IP6M_MINLEN) {
135                       /* too small */
136                       ip6stat.ip6s_toosmall++;
137                       /* 9.2 discard and SHOULD send ICMP Parameter Problem */
138                       icmp6_error(m, ICMP6_PARAM_PROB,
139                                  ICMP6_PARAMPROB_HEADER,
140                                  (caddr_t)&mh->ip6mh_len - (caddr_t)ip6);
141                       return (IPPROTO_DONE);
142               }
143
144               if (mh->ip6mh_proto != IPPROTO_NONE) {
....
147                       /* 9.2 discard and SHOULD send ICMP Parameter Problem */
148                       mip6stat.mip6s_payloadproto++;
149                       icmp6_error(m, ICMP6_PARAM_PROB,
150                                  ICMP6_PARAMPROB_HEADER,
151                                  (caddr_t)&mh->ip6mh_proto - (caddr_t)ip6);
152                       return (IPPROTO_DONE);
153               }
```
_____mobility6.c

133–152 This part performs some validation of the input packet. If the length of the Mobility Header is smaller than the minimal value (`IP6M_MINLEN` = 8), an Internet Control Message Protocol version 6 (ICMPv6) Parameter Problem error is sent and the packet is discarded. The problem pointer of the error packet is set to the length field, `ip6mh_len`. If the payload protocol number is not `IPPROTO_NONE`, no next header value, an ICMPv6 Parameter Problem error is sent and the packet is discarded. Currently, the Mobile IPv6 specification does not allow piggybacking any kind of upper layer headers after a Mobility Header. The problem pointer of the ICMPv6 error packet is set to the payload protocol field, `ip6mh_proto` in the `ip6_mh{}` structure.

Listing 13-4

_____mobility6.c
```
155               /*
156                * calculate the checksum
157                */
....
159               IP6_EXTHDR_CHECK(m, off, mhlen, IPPROTO_DONE);
160               mh = (struct ip6_mh *)(mtod(m, caddr_t) + off);
....
166               if ((sum = in6_cksum(m, IPPROTO_MH, off, mhlen)) != 0) {
....
```

```
171                         m_freem(m);
    ....
p173:                       return (IPPROTO_DONE);
174                 }
```
                                                                    ———mobility6.c

155–174 After header validation is finished, the checksum value is checked. At this point, memory contiguity is assured only for the ip6_mh{} structure. To compute the checksum value, we need to make sure the entire Mobility Header message is placed in contiguous memory. Before performing checksum validation, the IP6_EXTHDR_CHECK() macro is called to make sure that the input message is located in contiguous memory. The in6_cksum() function computes the checksum value. If the checksum field, ip6mh_cksum, contains the correct value, the returned value of in6_cksum() will be 0. If the checksum verification fails, the packet is discarded.

Listing 13-5
                                                                    ———mobility6.c
```
176             off += mhlen;
177
178             /* XXX sanity check. */
179
180             switch (mh->ip6mh_type) {
181             case IP6_MH_TYPE_HOTI:
182                     if (mip6_ip6mhi_input(m, (struct ip6_mh_home_test_init *)mh,
183                         mhlen) != 0)
184                             return (IPPROTO_DONE);
185                     break;
186
187             case IP6_MH_TYPE_COTI:
188                     if (mip6_ip6mci_input(m, (struct ip6_mh_careof_test_init *)mh,
189                         mhlen) != 0)
190                             return (IPPROTO_DONE);
191                     break;
192
193     #if defined(MIP6) && defined(MIP6_MOBILE_NODE)
194             case IP6_MH_TYPE_HOT:
195                     if (!MIP6_IS_MN)
196                             break;
197                     if (mip6_ip6mh_input(m, (struct ip6_mh_home_test *)mh,
198                         mhlen) != 0)
199                             return (IPPROTO_DONE);
200                     break;
201
202             case IP6_MH_TYPE_COT:
203                     if (!MIP6_IS_MN)
204                             break;
205                     if (mip6_ip6mc_input(m, (struct ip6_mh_careof_test *)mh,
206                         mhlen) != 0)
207                             return (IPPROTO_DONE);
208                     break;
209
210             case IP6_MH_TYPE_BRR:
211                     if (!MIP6_IS_MN)
212                             break;
213                     if (mip6_ip6mr_input(m, (struct ip6_mh_binding_request *)mh,
214                         mhlen) != 0)
215                             return (IPPROTO_DONE);
216                     break;
217
218             case IP6_MH_TYPE_BACK:
219                     if (!MIP6_IS_MN)
220                             break;
```

```
221                          if (mip6_ip6ma_input(m, (struct ip6_mh_binding_ack *)mh,
222                              mhlen) != 0)
223                                  return (IPPROTO_DONE);
224                          break;
225
226                  case IP6_MH_TYPE_BERROR:
227                          if (mip6_ip6me_input(m, (struct ip6_mh_binding_error *)mh,
228                              mhlen) != 0)
229                                  return (IPPROTO_DONE);
230                          break;
231          #endif /* MIP6 && MIP6_MOBILE_NODE */
232
233                  case IP6_MH_TYPE_BU:
234                          if (mip6_ip6mu_input(m, (struct ip6_mh_binding_update *)mh,
235                              mhlen) != 0)
236                                  return (IPPROTO_DONE);
237                          break;
```
                                                                            ───── mobility6.c

180–237 Each Mobility Header type has its own processing function. In this `switch` clause,
the corresponding functions are called based on the message type value. All processing
functions have three parameters: The first is a double pointer to the mbuf that contains
the input packet; the second is a pointer to the head of the Mobility Header message; and
the third is the length of the message. Table 13-2 shows the functions for each Mobility
Header type.

Listing 13-6
                                                                            ───── mobility6.c

```
239                  default:
240                          /*
241                           * if we receive a MH packet which type is unknown,
242                           * send a binding error message.
243                           */
244                          n = ip6_findaux(m);
245                          if (n) {
246                                  struct ip6aux *ip6a;
247                                  struct in6_addr src, home;
248
249                                  ip6a = (struct ip6aux *) (n + 1);
250                                  src = ip6->ip6_src;
251                                  if ((ip6a->ip6a_flags & IP6A_HASEEN) != 0) {
252                                          home = ip6->ip6_src;
253                                          if ((ip6a->ip6a_flags & IP6A_SWAP) != 0) {
254                                                  /*
255                                                   * HAO exists and swapped
256                                                   * already at this point.
257                                                   * send a binding error to CoA
258                                                   * of the sending node.
259                                                   */
260                                                  src = ip6a->ip6a_coa;
261                                          } else {
262                                                  /*
263                                                   * HAO exists but not swapped
264                                                   * yet.
265                                                   */
266                                                  home = ip6a->ip6a_coa;
267                                          }
268                                  } else {
269                                          /*
270                                           * if no HAO exists, the home address
271                                           * field of the binding error message
272                                           * must be an unspecified address.
273                                           */
274                                          home = in6addr_any;
```

```
275                                        }
276                                        (void)mobility6_send_be(&ip6->ip6_dst, &src,
277                                            IP6_MH_BES_UNKNOWN_MH, &home);
278                                    }
279                                mip6stat.mip6s_unknowntype++;
280                                break;
281                        }
```
———————————————————————————————————————mobility6.c

239–281 If a node receives a Mobility Header message that has a type value other than those listed in Table 13-2, the node sends a Binding Error message.

The destination address of a Binding Error message must be set to the IPv6 source address (which was seen on the wire) of the incoming packet. That is, if the packet contains a Home Address option, and the home and source addresses have been swapped at this point, the addresses have to be recovered before the swapping. The IP6A_HASEEN flag, which is in the ip6aux{} structure, indicates that the packet has a Home Address option and was processed before. Table 13-3 shows the flags extended for Mobile IPv6. The IP6A_SWAP flag is set when a Home Address option is processed and the source and home addresses have been swapped. The IP6A_SWAP flag is valid only if the IP6A_HASEEN flag is set. If the IP6A_SWAP flag is set, we need to use the address stored in the ip6aux{} structure, which was the source address while the packet was on the wire; otherwise, we can use the address stored in the source address field of the incoming IPv6 header. A Binding Error message also contains the home address of the

TABLE 13-2

Type	Function name
IP6_MH_TYPE_HOTI	mip6_ip6mhi_input()
IP6_MH_TYPE_COTI	mip6_ip6mci_input()
IP6_MH_TYPE_HOT	mip6_ip6mh_input()
IP6_MH_TYPE_COT	mip6_ip6mc_input()
IP6_MH_TYPE_BRR	mip6_ip6mr_input()
IP6_MH_TYPE_BACK	mip6_ip6ma_input()
IP6_MH_TYPE_BERROR	mip6_ip6me_input()
IP6_MH_TYPE_BU	mip6_ip6mu_input()

Input processing functions for Mobility Header messages.

TABLE 13-3

Name	Description
IP6A_HASEEN	The packet contains a Home Address option
IP6A_SWAP	The source and home addresses specified in a Home Address option have been swapped

Flags of ip6aux{} structure used for Mobile IPv6.

incoming packet if the packet has a Home Address option. The home address field of a Binding Error message, `ip6mhbe_homeaddr`, is set to an unspecified address if the incoming packet does not have a Home Address option.

The `mobility6_send_be()` function generates a Binding Error message. This function is described in Section 13.3.

Listing 13-7

```
282
283              /* deliver the packet to appropriate sockets */
284              if (mobility6_rip6_input(&m, *offp) == IPPROTO_DONE) {
285                      /* in error case, IPPROTO_DONE is returned. */
286                      return (IPPROTO_DONE);
287              }
288
289              *offp = off;
290
291              return (mh->ip6mh_proto);
292      }
```

283–292 All Mobility Header messages are delivered into user space through a raw socket. The `mobility6_rip6_input()` function sends the messages to every opened raw socket that is waiting for Mobility Header messages. Note that the comment on line 285 is wrong. Since we do not have any upper layer protocol headers after the Mobility Header, `mobility6_rip6_input()` always returns `IPPROTO_DONE`. We can finish packet processing after we have finished the Mobility Header processing.

13.3   Generating Binding Error Messages

When there is any error condition while processing Mobile IPv6 signaling messages, a Binding Error message is sent to the sender of the packet that caused the error. The `mobility6_send_be()` function generates a Binding Error packet and sends the packet to the source of the error.

Listing 13-8

```
297      int
298      mobility6_send_be(src, dst, status, home)
299              struct in6_addr *src;
300              struct in6_addr *dst;
301              u_int8_t status;
302              struct in6_addr *home;
303      {
304              struct mbuf *m;
305              struct ip6_pktopts opt;
306              int error = 0;
307
308              /* a binding message must be rate limited. */
309              if (mobility6_be_ratelimit(dst, home, status))
310                      return (0); /* rate limited. */
```

297–302 The `mobility6_send_be()` function has four parameters: the `src` and `dst` parameters are the source and destination addresses of the Binding Error message to be generated, the `status` parameter is a status code of the Binding Error message, and the home parameter is the home address that is set to the home address field of the Binding Error message. Table 10-6 in Chapter 10 shows the available codes.

309–310 Sending   Binding   Error   messages   must   be   rate   limited.   The `mobility6_be_ratelimit()` function returns whether we can send a Binding Error message or not, based on the rate-limitation algorithm we use. Function `mobility6_be_ratelimit()` is discussed in Section 13.4.

Listing 13-9

—— mobility6.c

```
312            ip6_initpktopts(&opt);
313
314            m = mip6_create_ip6hdr(src, dst, IPPROTO_NONE, 0);
315            if (m == NULL)
316                    return (ENOMEM);
317
318            error = mip6_ip6me_create(&opt.ip6po_mh, src, dst, status, home);
319            if (error) {
320                    m_freem(m);
321                    goto free_ip6pktopts;
322            }
323
324            /* output a binding missing message. */
....
326            error = ip6_output(m, &opt, NULL, 0, NULL, NULL
....
328                                    , NULL
....
330                                    );
331            if (error)
332                    goto free_ip6pktopts;
333
334    free_ip6pktopts:
335            if (opt.ip6po_mh)
336                    FREE(opt.ip6po_mh, M_IP6OPT);
337
338            return (error);
339    }
```

—— mobility6.c

314–322 The `mip6_create_ip6hdr()` function creates an IPv6 header based on the specified parameters. The `mip6_ip6me_create()` function creates a Binding Error message with the specified parameters and stores the message in the instance of the `ip6_pktopts{}` structure given as the first parameter, `opt.ip6po_mh` in this case. Function `mip6_ip6me_create()` is discussed in Section 13.5.

324–338 Since Mobility Header packets do not have any upper layer headers, we can simply call the `ip6_output()` function with a Mobility Header only as a packet option to send the packet. The Mobility Header message must be freed by the caller, since the `ip6_output()` function does not free packet option information.

13.4   Rate Limitation of Binding Error Messages

To avoid flooding the network with error messages, the sending rate of Binding Error messages is limited.

FIGURE 13-1

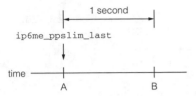

ip6me_pps_count = the number of packets sent between A and B

ppsratecheck() mechanism.

Listing 13-10

_____mobility6.c

```
341     static int
342     mobility6_be_ratelimit(dst, hoa, status)
343             const struct in6_addr *dst;      /* not used at this moment */
344             const struct in6_addr *hoa;      /* not used at this moment */
345             const int status;                /* not used at this moment */
346     {
347             int ret;
348
349             ret = 0;          /* okay to send */
350
351             /* PPS limit */
352             if (!ppsratecheck(&ip6me_ppslim_last, &ip6me_pps_count,
353                 ip6me_ppslim)) {
354                     /* The packet is subject to rate limit */
355                     ret++;
356             }
357
358             return ret;
359     }
```
_____mobility6.c

341–359 The `mobility6_be_ratelimit()` function decides whether we can send a Binding Error message, based on the system rate limit parameter. Currently, the rate limitation code does not take into account any information about the source or destination address of the Binding Error message to be sent. In the current algorithm, Binding Error messages are limited to the number specified by the `ip6me_ppslim` global variable per second. The `ppsratecheck()` function returns true if the number of packets has not reached the limit, otherwise it returns false. The `ip6me_ppslim_last` global variable keeps the start time of the one limitation unit time to count the number of error messages sent in 1. The `ip6me_pps_count` variable keeps the number of messages we sent after `ip6me_ppslim_last`. If `ip6me_pps_count` exceeds the `ip6me_ppslim` variable, we must not send the packet. The `ip6me_ppslim_last` and `ip6me_pps_count` variables are updated in the `ppsratecheck()` function. Figure 13-1 shows the mechanism.

13.5 Creation of Binding Error Message

A Binding Error message is created by calling the `mip6_ip6me_create()` function.

Listing 13-11

```
                                                                        mip6_cncore.c
2574    int
2575    mip6_ip6me_create(pktopt_mobility, src, dst, status, addr)
2576            struct ip6_mh **pktopt_mobility;
2577            struct in6_addr *src;
2578            struct in6_addr *dst;
2579            u_int8_t status;
2580            struct in6_addr *addr;
2581    {
2582            struct ip6_mh_binding_error *ip6me;
2583            int ip6me_size;
2584
2585            *pktopt_mobility = NULL;
2586
2587            ip6me_size = sizeof(struct ip6_mh_binding_error);
2588
2589            MALLOC(ip6me, struct ip6_mh_binding_error *,
2590                    ip6me_size, M_IP6OPT, M_NOWAIT);
2591            if (ip6me == NULL)
2592                    return (ENOMEM);
2593
2594            bzero(ip6me, ip6me_size);
2595            ip6me->ip6mhbe_proto = IPPROTO_NONE;
2596            ip6me->ip6mhbe_len = (ip6me_size >> 3) - 1;
2597            ip6me->ip6mhbe_type = IP6_MH_TYPE_BERROR;
2598            ip6me->ip6mhbe_status = status;
2599            ip6me->ip6mhbe_homeaddr = *addr;
2600            in6_clearscope(&ip6me->ip6mhbe_homeaddr);
2601
2602            /* calculate checksum. */
2603            ip6me->ip6mhbe_cksum = mip6_cksum(src, dst, ip6me_size, IPPROTO_MH,
2604                (char *)ip6me);
2605
2606            *pktopt_mobility = (struct ip6_mh *)ip6me;
2607
2608            return (0);
2609    }
                                                                        mip6_cncore.c
```

2574–2580 The `mip6_ip6me_create()` function has five parameters. The `pktopt_mobility` is a pointer to the `ip6po_mh` field of the `ip6_pktopts{}` structure, the `src` and `dst` parameters are the IPv6 addresses that are used as source and destination addresses of a Binding Error message, and the `status` and `addr` parameters are the status code and the home address included in the Binding Error message.

2587–2592 A Binding Error message is passed to the `ip6_output()` function as a packet option. Memory is allocated based on the packet format of a Binding Error message. If memory allocation fails, an error is returned.

2594–2608 The contents of a Binding Error message are filled. Care has to be taken when setting the length of the message since the length of an extension header is specified in units of 8 bytes. If the home address information passed by the caller has an embedded scope identifier, it must be cleared by calling the `in6_clearscope()` function. The checksum computation is done by the `mip6_cksum()` function, which serves the same function as the `in6_cksum()` function. The difference is that `in6_cksum()` takes the mbuf as a parameter while `mip6_cksum()` takes address information as an `in6_addr{}` structure. The created Binding Error message is set to the `ip6po_mh` field.

13.6 Mobility Header Message Delivery to Raw Sockets

All Mobility Header messages are delivered to raw sockets. A user space program can receive Mobility Header messages by opening a raw socket specifying IPPROTO_MH as a protocol. The delivery to raw sockets is implemented in the mobility6_rip6_input() function.

Listing 13-12

_____ mobility6.c

```
361     static int
362     mobility6_rip6_input(mp, off)
363             struct mbuf **mp;
364             int off;
365     {
366             struct mbuf *m = *mp;
367             struct ip6_hdr *ip6;
368             struct ip6_mh *mh;
369             struct sockaddr_in6 fromsa;
370             struct in6pcb *in6p;
371             struct in6pcb *last = NULL;
372             struct mbuf *opts = NULL;
373
374             ip6 = mtod(m, struct ip6_hdr *);
....
377             mh = (struct ip6_mh *)((caddr_t)ip6 + off);
....
386             /*
387              * XXX: the address may have embedded scope zone ID, which should be
388              * hidden from applications.
389              */
390             bzero(&fromsa, sizeof(fromsa));
391             fromsa.sin6_family = AF_INET6;
392             fromsa.sin6_len = sizeof(struct sockaddr_in6);
393             if (in6_recoverscope(&fromsa, &ip6->ip6_src, m->m_pkthdr.rcvif) != 0) {
394                     m_freem(m);
395                     return (IPPROTO_DONE);
396             }
```

_____ mobility6.c

361–364 The mobility6_rip6_input() function has two parameters. The mp parameter is a pointer to the packet that contains the Mobility Header message to be delivered and the off parameter is the offset from the head of the IPv6 header to the head of the Mobility Header.

377 Since the contiguous memory check has already been done in the initial part of the input processing, the pointer to the Mobility Header can be set to the mh variable without any validation.

386–396 The source address of the Mobility Header message is set to the variable fromsa. In the KAME implementation, the address used in the kernel may have an embedded scope identifier. The fromsa variable is used to communicate the source address to user space programs. This code makes sure that the scope identifier is cleared since we must not pass addresses that have embedded scope identifiers to user space programs.

Listing 13-13

_____ mobility6.c

```
399             LIST_FOREACH(in6p, &ripcb, inp_list)
....
408             {
....
```

```
410                         if ((in6p->inp_vflag & INP_IPV6) == 0)
411                                 continue;
....
414                         if (in6p->inp_ip_p != IPPROTO_MH)
....
418                                 continue;
419                     if (!IN6_IS_ADDR_UNSPECIFIED(&in6p->in6p_laddr) &&
420                         !IN6_ARE_ADDR_EQUAL(&in6p->in6p_laddr, &ip6->ip6_dst))
421                             continue;
422                     if (!IN6_IS_ADDR_UNSPECIFIED(&in6p->in6p_faddr) &&
423                         !IN6_ARE_ADDR_EQUAL(&in6p->in6p_faddr, &ip6->ip6_src))
424                             continue;
425                 if (last) {
426                         struct mbuf *n = NULL;
427
428                         /*
429                          * Recent network drivers tend to allocate a single
430                          * mbuf cluster, rather than to make a couple of
431                          * mbufs without clusters.  Also, since the IPv6 code
432                          * path tries to avoid m_pullup(), it is highly
433                          * probable that we still have an mbuf cluster here
434                          * even though the necessary length can be stored in an
435                          * mbuf's internal buffer.
436                          * Meanwhile, the default size of the receive socket
437                          * buffer for raw sockets is not so large.  This means
438                          * the possibility of packet loss is relatively higher
439                          * than before.  To avoid this scenario, we copy the
440                          * received data to a separate mbuf that does not use
441                          * a cluster, if possible.
442                          * XXX: it is better to copy the data after stripping
443                          * intermediate headers.
444                          */
445                         if ((m->m_flags & M_EXT) && m->m_next == NULL &&
446                             m->m_len <= MHLEN) {
447                                 MGET(n, M_DONTWAIT, m->m_type);
448                                 if (n != NULL) {
....
453                                         m_dup_pkthdr(n, m);
....
457                                         bcopy(m->m_data, n->m_data, m->m_len);
458                                         n->m_len = m->m_len;
459                                 }
460                         }
461                         if (n != NULL ||
462                             (n = m_copy(m, 0, (int)M_COPYALL)) != NULL) {
463                                 if (last->in6p_flags & IN6P_CONTROLOPTS)
464                                         ip6_savecontrol(last, n, &opts);
465                                 /* strip intermediate headers */
466                                 m_adj(n, off);
467                                 if (sbappendaddr(&last->in6p_socket->so_rcv,
468                                     (struct sockaddr *)&fromsa, n, opts)
469                                     == 0) {
470                                         /* should notify about lost packet */
471                                         m_freem(n);
472                                         if (opts) {
473                                                 m_freem(opts);
474                                         }
475                                 } else
476                                         sorwakeup(last->in6p_socket);
477                                 opts = NULL;
478                         }
479                 }
480                 last = in6p;
481         }
482         if (last) {
483                 if (last->in6p_flags & IN6P_CONTROLOPTS)
484                         ip6_savecontrol(last, m, &opts);
485                 /* strip intermediate headers */
```

```
486                        m_adj(m, off);
487
488                        /* avoid using mbuf clusters if possible (see above) */
489                        if ((m->m_flags & M_EXT) && m->m_next == NULL &&
490                            m->m_len <= MHLEN) {
491                                struct mbuf *n;
492
493                                MGET(n, M_DONTWAIT, m->m_type);
494                                if (n != NULL) {
....
499                                        m_dup_pkthdr(n, m);
....
503                                        bcopy(m->m_data, n->m_data, m->m_len);
504                                        n->m_len = m->m_len;
505
506                                        m_freem(m);
507                                        m = n;
508                                }
509                        }
510                        if (sbappendaddr(&last->in6p_socket->so_rcv,
511                            (struct sockaddr *)&fromsa, m, opts) == 0) {
512                                m_freem(m);
513                                if (opts)
514                                        m_freem(opts);
515                        } else
516                                sorwakeup(last->in6p_socket);
517                } else {
518                        m_freem(m);
519                        ip6stat.ip6s_delivered--;
520                }
521                return IPPROTO_DONE;
522        }
```
—— mobility6.c

399–424 The Protocol Control Block (PCB) entries for raw sockets are kept in the `ripcb` variable as a list of PCB entries. Each PCB entry is checked to determine whether the Mobility Header message should be delivered to the PCB entry. The following PCB entries are skipped because they have nothing to do with this processing:

- The version is not IPv6

- The protocol is not Mobility Header (`IPPROTO_MH`)

- The local address of the PCB entry does not match the destination address of the incoming Mobility Header packet

- The foreign address of the PCB entry does not match the source address of the incoming Mobility Header packet

445–460 If the incoming packet is stored in a cluster mbuf, but the size is smaller than the mbuf size of a noncluster mbuf, a new noncluster mbuf is created and the contents are copied to the noncluster mbuf to avoid exhaustion of cluster mbufs.

461–477 If the above code fails and a copy of the incoming packet has not been created, the `m_copy()` function is called to copy the mbuf that keeps the incoming packet. If the copy is created, it is delivered to the raw socket of the current PCB entry that is being processed. The `ip6_savecontrol()` function extracts packet information and stores it in the `opts` variable. A user needs to specify required information using the socket application programming interfaces (API) if the user wants to use packet information in the program. The `ip6_savecontrol()` function is discussed in chapter 7 of *IPv6 Core Protocols*

Implementation, "Socket API Extensions." Before putting the Mobility Header message in the socket with the `sbappendaddr()` function, the IPv6 header and other extension headers have to be removed by the `m_adj()` function if it exists. The `sbappendaddr()` function appends the Mobility Header message to the tail of the socket buffer that is bound to this PCB entry. To give a user program the source address of a packet, the value of the `fromsa` variable is passed. In addition, the saved packet information, `opt`, is passed to function `sbappendaddr()`. If the call to function `sbappendaddr()` succeeds, the `sorwakeup()` function is called to notify the socket that there is unreceived data in the socket buffer. If the call fails, the copied Mobility Header message and packet information are freed.

482–517 This code is almost the same as the code written on lines 461–477 but handles the last PCB entry in the raw PCB entry list. The difference from the previous code is that we need not copy the incoming packet since we do not have any other PCB entry to be processed. We can directly modify the original packet to deliver the data.

518–521 If there are no raw PCB entries in the kernel, the incoming mbuf is simply freed. After all the PCB entries have been processed, `IPPROTO_DONE` is returned to indicate that the delivery has finished. There is no need to return a value other than `IPPROTO_DONE` since the Mobility Header does not have any upper layer protocols. Incoming packet processing can be finished after Mobility Header processing is finished.

14

Home Agent and Correspondent Node

A home agent is a special router located on a home network to support mobile nodes. A home agent has the following capabilities:

- Maintaining binding information for home registration
- Intercepting and forwarding packets sent to the home addresses of mobile nodes using the Internet Protocol version 6 (IPv6) in IPv6 encapsulating mechanism
- Forwarding packets sent from mobile nodes using the IPv6 in IPv6 encapsulating mechanism to the final destination of those packets

A correspondent node may have the capability to perform the route optimization procedure described in Chapter 5. Such a node has the following capabilities:

- Receiving binding information for route optimized communication
- Performing direct communication with mobile nodes using the Home Address Option and the Type 2 Routing Header

A home agent also has the capabilities to perform route optimization. The above functions are implemented as the binding cache mechanism that keeps information about mobile nodes. We describe the mechanism in this section for both a home agent and a correspondent node that supports route optimization at the same time, since those two types of nodes share pieces of code.

14.1 Files

Table 14-1 shows the files used by a home agent and route optimization functions.

TABLE 14-1

File	*Description*
`${KAME}/kame/sys/netinet/icmp6.h`	Dynamic Home Agent Address Discovery and Mobile Prefix Solicitation/Advertisement structures
`${KAME}/kame/sys/netinet/ip6.h`	Home Address option structure
`${KAME}/kame/sys/netinet/ip6mh.h`	Mobility Header structures
`${KAME}/kame/sys/netinet6/dest6.c`	Processing code of the Home Address option
`${KAME}/kame/sys/netinet6/mip6_var.h`	All structures that are used in the Mobile IPv6 stack
`${KAME}/kame/sys/netinet6/mip6_cncore.c`	Implementation of correspondent node functions
`${KAME}/kame/sys/netinet6/mip6_hacore.c`	Implementation of home agent functions
`${KAME}/kame/sys/netinet6/mip6_icmp6.c`	Implementation of ICMPv6 message-related processing
`${KAME}/kame/sys/netinet6/ip6_forward.c`	Sending packets to the bidirectional tunnel on a home agent side
`${KAME}/kame/sys/netinet6/ip6_output.c`	Insertion of extension headers for Mobile IPv6 signaling
`${KAME}/kame/kame/had/halist.h`	Home agent information structure used by the home agent side
`${KAME}/kame/kame/had/halist.c`	Home Agent information management
`${KAME}/kame/kame/had/haadisc.c`	Implementation of the Dynamic Home Agent Address Discovery mechanism of the home agent side
`${KAME}/kame/kame/had/mpa.c`	Implementation of the Mobile Prefix Solicitation and Advertisement mechanism of the home agent side

Files for home agent and route optimization functions.

14.2   Binding Update Message Input

A mobile node sends its binding information via the Binding Update message, which is a Mobility Header message. A Binding Update message is dispatched to the `mip6_ip6mu_input()` function from the `mobility6_input()` function.

Listing 14-1

——— mip6_cncore.c

```
2002    #define IS_REQUEST_TO_CACHE(lifetime, hoa, coa) \
2003            (((lifetime) != 0) &&                   \
2004             (!IN6_ARE_ADDR_EQUAL((hoa), (coa))))
2005    int
2006    mip6_ip6mu_input(m, ip6mu, ip6mulen)
2007            struct mbuf *m;
2008            struct ip6_mh_binding_update *ip6mu;
```

```
2009            int ip6mulen;
2010   {
2011            struct ip6_hdr *ip6;
2012            struct m_tag *mtag;
2013            struct ip6aux *ip6a = NULL;
2014            u_int8_t isprotected = 0;
2015            struct mip6_bc *mbc;
2016
2017            int error = 0;
2018            u_int8_t bu_safe = 0;    /* To accept bu always without authentication,
    this value is set to non-zero */
2019            struct mip6_mobility_options mopt;
2020            struct mip6_bc bi;
2021
2022            mip6stat.mip6s_bu++;
2023            bzero(&bi, sizeof(bi));
2024            bi.mbc_status = IP6_MH_BAS_ACCEPTED;
2025            /*
2026             * we send a binding ack immediately when this binding update
2027             * is not a request for home registration and has an ACK bit
2028             * on.
2029             */
2030            bi.mbc_send_ba = ((ip6mu->ip6mhbu_flags & IP6MU_ACK)
2031                && !(ip6mu->ip6mhbu_flags & IP6MU_HOME));
```
―――――――――――――――――――――――――――――――――――――― *mip6_cncore.c*

Note: Line 2018 is broken here for layout reasons. However, it is a single line of code.

2006–2031 The `mip6_ip6mu_input()` function has three parameters. The m parameter is a
pointer to the mbuf that contains a Binding Update message, the `ip6mu` and `ip6mulen`
parameters are pointers to the head of the Binding Update message and the length of the
message, respectively.

The variable `bi`, which is an instance of the `mip6_bc{}` structure, is used as a tempo-
rary buffer for the binding information that will be stored in the node. The `mbc_send_ba`
variable is a flag that indicates whether the node needs to reply to the mobile node that
has sent this Binding Update message with a Binding Acknowledgment message in this
function. A Binding Update message for home registration must have the A (Acknowledge)
flag set. If the input message has the H (Home registration) flag set but does not have the
A flag set, the `mbc_send_ba` variable is set to true to send a Binding Acknowledgment
message to the sender to indicate an error.

Listing 14-2
―――――――――――――――――――――――――――――――――――――― *mip6_cncore.c*
```
2045            ip6 = mtod(m, struct ip6_hdr *);
2046            bi.mbc_addr = ip6->ip6_dst;
2047
2048            /* packet length check. */
2049            if (ip6mulen < sizeof(struct ip6_mh_binding_update)) {
....
2054                    ip6stat.ip6s_toosmall++;
2055                    /* send ICMP parameter problem. */
2056                    icmp6_error(m, ICMP6_PARAM_PROB, ICMP6_PARAMPROB_HEADER,
2057                        (caddr_t)&ip6mu->ip6mhbu_len - (caddr_t)ip6);
2058                    return (EINVAL);
2059            }
2060
2061            bi.mbc_flags = ip6mu->ip6mhbu_flags;
2062
2063   #ifdef M_DECRYPTED       /* not openbsd */
2064            if (((m->m_flags & M_DECRYPTED) != 0)
2065                || ((m->m_flags & M_AUTHIPHDR) != 0)) {
```

```
2066                     isprotected = 1;
2067            }
2068    #endif
```
_____ mip6_cncore.c

2045–2059 The length of the incoming packet must be greater than the size of an `ip6_mh_binding_update{}` structure, otherwise an Internet Control Message Protocol version 6 (ICMPv6) Parameter Problem message will be sent. The code value is set to `ICMP6_PARAMPROB_HEADER` to indicate that the header is invalid and the problem pointer points to the length field of the Binding Update message.

2064–2067 The KAME IPsec stack adds the `M_DECRYPTED` mbuf flag if the packet was encrypted by the ESP mechanism. The stack also adds the `M_AUTHIPHDR` mbuf flag if the packet was protected by an AH. The Mobile IPv6 specification requires that Binding Update messages for home registration must be protected by ESP or AH. The `isprotected` variable is used later when we perform the home registration procedure.

Listing 14-3

_____ mip6_cncore.c

```
2075            bi.mbc_pcoa = ip6->ip6_src;
2076            mtag = ip6_findaux(m);
2077            if (mtag == NULL) {
2078                    m_freem(m);
2079                    return (EINVAL);
2080            }
2081            ip6a = (struct ip6aux *) (mtag + 1);
2082            if (((ip6a->ip6a_flags & IP6A_HASEEN) != 0) &&
2083                ((ip6a->ip6a_flags & IP6A_SWAP) != 0)) {
2084                    bi.mbc_pcoa = ip6a->ip6a_coa;
2085            }
```
_____ mip6_cncore.c

2075–2085 The care-of address of the mobile node is extracted from the incoming Binding Update packet and is stored in the Home Address Option at this point. We can copy the address from the auxiliary mbuf since the address has already been copied to the auxiliary mbuf during the Destination Option processing discussed in Section 14.15. The exception is a Binding Update message for home deregistration. In the deregistration case, the Home Address option may not exist, in which case the source address of the IPv6 header is considered a care-of address.

Listing 14-4

_____ mip6_cncore.c

```
2087            if (!mip6ctl_use_ipsec && (bi.mbc_flags & IP6MU_HOME)) {
2088                    bu_safe = 1;
2089                    goto accept_binding_update;
2090            }
2091
2092            if (isprotected) {
2093                    bu_safe = 1;
2094                    goto accept_binding_update;
2095            }
2096            if ((bi.mbc_flags & IP6MU_HOME) == 0)
2097                    goto accept_binding_update;     /* Must be checked its safety
2098                                                     * with RR later */
2099
2100            /* otherwise, discard this packet. */
```

```
2101              m_freem(m);
2102              mip6stat.mip6s_haopolicy++;
2103              return (EINVAL);
```
_____mip6_cncore.c

2087–2090 The `mip6ctl_use_ipsec` variable is a configurable variable, which can be set using the **sysctl** program. A Binding Update message is accepted even if it is not protected by IPsec when the `mip6ctl_use_ipsec` variable is set to false.

2092–2097 If the Binding Update message is protected by the IPsec mechanism, the message is accepted, otherwise, the validity is checked later by return routability information if the message is for route optimization and is not for home registration.

2100–2103 Any other Binding Update messages are silently discarded.

Listing 14-5
_____mip6_cncore.c

```
2105    accept_binding_update:
2106
2107            /* get home address. */
2108            bi.mbc_phaddr = ip6->ip6_src;
2109
2110            if ((error = mip6_get_mobility_options((struct ip6_mh *)ip6mu,
2111                    sizeof(*ip6mu), ip6mulen, &mopt)))) {
2112                    /* discard. */
2113                    m_freem(m);
2114                    mip6stat.mip6s_invalidopt++;
2115                    return (EINVAL);
2116            }
2117
2118            if (mopt.valid_options & MOPT_ALTCOA)
2119                    bi.mbc_pcoa = mopt.mopt_altcoa;
2120
2121            if (IN6_IS_ADDR_MULTICAST(&bi.mbc_pcoa) ||
2122                    IN6_IS_ADDR_UNSPECIFIED(&bi.mbc_pcoa) ||
2123                    IN6_IS_ADDR_V4MAPPED(&bi.mbc_pcoa) ||
2124                    IN6_IS_ADDR_V4COMPAT(&bi.mbc_pcoa) ||
2125                    IN6_IS_ADDR_LOOPBACK(&bi.mbc_pcoa)) {
2126                    /* discard. */
2127                    m_freem(m);
2128                    mip6stat.mip6s_invalidcoa++;
2129                    return (EINVAL);
2130            }
```
_____mip6_cncore.c

2108 The home address of a mobile node is stored in the source address field of the IPv6 header. As already described, the address may be the same as the care-of address when the Binding Update message is for home deregistration.

2110–2130 A mobile node may specify an alternate care-of address when it wants to use an address other than the address specified in the source address field of a Binding Update message. The source address is taken as a care-of address unless the mobile node explicitly specifies it is not. The alternate address is carried in the option field of the Binding Update message. The `mip6_get_mobility_options()` function extracts all options contained in a Mobility message. The function is discussed in Section 14.4. If the message has an Alternate Care-of Address mobility option, the specified address in the option is set as a care-of address. The care-of address must be a global unicast address. We discard the packet if the care-of address is not a global unicast address.

Listing 14-6

```
_____mip6_cncore.c
2132            if ((mopt.valid_options & MOPT_AUTHDATA) &&
2133                 ((mopt.mopt_auth + IP6MOPT_AUTHDATA_SIZE) -
(caddr_t)ip6mu < ip6mulen)) {
2134                    /* Auth. data options is not the last option */
2135                    /* discard. */
2136                    m_freem(m);
2137                    /* XXX Statistics */
2138                    return (EINVAL);
2139            }
2140
2141
2142            if ((mopt.valid_options & (MOPT_AUTHDATA | MOPT_NONCE_IDX)) &&
2143                (ip6mu->ip6mhbu_flags & IP6MU_HOME)) {
2144                    /* discard. */
2145                    m_freem(m);
....
2147                    return (EINVAL);
2148            }
_____mip6_cncore.c
```

Note: Line 2133 is broken here for layout reasons. However, it is a single line of code.

2132–2148 When a mobile node tries to optimize the path to a correspondent node, the node
needs to send a Binding Update message with an Authentication Data mobility option.
[RFC3775] says that the option must be the last option in the message. If the option is not
located at the end of options, the packet is discarded.

   The packet is discarded if a mobile node uses an Authentication Data mobility option
in a home registration request.

Listing 14-7

```
_____mip6_cncore.c
2150            bi.mbc_seqno = ntohs(ip6mu->ip6mhbu_seqno);
2151            bi.mbc_lifetime = ntohs(ip6mu->ip6mhbu_lifetime) << 2;
/* units of 4 secs */
2152            /* XXX Should this check be done only when this bu is confirmed
with RR ? */
2153            if (bi.mbc_lifetime > MIP6_MAX_RR_BINDING_LIFE)
2154                    bi.mbc_lifetime = MIP6_MAX_RR_BINDING_LIFE;
2155
2156            if (IS_REQUEST_TO_CACHE(bi.mbc_lifetime, &bi.mbc_phaddr, &bi.mbc_pcoa)
2157                && mip6_bc_list_find_withphaddr(&mip6_bc_list, &bi.mbc_pcoa)) {
2158                    /* discard */
2159                    m_freem(m);
2160                    mip6stat.mip6s_circularrefered++;        /* XXX */
2161                    return (EINVAL);
2162            }
2163            if (!bu_safe &&
2164                mip6_is_valid_bu(ip6, ip6mu, ip6mulen, &mopt, &bi.mbc_phaddr,
2165                    &bi.mbc_pcoa, IS_REQUEST_TO_CACHE(bi.mbc_lifetime,
2166                        &bi.mbc_phaddr, &bi.mbc_pcoa), &bi.mbc_status)) {
....
2170                    /* discard. */
2171                    m_freem(m);
2172                    mip6stat.mip6s_rrauthfail++;
2173                    if (bi.mbc_status >= IP6_MH_BAS_HOME_NI_EXPIRED &&
2174                        bi.mbc_status <= IP6_MH_BAS_NI_EXPIRED) {
2175                            bi.mbc_send_ba = 1;
2176                            error = EINVAL;
2177                            goto send_ba;
2178                    }
2179                    return (EINVAL);
2180            }
_____mip6_cncore.c
```

Note: Lines 2151 and 2152 are broken here for layout reasons. However, they are each a single line of code.

2150–2151 A sequence number and lifetime are extracted from the incoming Binding Update message. Both values are stored in network byte order. We need to shift the lifetime value by 2 bits since the value is specified in units of 4 seconds.

2153–2162 The KAME implementation restricts the lifetime of all binding cache information to `MIP6_MAX_RR_BINDING_LIFE` seconds. As the macro name says, the lifetime limitation should be applied to the binding cache entries created by the return routability procedure. In the home registration case, the lifetime of a binding cache can be larger than `MIP6_MAX_RR_BINDING_LIFE`; however, the current implementation does not consider this case and limits the lifetime of all entries to `MIP6_MAX_RR_BINDING_LIFE`.

2156–2162 A care-of address cannot also be a home address of other binding information. Such a binding may cause unwanted loop conditions.

2163–2180 The `mip6_is_valid_bu()` function checks if the Binding Update message is protected by the return routability procedure discussed in Section 15.23 of Chapter 15. If the message is not protected by the IPsec mechanism or the `use_ipsec` sysctl switch is turned off (`bu_safe` is false), we call `mip6_is_valid_bu()`. The function returns a nonzero value if the message is not acceptable. In this case, the node returns a Binding Acknowledgment message with an error status. The error status is decided by the `mip6_is_valid_bu()` function.

Listing 14-8

_____mip6_cncore.c

```
2182                /* ip6_src and HAO has been already swapped at this point. */
2183                mbc = mip6_bc_list_find_withphaddr(&mip6_bc_list, &bi.mbc_phaddr);
2184                if (mbc != NULL) {
2185                        /* check a sequence number. */
2186                        if (MIP6_LEQ(bi.mbc_seqno, mbc->mbc_seqno)) {
....
2192                                /*
2193                                 * the seqno of this binding update is smaller than the
2194                                 * corresponding binding cache.  we send TOO_SMALL
2195                                 * binding ack as an error.  in this case, we use the
2196                                 * coa of the incoming packet instead of the coa
2197                                 * stored in the binding cache as a destination
2198                                 * addrress.  because the sending mobile node's coa
2199                                 * might have changed after it had registered before.
2200                                 */
2201                                bi.mbc_status = IP6_MH_BAS_SEQNO_BAD;
2202                                bi.mbc_seqno = mbc->mbc_seqno;
2203                                bi.mbc_send_ba = 1;
2204                                error = EINVAL;
2205
2206                                /* discard. */
2207                                m_freem(m);
2208                                mip6stat.mip6s_seqno++;
2209                                goto send_ba;
2210                        }
```

_____mip6_cncore.c

2182–2210 A Binding Update message has a sequence number field to detect an out-of-sequence packet. We need to check to see if the sequence number of the incoming Binding Update message is greater than the sequence number of the existing binding cache entry. To compare the sequence numbers, the `MIP6_LEQ()` macro is used. The comparison algorithm is described in Figure 4-3 of Chapter 4. If the sequence number of the incoming message is equal to or smaller than the existing one, the incoming Binding

Update message is discarded. The node needs to reply to the message sender with a
Binding Acknowledgment message with error code IP6_MH_BAS_SEQNO_BAD so that
the message sender can catch up to the latest sequence number and resend a Binding
Update message with a valid sequence number.

Listing 14-9

```
                                                                    mip6_cncore.c
2211                    if ((bi.mbc_flags & IP6MU_HOME) ^ (mbc->mbc_flags &
    IP6MU_HOME)) {
2212                            /* 9.5.1 */
2213                            bi.mbc_status = IP6_MH_BAS_REG_NOT_ALLOWED;
2214                            bi.mbc_send_ba = 1;
2215                            error = EINVAL;
2216
2217                            /* discard. */
2218                            m_freem(m);
2219                            goto send_ba;
2220                    }
2221            }
                                                                    mip6_cncore.c
```
Note: Line 2211 is broken here for layout reasons. However, it is a single line of code.

2211–2221 A mobile node cannot change the home registration flag (the H flag) of the existing
binding cache entry. If there is a binding cache entry that corresponds to the incoming
Binding Update message and the H flag of the message differs from the H flag of the
binding cache entry, the message is discarded. In this case, the node replies a Binding
Acknowledgment message with the error code IP6_MH_BAS_REG_NOT_ALLOWED.

Listing 14-10

```
                                                                    mip6_cncore.c
2223                if (ip6mu->ip6mhbu_flags & IP6MU_HOME) {
2224                        /* request for the home (un)registration. */
2225                        if (!MIP6_IS_HA) {
2226                                /* this is not a homeagent. */
2227                                /* XXX */
2228                                bi.mbc_status = IP6_MH_BAS_HA_NOT_SUPPORTED;
2229                                bi.mbc_send_ba = 1;
2230                                goto send_ba;
2231                        }
2232
2233    #ifdef MIP6_HOME_AGENT
2234                        /* limit the max duration of bindings. */
2235                        if (mip6ctl_hrbc_maxlifetime > 0 &&
2236                            bi.mbc_lifetime > mip6ctl_hrbc_maxlifetime)
2237                                bi.mbc_lifetime = mip6ctl_hrbc_maxlifetime;
2238
2239                        if (IS_REQUEST_TO_CACHE(bi.mbc_lifetime, &bi.mbc_phaddr, &
    bi.mbc_pcoa)) {
2240                                if (mbc != NULL && (mbc->mbc_flags & IP6MU_CLONED)) {
....
2244                                        /* XXX */
2245                                }
2246                                if (mip6_process_hrbu(&bi)) {
....
2250                                        /* continue. */
2251                                }
2252                        } else {
2253                                if (mbc == NULL || (mbc->mbc_flags & IP6MU_CLONED)) {
2254                                        bi.mbc_status = IP6_MH_BAS_NOT_HA;
2255                                        bi.mbc_send_ba = 1;
```

```
2256                                    goto send_ba;
2257                            }
....
2265                            if (mip6_process_hurbu(&bi)) {
....
2269                                    /* continue. */
2270                            }
2271                    }
2272     #endif /* MIP6_HOME_AGENT */
```
——————————————————————————————————————— mip6_cncore.c

Note: Line 2239 is broken here for layout reasons. However, it is a single line of code.

2223–2231 If the incoming Binding Update message has the H flag on, the message is for home registration or home deregistration. A node must return an error with IP6_MH_BAS_HA_NOT_SUPPORTED status if the node is not acting as a home agent.

2235–2272 The lifetime of a binding cache entry for home registration cannot be longer than the value stored in the mip6ctl_hrbc_maxlifetime variable. The mip6ctl_hrbc_maxlifetime variable is tunable; a user can change the value with the **sysctl** command. A mobile node uses a Binding Update message for both home registration and deregistration. We can check to see if the message is for registration or deregistration with the following two conditions.

- If the lifetime requested is 0, the message is for deregistration

- If the home address and the care-of address that are included in the message are the same, the message is for deregistration

If the message is for home registration, the mip6_process_hrbu() function is called, which performs the home registration procedure. The code fragment from 2240 to 2245 is intended for error handling when there is a binding cache entry for the same mobile node and the entry has the CLONED flag set. There is no error processing code right now. If the message is for home deregistration, the mip6_process_hurbu() function is called, which performs the home deregistration procedure. If there is a binding cache entry for the mobile node that sent the incoming Binding Update message and the cache entry has the CLONED flag set, a Binding Acknowledgment message with IP6_MH_BAS_NOT_HA status is sent.

Listing 14-11
——————————————————————————————————————— mip6_cncore.c

```
2273              } else {
2274                      /* request to cache/remove a binding for CN. */
2275                      if (IS_REQUEST_TO_CACHE(bi.mbc_lifetime, &bi.mbc_phaddr, &
         bi.mbc_pcoa)) {
2276                              int bc_error;
2277
2278                              if (mbc == NULL)
2279                                      bc_error = mip6_bc_register(&bi.mbc_phaddr,
2280                                                        &bi.mbc_pcoa,
2281                                                        &bi.mbc_addr,
2282                                                        ip6mu->ip6mhbu_flags,
2283                                                        bi.mbc_seqno,
2284                                                        bi.mbc_lifetime);
2285                              else
2286                                /* Update a cache entry */
2287                                      bc_error = mip6_bc_update(mbc, &bi.mbc_pcoa,
2288                                                        &bi.mbc_addr,
```

```
2289                                                        ip6mu->ip6mhbu_flags,
2290                                                        bi.mbc_seqno,
2291                                                        bi.mbc_lifetime);
2292                     } else {
2293                             mip6_bc_delete(mbc);
2294                     }
2295             }
```
—— mip6_cncore.c

Note: Line 2275 is broken here for layout reasons. However, it is a single line of code.

2274–2291 If the H flag is not set in the incoming Binding Update message, the request is sent to a correspondent node. If the message is a request to register the binding information, then the `mip6_bc_register()` or `mip6_bc_update()` function is called. The `mip6_bc_register()` function creates a new binding cache entry and is called when there is no binding cache entry corresponding to the incoming Binding Update message. The `mip6_bc_update()` function is used to update the existing binding cache entry.

2293 If the incoming message is a request to remove the existing binding cache information, the `mip6_bc_delete()` function is called to remove the corresponding binding cache entry.

Listing 14-12

—— mip6_cncore.c
```
2297    send_ba:
2298            if (bi.mbc_send_ba) {
2299                    int ba_error;
2300
2301                    ba_error = mip6_bc_send_ba(&bi.mbc_addr, &bi.mbc_phaddr,
2302                            &bi.mbc_pcoa, bi.mbc_status, bi.mbc_seqno,
2303                            bi.mbc_lifetime, bi.mbc_refresh, &mopt);
2304                    if (ba_error) {
.... (log the error)
2308                    }
2309            }
2310
2311            return (error);
2312    }
```
—— mip6_cncore.c

2297–2311 We need to send back a Binding Acknowledgment message in certain cases:

- There is an error in processing a Binding Update message
- The Binding Update message is for home registration/deregistration

If one of the conditions listed above matches, the `mbc_send_ba` variable is set to 1. The `mip6_bc_send_ba()` function creates a Binding Acknowledgment message and sends it to the appropriate destination. If the sending process fails, an error is logged and packet processing is continued.

14.3   Binding Cache Entry Management

A binding cache entry is represented by the `mip6_bc{}` structure. In an IPv6 node, binding cache entries are managed as a list of `mip6_bc{}` entries. The Mobile IPv6 stack provides several access methods for the structure and list. Table 14-2 shows the list of access functions.

TABLE 14-2

Name	Description
mip6_bc_create()	Create an mip6_bc structure
mip6_bc_delete()	Delete an mip6_bc structure
mip6_bc_list_insert()	Insert an mip6_bc structure to the list
mip6_bc_list_remove()	Remove an mip6_bc structure from the list
mip6_bc_list_find_withphaddr()	Search an mip6_bc structure from the list that has the specified home address
mip6_bc_settimer()	Set a timer to call the timeout function
mip6_bc_timer()	The function called when a timer expires

Binding cache entry management functions.

14.3.1 Creating a `mip6_bc{}` Structure

A mip6_bc{} structure is created by the mip6_bc_create() function.

Listing 14-13

_____ mip6_cncore.c

```
750     struct mip6_bc *
751     mip6_bc_create(phaddr, pcoa, addr, flags, seqno, lifetime, ifp)
752             struct in6_addr *phaddr;
753             struct in6_addr *pcoa;
754             struct in6_addr *addr;
755             u_int8_t flags;
756             u_int16_t seqno;
757             u_int32_t lifetime;
758             struct ifnet *ifp;
759     {
760             struct mip6_bc *mbc;
....
765             MALLOC(mbc, struct mip6_bc *, sizeof(struct mip6_bc),
766                     M_TEMP, M_NOWAIT);
767             if (mbc == NULL) {
....
771                     return (NULL);
772             }
773             bzero(mbc, sizeof(*mbc));
774
775             mbc->mbc_phaddr = *phaddr;
776             mbc->mbc_pcoa = *pcoa;
777             mbc->mbc_addr = *addr;
778             mbc->mbc_flags = flags;
779             mbc->mbc_seqno = seqno;
780             mbc->mbc_lifetime = lifetime;
781             mbc->mbc_state = MIP6_BC_FSM_STATE_BOUND;
782             mbc->mbc_mpa_exp = time_second; /* set to current time to send mpa as
    soon as created it */
783             mbc->mbc_ifp = ifp;
784             mbc->mbc_llmbc = NULL;
785             mbc->mbc_refcnt = 0;
....
789             callout_init(&mbc->mbc_timer_ch);
....
793             mbc->mbc_expire = time_second + lifetime;
794             /* sanity check for overflow */
```

```
795              if (mbc->mbc_expire < time_second)
796                      mbc->mbc_expire = 0x7fffffff;
797              mip6_bc_settimer(mbc, mip6_brr_time(mbc));
798
799              if (mip6_bc_list_insert(&mip6_bc_list, mbc)) {
800                      FREE(mbc, M_TEMP);
801                      return (NULL);
802              }
803
804              return (mbc);
805      }
```
                                                                        ———————mip6_cncore.c

Note: Line 782 is broken here for layout reasons. However, it is a single line of code.

750–759 The `mip6_bc_create()` function has seven parameters. The `phaddr` and `pcoa` parameters are, respectively, the home address and the care-of address of the communication peer; the `addr` parameter is the IPv6 address of this node; the `flags`, `seqno`, and `lifetime` parameters are information that is included in the incoming Binding Update message. The `ifp` parameter is the network interface of this node on which the Binding Update message has been received.

765–796 Memory is allocated for the `mip6_bc{}` structure and each member variable is filled with parameters specified in the parameter list of the function. The `mbc_state` variable is initialized with `MIP6_BC_FSM_STATE_BOUND`, which means the lifetime left is long enough. The `mbc_mpa_exp` variable is not currently used. The `callout_init()` function initializes the timer routine related to the timer handler `mbc_timer_ch`. Each entry has its own timer handler to manage its timeouts. The `mbc_expire` variable is set to the time when the lifetime of this entry expires. The variable is set to 0x7fffffff if the value overflows the limitation of the `time_t` type.

797 The `mip6_bc_settimer()` function schedules the next timeout for the entry. The `mip6_brr_time()` function returns the time when the node should send a Binding Refresh Request message to the mobile node of this entry so that the entry does not expire. The timer function for a binding cache entry is called when the time specified to send a Binding Refresh Request message has passed.

799–804 The `mip6_bc_list_insert()` function inserts the `mip6_bc{}` instance in the list specified by the first parameter of the `mip6_bc_list_insert()` function. NULL will be returned if the insertion fails, otherwise, the pointer to the new instance of `mip6_bc{}` structure is returned.

14.3.2   Deleting the `mip6_bc{}` Structure

The `mip6_bc_delete()` function removes an instance of the `mip6_bc{}` structure from `mip6_bc_list`.

Listing 14-14
                                                                        ———————mip6_cncore.c
```
854      static int
855      mip6_bc_delete(mbc)
856              struct mip6_bc *mbc;
857      {
858              int error;
859
```

```
860                   /* a request to delete a binding. */
861              if (mbc) {
862                      error = mip6_bc_list_remove(&mip6_bc_list, mbc);
863                      if (error) {
....
867                              return (error);
868                      }
869              } else {
870                      /* There was no Binding Cache entry */
871                      /* Is there something to do ? */
872              }
873
874              return (0);
875      }
```
——— mip6_cncore.c

854–857 The mip6_bc_delete() function takes a pointer of the mip6_bc{} instance as the parameter mbc.

861–874 The mip6_bc_list_remove() function is called to remove the entry if mbc is not a NULL pointer, otherwise nothing happens. The mip6_bc_delete() function returns to 0 when the removal succeeded or returns an error code of the mip6_bc_list_remove() function if any error occurs when removing the entry.

14.3.3 Inserting the mip6_bc{} Structure to List

The mip6_bc_list_insert() function inserts a newly created instance of the mip6_bc{} structure to the mip6_bc_list.

Listing 14-15

——— mip6_cncore.c
```
877      static int
878      mip6_bc_list_insert(mbc_list, mbc)
879              struct mip6_bc_list *mbc_list;
880              struct mip6_bc *mbc;
881      {
882              int id = MIP6_BC_HASH_ID(&mbc->mbc_phaddr);
883
884              if (mip6_bc_hash[id] != NULL) {
885                      LIST_INSERT_BEFORE(mip6_bc_hash[id], mbc, mbc_entry);
886              } else {
887                      LIST_INSERT_HEAD(mbc_list, mbc, mbc_entry);
888              }
889              mip6_bc_hash[id] = mbc;
890
891              mbc->mbc_refcnt++;
892
893              return (0);
894      }
```
——— mip6_cncore.c

877–894 The mip6_bc_list_insert() function has two parameters. The mbc_list parameter is a pointer to the list of mip6_bc{} structures and usually points to the mip6_bc_list global variable. The mbc parameter is a pointer to the newly created instance of the mip6_bc structure. The list of mip6_bc{} structures is maintained as a hashed list. The MIP6_BC_HASH_ID() macro computes the hash ID of the mip6_bc{} instance based on the peer home address, mbc_phaddr. If there is an mip6_bc{} instance that has the same hash ID in the mip6_bc_list already, the new instance

FIGURE 14-1

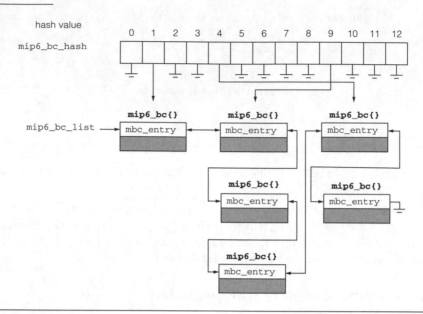

The hashed list structure of the `mip6_bc{}` *structure.*

is inserted before the existing instance; otherwise, the new instance is inserted at the head of the `mip6_bc_list`. Figure 14-1 shows the structure of the hashed list. In the figure, three different hash IDs (1, 4, 9) are active. The Hash Index 1 points to the first `mip6_bc{}` instance in the `mip6_bc_list`. The Hash Index 9 points to the second `mip6_bc{}` instance. The third and fourth instances have the same hash value with the second instance. The Hash Index 4 points to the fifth instance and the last instance also has the same hash value.

14.3.4   Removing the `mip6_bc{}` Structure from List

The `mip6_bc_list_remove()` function removes the specified `mip6_bc{}` instance from the `mip6_bc_list`.

Listing 14-16

—— `mip6_cncore.c`

```
896      int
897      mip6_bc_list_remove(mbc_list, mbc)
898           struct mip6_bc_list *mbc_list;
899           struct mip6_bc *mbc;
900      {
901           int error = 0;
902           int id;
903
904           if ((mbc_list == NULL) || (mbc == NULL)) {
905                return (EINVAL);
906           }
907
```

```
908                 id = MIP6_BC_HASH_ID(&mbc->mbc_phaddr);
909                 if (mip6_bc_hash[id] == mbc) {
910                         struct mip6_bc *next = LIST_NEXT(mbc, mbc_entry);
911                         if (next != NULL &&
912                             id == MIP6_BC_HASH_ID(&next->mbc_phaddr)) {
913                                 mip6_bc_hash[id] = next;
914                         } else {
915                                 mip6_bc_hash[id] = NULL;
916                         }
917                 }
```
                                                              _____mip6_cncore.c

896–899 The `mip6_bc_list_remove()` function has two parameters. The `mbc_list` parameter is a pointer to the list of `mip6_bc{}` structures and usually points to the `mip6_bc_list` global variable. The `mbc` parameter is a pointer to the instance of the `mip6_bc{}` structure to be removed.

908–917 We need to rehash the hash table of the `mip6_bc{}` list since removing an entry may cause inconsistency in the hash table.

Listing 14-17
                                                              _____mip6_cncore.c

```
919                 mbc->mbc_refcnt--;
920                 if (mbc->mbc_flags & IP6MU_CLONED) {
921                         if (mbc->mbc_refcnt > 1)
922                                 return (0);
923                 } else {
924                         if (mbc->mbc_refcnt > 0)
925                                 return (0);
926                 }
927                 mip6_bc_settimer(mbc, -1);
928                 LIST_REMOVE(mbc, mbc_entry);
```
                                                              _____mip6_cncore.c

920–926 The `IP6MU_CLONED` flag is a special internal flag, which indicates that the `mip6_bc{}` instance is cloned as a result of a Binding Update message with the link-local compatibility bit (the L bit). Only home agents have such entries that have the L bit on.

Every `mip6_bc{}` instance has a reference count. The entry is released only when there are no references from other instances. The exception is a cloned entry. A cloned entry has two references initially. One is a reference as a list entry, the other is a reference from the original entry that clones the entry. An entry is released when the reference counter goes to 1 if `IP6MU_CLONED` is set. Otherwise, it is released when the reference count goes to 0.

927–928 The `mip6_bc_settimer()` function controls the timer for the entry. Specifying -1 stops the timer function. After stopping the timer, the entry is removed from the list. At this point, the entry itself is not released.

Listing 14-18
                                                              _____mip6_cncore.c

```
929     #ifdef MIP6_HOME_AGENT
930             if (mbc->mbc_flags & IP6MU_HOME) {
931                     if (MIP6_IS_BC_DAD_WAIT(mbc)) {
932                             mip6_dad_stop(mbc);
933                     } else {
934                             error = mip6_bc_proxy_control(&mbc->mbc_phaddr,
```

```
935                                  &mbc->mbc_addr, RTM_DELETE);
936                              if (error) {
 ....
942                              }
943                              error = mip6_tunnel_control(MIP6_TUNNEL_DELETE,
944                                  mbc, mip6_bc_encapcheck, &mbc->mbc_encap);
945                              if (error) {
 ....
951                              }
952                          }
953                      }
954      #endif /* MIP6_HOME_AGENT */
955              FREE(mbc, M_TEMP);
956
957              return (error);
958      }
```
_____ mip6_cncore.c

929–954 If a node is acting as a home agent, then the proxy Neighbor Discovery and tunneling for the mobile node needs to be stopped. The `MIP6_IS_BC_DAD_WAIT()` macro checks to see if the entry has finished the Duplicate Address Detection (DAD) procedure. If the DAD procedure of the entry is incomplete, the entry can simply be released since the proxy/tunnel service has not started yet. If the DAD procedure has finished, the proxy service has to be stopped by calling the `mip6_bc_proxy_control()` function and the tunneling service has to be stopped by calling the `mip6_tunnel_control()` function.

955–957 If an error occurs, the memory used for the entry is released and the error code is returned.

14.3.5   Looking Up the `mip6_bc{}` Structure

The `mip6_bc_list_find_withphaddr()` function is used when we need to find a certain `mip6_bc{}` instance from the `mip6_bc_list` by its peer home address.

Listing 14-19
_____ mip6_cncore.c

```
960      struct mip6_bc *
961      mip6_bc_list_find_withphaddr(mbc_list, haddr)
962              struct mip6_bc_list *mbc_list;
963              struct in6_addr *haddr;
964      {
965              struct mip6_bc *mbc;
966              int id = MIP6_BC_HASH_ID(haddr);
967
968              for (mbc = mip6_bc_hash[id]; mbc;
969                  mbc = LIST_NEXT(mbc, mbc_entry)) {
970                      if (MIP6_BC_HASH_ID(&mbc->mbc_phaddr) != id)
971                              return NULL;
972                      if (IN6_ARE_ADDR_EQUAL(&mbc->mbc_phaddr, haddr))
973                              break;
974              }
975
976              return (mbc);
977      }
```
_____ mip6_cncore.c

960–963 The `mip6_bc_list_find_withphaddr()` function has two parameters. The `mbc_list` parameter is a pointer to the list of `mip6_bc{}` structures and usually points to the `mip6_bc_list` global variable. The `haddr` parameter is the home address of the `mip6_bc{}` entry we are looking for.

968–976 As described in Listing 14-15, the `mip6_bc_list` variable is maintained as a hashed list. The hash ID is computed from the home address specified as the second parameter, and the hash list of binding cache entries is checked to find the target entry by comparing `haddr` and `mbc_phaddr` variables.

14.3.6 Timer Processing of the `mip6_bc{}` Structure

The `mip6_bc_settimer()` sets the next timeout of an `mip6_bc{}` instance. The `mip6_bc_timer()` function will be called when the timer of the entry expires.

Listing 14-20
_____ mip6_cncore.c

```
815     void
816     mip6_bc_settimer(mbc, t)
817             struct mip6_bc *mbc;
818             int t;   /* unit: second */
819     {
820             long tick;
821             int s;
822
....
826             s = splnet();
....
828
829             if (t != 0) {
830                     tick = t * hz;
831                     if (t < 0) {
....
833                             callout_stop(&mbc->mbc_timer_ch);
....
839                     } else {
....
841                             callout_reset(&mbc->mbc_timer_ch, tick,
842                                 mip6_bc_timer, mbc);
....
848                     }
849             }
850
851             splx(s);
852     }
```
_____ mip6_cncore.c

815–818 The `mip6_bc_settimer()` has two parameters. The `mbc` parameter is a pointer to the `mip6_bc{}` instance whose timer is set, and the `t` parameter is the time until the next timeout event in seconds.

829–849 The `t` parameter can be a negative value, in which case, the timer configuration is cleared by calling the `callout_stop()` function. Otherwise, the next timeout event is set by calling the `callout_reset()` function.

Listing 14-21
_____ mip6_cncore.c

```
1256    static void
1257    mip6_bc_timer(arg)
1258            void *arg;
```

```
1259    {
1260            int s;
1261            u_int brrtime;
1262            struct mip6_bc *mbc = arg;
....
1270            s = splnet();
....
1272
1273            switch (mbc->mbc_state) {
1274            case MIP6_BC_FSM_STATE_BOUND:
1275                    mbc->mbc_state = MIP6_BC_FSM_STATE_WAITB;
1276                    mbc->mbc_brr_sent = 0;
1277                    /* No break; */
1278            case MIP6_BC_FSM_STATE_WAITB:
1279                    if (mip6_bc_need_brr(mbc) &&
1280                        (mbc->mbc_brr_sent < mip6_brr_maxtries)) {
1281                            brrtime = mip6_brr_time(mbc);
1282                            if (brrtime == 0) {
1283                                    mbc->mbc_state = MIP6_BC_FSM_STATE_WAITB2;
1284                            } else {
1285                                    mip6_bc_send_brr(mbc);
1286                            }
1287                            mip6_bc_settimer(mbc, mip6_brr_time(mbc));
1288                            mbc->mbc_brr_sent++;
1289                    } else {
1290                            mbc->mbc_state = MIP6_BC_FSM_STATE_WAITB2;
1291                            mip6_bc_settimer(mbc, mbc->mbc_expire - time_second);
1292                    }
1293                    break;
1294            case MIP6_BC_FSM_STATE_WAITB2:
1295    #ifdef MIP6_HOME_AGENT
1296                    if (mbc->mbc_flags & IP6MU_CLONED) {
1297                            /*
1298                             * cloned entry is removed
1299                             * when the last referring mbc
1300                             * is removed.
1301                             */
1302                            break;
1303                    }
1304                    if (mbc->mbc_llmbc != NULL) {
1305                            /* remove a cloned entry. */
1306                            if (mip6_bc_list_remove(
1307                                &mip6_bc_list, mbc->mbc_llmbc) != 0) {
....
1312                            }
1313                    }
1314    #endif /* MIP6_HOME_AGENT */
1315                    mip6_bc_list_remove(&mip6_bc_list, mbc);
1316                    break;
1317            }
1318
1319            splx(s);
1320    }
```

_____ mip6_cncore.c

1256–1258 The `mip6_bc_timer()` function has one parameter that specifies a pointer to the `mip6_bc{}` instance whose timer expired.

1273–1277 An `mip6_bc{}` structure has three states as described in Table 10-11 of Chapter 10. When the state is BOUND, it is changed to WAITB when the timer expires.

1278–1286 When the state is WAITB, a Binding Refresh Request message is sent. The `mip6_bc_need_brr()` function checks the current Transmission Control Protocol (TCP) connections to see if there is any TCP connection related to this `mip6_bc{}` entry.

The `mip6_brr_maxtries` variable is an upper limit on the number of Binding Refresh Request messages to be sent; its default value is 2. If `mip6_bc_need_brr()` returns true and `mbc_brr_sent` is smaller than `mip6_brr_maxtries`, then a Binding Refresh Request message is sent: the `mip6_brr_time()` function returns the time when the next Binding Refresh Request message should be sent. If `mip6_brr_time()` returns 0, the state is changed to `WAITB2`; otherwise, a Binding Refresh Request message is sent by the `mip6_bc_send_brr()` function.

1287–1288 The next timeout event is scheduled by calling the `mip6_bc_settimer()` function and a counter that tracks the number of Binding Refresh Request messages sent is incremented.

1290–1291 If `mip6_bc_need_brr()` returns false, the state is changed to `WAITB2` and the next timeout is set to the time when the entry expires.

1294–1316 If the state is `WAITB2`, the entry is removed. When removing an entry, cloned entries must also be considered. These entries must be removed after all references from other `mip6_bc{}` instances are removed. A cloned entry is referenced to by the `mbc_llmbc` member variable of the `mip6_bc{}` structure. If the entry is not a cloned entry and the entry has a valid `mbc_llmbc` pointer, the cloned entry is removed or the reference counter of the cloned entry is decremented as discussed in Section 14.3.

Figure 14-2 shows the state transition diagram of a binding cache entry.

FIGURE 14-2

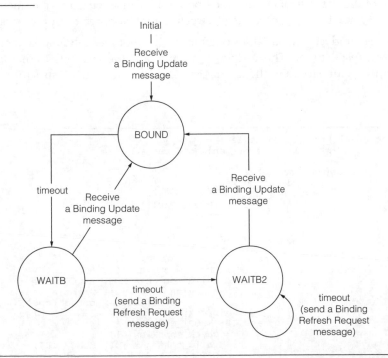

The state transition diagram of a binding cache entry.

14.4   Mobility Options Processing

Some mobility messages have optional parameters at the end of the message. These options are called mobility options. The options are encoded in the Type-Length-Value (TLV) format. The `mip6_get_mobility_options()` parses the options and sets the values of each option in the `mip6_mobility_options{}` structure.

Listing 14-22

```
                                                                    mip6_cncore.c
2611    int
2612    mip6_get_mobility_options(ip6mh, hlen, ip6mhlen, mopt)
2613            struct ip6_mh *ip6mh;
2614            int hlen, ip6mhlen;
2615            struct mip6_mobility_options *mopt;
2616    {
2617            u_int8_t *mh, *mhend;
2618            u_int16_t valid_option;
2619
2620            mh = (caddr_t)(ip6mh) + hlen;
2621            mhend = (caddr_t)(ip6mh) + ip6mhlen;
2622            mopt->valid_options = 0;
                                                                    mip6_cncore.c
```

2611–2616 The `mip6_get_mobility_options()` function has four parameters. The `ip6mh` parameter is a pointer to the head of the incoming Mobility Header message; the `hlen` parameter is the size of the message excluding the option area; the `ip6mhlen` parameter is the size of the message including the options area; and the `mopt` parameter is a pointer to the `mip6_mobility_options{}` structure that stores the result of this function.

2620–2622 The `mh` and `mhend` variables point to the head of the option area and the tail of the option area, respectively. The `valid_options` variable is a bitfield that indicates what kind of options are stored in the `mip6_mobility_options{}` structure.

Listing 14-23

```
                                                                    mip6_cncore.c
2624    #define check_mopt_len(mopt_len)             \
2625            if (*(mh + 1) != mopt_len) goto bad;
2626
2627            while (mh < mhend) {
2628                    valid_option = 0;
2629                    switch (*mh) {
2630                            case IP6_MHOPT_PAD1:
2631                                    mh++;
2632                                    continue;
2633                            case IP6_MHOPT_PADN:
2634                                    break;
2635                            case IP6_MHOPT_ALTCOA:
2636                                    check_mopt_len(16);
2637                                    valid_option = MOPT_ALTCOA;
2638                                    bcopy(mh + 2, &mopt->mopt_altcoa,
2639                                            sizeof(mopt->mopt_altcoa));
2640                                    break;
2641                            case IP6_MHOPT_NONCEID:
2642                                    check_mopt_len(4);
2643                                    valid_option = MOPT_NONCE_IDX;
2644                                    GET_NETVAL_S(mh + 2, mopt->mopt_ho_nonce_idx);
2645                                    GET_NETVAL_S(mh + 4, mopt->mopt_co_nonce_idx);
2646                                    break;
2647                            case IP6_MHOPT_BAUTH:
```

```
2648                                      valid_option = MOPT_AUTHDATA;
2649                                      mopt->mopt_auth = mh;
2650                                      break;
2651                          case IP6_MHOPT_BREFRESH:
2652                                      check_mopt_len(2);
2653                                      valid_option = MOPT_REFRESH;
2654                                      GET_NETVAL_S(mh + 2, mopt->mopt_refresh);
2655                                      break;
2656                          default:
2657                                      /*        '... MUST quietly ignore ... (6.2.1)'
    ....
2661                                       */
2662                                      break;
2663                          }
2664
2665                  mh += *(mh + 1) + 2;
2666                  mopt->valid_options |= valid_option;
2667          }
2668
2669  #undef check_mopt_len
2670
2671          return (0);
2672
2673   bad:
2674          return (EINVAL);
2675  }
```
—————————————————————————————————— mip6_cncore.c

2624–2625 The check_mopt_len() macro validates the option length. This macro terminates the mip6_get_mobility_options() function if the received option length does not match the length as specified in [RFC3775].

2627–2663 Each option is parsed based on its option type number stored in the first byte of the option. If the option type is one of IP6_MHOPT_ALTCOA, IP6_MHOPT_NONCEID or IP6_MHOPT_BREFRESH, the option values are copied to the corresponding member fields of the mip6_mobility_options{} structure. The macro GET_NETVAL_S() provides a safe operation to copy 2 bytes of data even when the data is not aligned as the processor architecture permits. If the option type is IP6_MHOPT_BAUTH, the address of the option is stored in the mopt_auth member variable. Unknown options are ignored.

2665–2666 The pointer is incremented as specified in the option length field to proceed to the next mobility option, and the valid_options variable is updated to indicate that the structure has a valid option value.

14.5   Validation of Binding Update Message for Correspondent Node

When a correspondent node receives a Binding Update message, the node needs to validate whether or not the message is acceptable. The validation is done by the mip6_is_valid_bu() function.

Listing 14-24
—————————————————————————————————— mip6_cncore.c
```
1527   int
1528   mip6_is_valid_bu(ip6, ip6mu, ip6mulen, mopt, hoa, coa, cache_req, status)
1529          struct ip6_hdr *ip6;
1530          struct ip6_mh_binding_update *ip6mu;
1531          int ip6mulen;
```

```
1532              struct mip6_mobility_options *mopt;
1533              struct in6_addr *hoa, *coa;
1534              int cache_req;  /* true if this request is cacheing */
1535              u_int8_t *status;
1536       {
```
_____mip6_cncore.c

1527–1535 The mip6_is_valid_bu() function has eight parameters. The ip6 and ip6mu
parameters are pointers to the head of the incoming IPv6 packet and Binding Update
message, respectively; the ip6mulen parameter is the size of the Binding Update message;
the mopt parameter is a pointer to the instance of the mip6_mobility_option{}
structure that holds received options; the hoa and coa parameters indicate the home
and care-of addresses of the mobile node, which sent this Binding Update message; the
cache_req parameter is a boolean value that indicates that this message is for registration
or for deregistration; and the status parameter is space to store the status code that is
used when sending a Binding Acknowledgment message.

Listing 14-25
_____mip6_cncore.c
```
1537              u_int8_t key_bm[MIP6_KBM_LEN]; /* Stated as 'Kbm' in the spec */
1538              u_int8_t authdata[SHA1_RESULTLEN];
1539              u_int16_t cksum_backup;
1540
1541              *status = IP6_MH_BAS_ACCEPTED;
1542              /* Nonce index & Auth. data mobility options are required */
1543              if ((mopt->valid_options & (MOPT_NONCE_IDX | MOPT_AUTHDATA))
1544                  != (MOPT_NONCE_IDX | MOPT_AUTHDATA)) {
 ....
1549                  return (EINVAL);
1550              }
1551              if ((*status = mip6_calculate_kbm_from_index(hoa, coa,
1552                  mopt->mopt_ho_nonce_idx, mopt->mopt_co_nonce_idx,
1553                  !cache_req, key_bm))) {
1554                  return (EINVAL);
1555              }
1556
1557              cksum_backup = ip6mu->ip6mhbu_cksum;
1558              ip6mu->ip6mhbu_cksum = 0;
1559              /* Calculate authenticator */
1560              if (mip6_calculate_authenticator(key_bm, authdata, coa, &ip6->ip6_dst,
1561                  (caddr_t)ip6mu, ip6mulen,
1562                  (u_int8_t *)mopt->mopt_auth + sizeof(struct ip6_mh_opt_auth_data)
   - (u_int8_t *)ip6mu,
1563                  MOPT_AUTH_LEN(mopt) + 2)) {
1564                  return (EINVAL);
1565              }
1566
1567              ip6mu->ip6mhbu_cksum = cksum_backup;
1568
1569              return (bcmp(mopt->mopt_auth + 2, authdata, MOPT_AUTH_LEN(mopt)));
1570       }
```
_____mip6_cncore.c
Note: Line 1562 is broken here for layout reasons. However, it is a single line of code.

1541–1555 A Binding Update message to a correspondent node must include the nonce index
and authentication data mobility options. An error is returned if the incoming packet does
not have these options. The mip6_calculate_kbm_from_index() function computes the K_{bm} value from home address, care-of address, home nonce, and care-of nonce.
The result is stored in the key_mb variable.

1557–1565 The authentication data value is verified. The `mip6_calculate_ authenticator()` function computes the value of the authentication data field using the K_{bm} computed previously. The result will be stored in `authdata`, the second parameter of the function. Note that we need to clear the checksum field of a Mobility Header message before computing the authentication data value, since the sender of a Binding Update message clears the field when computing the authentication value.

1567–1569 After finishing the validation of authentication data, the function recovers the checksum field and returns the result of the comparison of the authentication data value sent as an option value (the `mopt_auth` member variable) and the verified value (the `authdata` variable).

14.6 K_{bm} and Authorization Data Computation

As described in Chapter 5, a Binding Update message and a Binding Acknowledgment message must have an authentication data mobility option to protect them. The content of the authentication data is a hash value based on HMAC-SHA1. The key of the hash function is computed by the `mip6_calculate_kbm_from_index()` function and the value is computed by the `mip6_calculate_authenticator()` function.

Listing 14-26

`mip6_cncore.c`

```
1572    int
1573    mip6_calculate_kbm_from_index(hoa, coa, ho_nonce_idx, co_nonce_idx,
      ignore_co_nonce, key_bm)
1574            struct in6_addr *hoa;
1575            struct in6_addr *coa;
1576            u_int16_t ho_nonce_idx; /* Home Nonce Index */
1577            u_int16_t co_nonce_idx; /* Care-of Nonce Index */
1578            int ignore_co_nonce;
1579            u_int8_t *key_bm;       /* needs at least MIP6_KBM_LEN bytes */
1580    {
```

`mip6_cncore.c`

Note: Line 1573 is broken here for layout reasons. However, it is a single line of code.

1572–1579 The `mip6_calculate_kbm_from_index()` function has six parameters. The `hoa` and `coa` parameters are the home and care-of addresses of a Binding Update message; the `ho_nonce_idx` and `co_nonce_idx` parameters are indices that indicate the home and care-of nonces that are used to compute the key (K_{bm}). The `ignore_co_nonce` parameter is a flag that is set when deregistering from home network. When a mobile node is home, the care-of nonce is ignored since the home and care-of addresses are the same address. The variable `key_bm` points to the address where the key value is stored.

Listing 14-27

`mip6_cncore.c`

```
1581            int stat = IP6_MH_BAS_ACCEPTED;
1582            mip6_nonce_t home_nonce, careof_nonce;
1583            mip6_nodekey_t home_nodekey, coa_nodekey;
1584            mip6_home_token_t home_token;
1585            mip6_careof_token_t careof_token;
```

```
1586
1587            if (mip6_get_nonce(ho_nonce_idx, &home_nonce) != 0) {
....
1592                    stat =IP6_MH_BAS_HOME_NI_EXPIRED;
1593            }
1594            if (!ignore_co_nonce &&
1595                mip6_get_nonce(co_nonce_idx, &careof_nonce) != 0){
....
1600                    stat = (stat == IP6_MH_BAS_ACCEPTED) ?
1601                        IP6_MH_BAS_COA_NI_EXPIRED : IP6_MH_BAS_NI_EXPIRED;
1602            }
1603            if (stat != IP6_MH_BAS_ACCEPTED)
1604                    return (stat);
```
————————————————————————————————————— mip6_cncore.c

1587–1604 The `mip6_get_nonce()` function finds a nonce value based on the index number provided as the first parameter. The function returns true when the index has already been expired and is invalid. `IP6_MH_BAS_HOME_NI_EXPIRED` is set as a status value if the home nonce index is invalid.

    If the message is for deregistration, the `ignore_co_nonce` variable is set to true and the care-of nonce can be ignored. If the `ignore_co_nonce` variable is false, the `mip6_get_nonce()` function is called to check whether the care-of nonce index is valid. If both home and care-of nonce indices are invalid, `IP6_MH_BA_NI_EXPIRED` is set as a status value. If only the care-of nonce index is invalid, `IP6_MH_BA_COA_NI_EXPIRED` is set as a status value.

    `IP6_MH_BAS_ACCEPTED` is set as a status value if both nonce indices are valid.

Listing 14-28

————————————————————————————————————— mip6_cncore.c
```
1610            if ((mip6_get_nodekey(ho_nonce_idx, &home_nodekey) != 0) ||
1611                (!ignore_co_nonce &&
1612                    (mip6_get_nodekey(co_nonce_idx, &coa_nodekey) != 0))) {
....
1617                    return (IP6_MH_BAS_NI_EXPIRED);
1618            }
....
1624            /* Calculate home keygen token */
1625            if (mip6_create_keygen_token(hoa, &home_nodekey, &home_nonce, 0,
1626                &home_token)) {
....
1631                    return (IP6_MH_BAS_UNSPECIFIED);
1632            }
....
1637            if (!ignore_co_nonce) {
1638                    /* Calculate care-of keygen token */
1639                    if (mip6_create_keygen_token(coa, &coa_nodekey, &careof_nonce,
1640                        1, &careof_token)) {
....
1645                            return (IP6_MH_BAS_UNSPECIFIED);
1646                    }
....
1650            }
1651
1652            /* Calculate K_bm */
1653            mip6_calculate_kbm(&home_token, ignore_co_nonce ? NULL : &careof_token,
1654                key_bm);
....
1658
1659            return (IP6_MH_BAS_ACCEPTED);
1660    }
```
————————————————————————————————————— mip6_cncore.c

1610–1618 The `mip6_get_nodekey()` function finds a nodekey value based on the index value specified as the first parameter. Nonce and nodekey management is discussed in Section 14.14. Nodekeys for the home and care-of addresses are stored in the `home_nodekey` and the `coa_nodekey` variables. If any error occurs while getting nodekey values, `IP6_MH_BAS_NI_EXPIRED` is returned.

1625–1650 The `mip6_create_keygen_token()` function computes a keygen token based on home and care-of addresses, nodekey and nonce values. The function is described later in this section. If computing these tokens meets any error, then `IP6_MH_BAS_UNSPECIFIED` is returned.

1653–1659 The K_{bm} value is computed from keygen tokens by calling the `mip6_calculate_kbm()` function. This function is described later in this section.

Listing 14-29

mip6_cncore.c

```
1429    int
1430    mip6_create_keygen_token(addr, nodekey, nonce, hc, token)
1431            struct in6_addr *addr;
1432            mip6_nodekey_t *nodekey;
1433            mip6_nonce_t *nonce;
1434            u_int8_t hc;
1435            void *token;              /* 64 bit */
1436    {
1437            /* keygen token = HMAC_SHA1(Kcn, addr | nonce | hc) */
1438            HMAC_CTX hmac_ctx;
1439            u_int8_t result[HMACSIZE];
1440
1441            hmac_init(&hmac_ctx, (u_int8_t *)nodekey,
1442                      sizeof(mip6_nodekey_t), HMAC_SHA1);
1443            hmac_loop(&hmac_ctx, (u_int8_t *)addr, sizeof(struct in6_addr));
1444            hmac_loop(&hmac_ctx, (u_int8_t *)nonce, sizeof(mip6_nonce_t));
1445            hmac_loop(&hmac_ctx, (u_int8_t *)&hc, sizeof(hc));
1446            hmac_result(&hmac_ctx, result, sizeof(result));
1447            /* First64 */
1448            bcopy(result, token, 8);
1449
1450            return (0);
1451    }
```

mip6_cncore.c

1430–1435 The `mip6_create_keygen_token()` function has five parameters. The `addr` parameter is an IPv6 address, which is either the home or the care-of address of a Mobility Header message, and the `hc` parameter is a decimal value either 0 or 1. 0, which is specified if `addr` is a home address and 1 is specified if `addr` is a care-of address. The parameter `token` points to the address in which the computed keygen token is stored.

1438–1450 Keygen token is a hash value computed over the concatenation of a nodekey, a home or care-of address, a nonce value and 0 or 1 based on the kind of address. The algorithm (HMAC-SHA1) used to compute the hash value generates 128-bit data. Only the first 64 bits of data is used for a keygen token since it needs only 64-bit data.

Listing 14-30

mip6_cncore.c

```
1662    void
1663    mip6_calculate_kbm(home_token, careof_token, key_bm)
```

```
1664            mip6_home_token_t *home_token;
1665            mip6_careof_token_t *careof_token;        /* could be null */
1666            u_int8_t *key_bm;        /* needs at least MIP6_KBM_LEN bytes */
1667    {
1668            SHA1_CTX sha1_ctx;
1669            u_int8_t result[SHA1_RESULTLEN];
1670
1671            SHA1Init(&sha1_ctx);
1672            SHA1Update(&sha1_ctx, (caddr_t)home_token, sizeof(*home_token));
1673            if (careof_token)
1674                    SHA1Update(&sha1_ctx, (caddr_t)careof_token,
    sizeof(*careof_token));
1675            SHA1Final(result, &sha1_ctx);
1676            /* First 128 bit */
1677            bcopy(result, key_bm, MIP6_KBM_LEN);
1678    }
```
── mip6_cncore.c

Note: Line 1674 is broken here for layout reasons. However, it is a single line of code.

1663–1666 The `mip6_calculate_kbm()` function has three parameters. The `home_token` and `careof_token` parameters are the keygen tokens for the home and the care-of addresses, and the `key_bm` parameter stores the computed K_{bm} value.

1668–1677 The K_{bm} value is computed using the SHA1 hash algorithm over the concatenated data of a home keygen token and a care-of keygen token. Note that a care-of keygen token may not be specified when performing deregistering from the home network. The first 128 bits of the computed data (which are 196 bits long) are copied as a K_{bm} value.

Listing 14-31

── mip6_cncore.c

```
1690    int
1691    mip6_calculate_authenticator(key_bm, result, addr1, addr2, data, datalen,
1692        exclude_offset, exclude_data_len)
1693            u_int8_t *key_bm;                /* Kbm */
1694            u_int8_t *result;
1695            struct in6_addr *addr1, *addr2;
1696            caddr_t data;
1697            size_t datalen;
1698            int exclude_offset;
1699            size_t exclude_data_len;
1700    {
1701            HMAC_CTX hmac_ctx;
1702            int restlen;
1703            u_int8_t sha1_result[SHA1_RESULTLEN];
1704
1705            /* Calculate authenticator (5.5.6) */
1706            /* MAC_Kbm(addr1, | addr2 | (BU|BA) ) */
1707            hmac_init(&hmac_ctx, key_bm, MIP6_KBM_LEN, HMAC_SHA1);
1708            hmac_loop(&hmac_ctx, (u_int8_t *)addr1, sizeof(*addr1));
....
1712            hmac_loop(&hmac_ctx, (u_int8_t *)addr2, sizeof(*addr2));
....
1716            hmac_loop(&hmac_ctx, (u_int8_t *)data, exclude_offset);
....
1721            /* Exclude authdata field in the mobility option to calculate authdata
1722                But it should be included padding area */
1723            restlen = datalen - (exclude_offset + exclude_data_len);
1724            if (restlen > 0) {
1725                    hmac_loop(&hmac_ctx,
1726                            data + exclude_offset + exclude_data_len,
1727                            restlen);
....
1732            }
```

```
1733                    hmac_result(&hmac_ctx, sha1_result, sizeof(sha1_result));
1734                    /* First(96, sha1_result) */
1735                    bcopy(sha1_result, result, MIP6_AUTHENTICATOR_LEN);
    ....
1739
1740                    return (0);
1741        }
```
                                                                                    ———— mip6_cncore.c

1690–1699 The `mip6_calculate_authenticator()` function has eight parameters. The `key_bm` parameter is the K_{bm} value computed by the `mip6_calculate_kbm()` function; the `result` parameter points to the memory in which the computed authentication data is stored; the `addr1` and `addr2` parameters are the care-of and the destination addresses of the Mobility Header message that will be protected; the `data` and `datalen` parameters specify the address of the Mobility Header message and size, respectively; and the `exclude_offset` and `exclude_data_len` parameters specify the region of the authentication data itself located in the Mobility Header message. The region must not be included in computation. Figure 14-3 shows the meanings of each parameter.

1707–1716 The computation algorithm we use is HMAC-SHA1. We first compute two addresses specified as parameters and the Mobility Header message before the region that is excluded.

1723–1740 The specification says that the authentication data should be placed as the last mobility option; however, there is a possibility that there is a padding option after the authentication data. The `restlen` variable indicates the length to the end of the Mobility Header message after the authentication data. The length of the authentication data is 96 bits, whereas the result of HMAC-SHA1 computation will be 128 bits. The first 96 bits are taken as the authentication data value.

14.7 Managing Binding Cache Entry as Correspondent Node

After receiving a valid Binding Update message from a mobile node, a correspondent node will create a binding cache entry to perform the route optimization. The procedure for registering the binding information is implemented in the `mip6_bc_register()` function, and the procedure for updating is implemented in the `mip6_bc_update()` function.

FIGURE 14-3

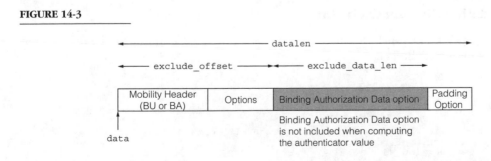

Calculation of the authenticator value.

14.7.1   Adding Binding Cache Entry

The mip6_bc_register() function is used by a correspondent node to create a new binding cache entry.

Listing 14-32

```
                                                                    mip6_cncore.c
979     static int
980     mip6_bc_register(hoa, coa, dst, flags, seqno, lifetime)
981             struct in6_addr *hoa;
982             struct in6_addr *coa;
983             struct in6_addr *dst;
984             u_int16_t flags;
985             u_int16_t seqno;
986             u_int32_t lifetime;
987     {
988             struct mip6_bc *mbc;
989
990             /* create a binding cache entry. */
991             mbc = mip6_bc_create(hoa, coa, dst, flags, seqno, lifetime,
992                 NULL);
993             if (mbc == NULL) {
 ....
998                     return (ENOMEM);
999             }
1000
1001            return (0);
1002    }
                                                                    mip6_cncore.c
```

979–987 The mip6_bc_register() function has six parameters. The hoa and coa parameters are the home and care-of addresses of the mobile node that sent the Binding Update message; the dst parameter is the IPv6 address of the correspondent node; the flags and seqno parameters are copies of the flags and seqno fields of the incoming Binding Update message; and the lifetime parameter is the lifetime of the binding cache entry. As discussed in Section 10.8 of Chapter 10, a correspondent node may not always create a binding cache entry with the lifetime requested by the mobile node. The actual lifetime may be shortened by the correspondent node.

991–1001 The mip6_bc_register() function simply calls the mip6_bc_create() function to create a binding cache entry. It returns ENOMEM error when mip6_bc_create() cannot create a new binding cache entry; otherwise, it returns to 0.

14.7.2   Updating Binding Cache Entry

The mip6_bc_update() function is used by a correspondent node to update the information of an existing binding cache entry.

Listing 14-33

```
                                                                    mip6_cncore.c
1004    static int
1005    mip6_bc_update(mbc, coa, dst, flags, seqno, lifetime)
1006            struct mip6_bc *mbc;
1007            struct in6_addr *coa;
1008            struct in6_addr *dst;
1009            u_int16_t flags;
```

```
1010              u_int16_t seqno;
1011              u_int32_t lifetime;
1012     {
  ....
1014              struct timeval mono_time;
  ....
1018              microtime(&mono_time);
  ....
1020              /* update a binding cache entry. */
1021              mbc->mbc_pcoa = *coa;
1022              mbc->mbc_flags = flags;
1023              mbc->mbc_seqno = seqno;
1024              mbc->mbc_lifetime = lifetime;
1025              mbc->mbc_expire = mono_time.tv_sec + mbc->mbc_lifetime;
1026              /* sanity check for overflow */
1027              if (mbc->mbc_expire < mono_time.tv_sec)
1028                      mbc->mbc_expire = 0x7fffffff;
1029              mbc->mbc_state = MIP6_BC_FSM_STATE_BOUND;
1030              mip6_bc_settimer(mbc, -1);
1031              mip6_bc_settimer(mbc, mip6_brr_time(mbc));
1032
1033              return (0);
1034     }
```
——mip6_cncore.c

1004–1011 The `mip6_bc_update()` function has six parameters, most of which have the same meanings as the `mip6_bc_register()` function. The home address of a mobile node never changes when updating other information. The first parameter of the function is a pointer to the address of the related binding cache entry. The rest of parameters are the same as the `mip6_bc_register()` function.

1018 The `microtime()` function returns the current time in the first parameter. This value is used to compute the expiration time of the binding cache entry.

1021–1029 Based on the parameters of the `mip6_bc_update()` function, the information stored in the existing binding cache entry is updated. Note that we need to take care of the overflow of the `mbc_expire` field since it is a 32-bit signed integer. The `mip6_bc_update()` function is called when a correspondent node receives a valid Binding Update message. The `mbc_state` field is set to `MIP6_BC_FSM_STATE_BOUND`, which indicates that the entry is valid and usable.

1030–1031 When a node receives a Binding Update message, it needs to reset the pending timer event. The next timeout is set to the time when the node needs to send a Binding Refresh Request message.

14.7.3 Calculating Next Timeout of Binding Refresh Request Message

The `mip6_brr_time()` function returns the time at which a node needs to send a Binding Refresh Request message.

Listing 14-34
——mip6_cncore.c

```
1228     u_int
1229     mip6_brr_time(mbc)
1230              struct mip6_bc *mbc;
1231     {
  ....
```

```
1236              switch (mbc->mbc_state) {
1237              case MIP6_BC_FSM_STATE_BOUND:
1238                  if (mip6_brr_mode == MIP6_BRR_SEND_EXPONENT)
1239                      return ((mbc->mbc_expire - mbc->mbc_lifetime / 2)
    - time_second);
1240                  else
1241                      return ((mbc->mbc_expire -
1242                              mip6_brr_tryinterval * mip6_brr_maxtries)
    - time_second);
1243                  break;
1244              case MIP6_BC_FSM_STATE_WAITB:
1245                  if (mip6_brr_mode == MIP6_BRR_SEND_EXPONENT)
1246                      return (mbc->mbc_expire - time_second) / 2;
1247                  else
1248                      return (mip6_brr_tryinterval < mbc->mbc_expire
    - time_second
1249                              ? mip6_brr_tryinterval : mbc->mbc_expire
    - time_second);
1250                  break;
1251              }
1252
1253              return (0); /* XXX; not reach */
1254      }
```
── mip6_cncore.c

Note: Lines 1239, 1242, 1248, and 1249 are broken here for layout reasons. However, they are each a single line of code.

1228–1230 The function has only one parameter mbc, which points to the address of the binding cache entry to be updated by the Binding Refresh Request message.

1236–1250 A binding cache entry has a state field that indicates the current state as described in Table 10-11 of Chapter 10. The computation is slightly different depending on the state and the mode specified as the mip6_brr_mode variable. Table 14-3 shows the available values for the mip6_brr_mode global variable. The default value of the mip6_brr_mode variable is hard-coded to MIP6_BRR_SEND_EXPONENT and we need to modify mip6_cncore.c if we want to change the mode. When mip6_brr_mode is set to MIP6_BRR_SEND_EXPONENT, a correspondent node will try to send Binding Refresh Request messages repeatedly with an exponential timeout. When mip6_brr_mode is set to MIP6_BRR_SEND_LINER, the message will be sent at a constant interval. Figure 14-4 shows the calculation algorithm. The interval between each try when the mode is MIP6_BRR_SEND_LINER is specified as the mip6_brr_tryinterval variable, which is 10 s by default. Note that Figure 14-4 describes four retries to make it easy to understand the difference between the MIP6_BRR_SEND_EXPONENT mode and the MIP6_BRR_SEND_LINER mode; however, the KAME implementation only tries twice since mip6_brr_maxtries is set to 2.

TABLE 14-3

Name	Description
MIP6_BRR_SEND_EXPONENT	Try to send a Binding Refresh Request message exponentially (e.g., when 1/2, 3/4, 7/8, … of lifetime is elapsed)
MIP6_BRR_SEND_LINER	Try to send a Binding Refresh Request message at a constant interval

The mode of mip6_brr_mode.

FIGURE 14-4

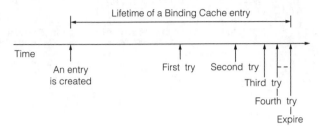

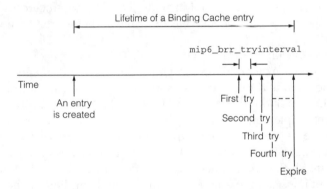

The calculation of time for when to send a Binding Refresh Request message.

14.8 Sending Binding Refresh Request Message

A correspondent node can send a Binding Refresh Request message to extend the lifetime of current binding information between a mobile node and the correspondent node. In the KAME Mobile IPv6, a correspondent node sends the message only when there is more than one TCP connection between those nodes. This means that the binding information between those nodes is kept as long as there is communication using TCP; otherwise, the binding information will be removed when the lifetime of the binding information expires.

Listing 14-35

mip6_cncore.c

```
1187    static int
1188    mip6_bc_need_brr(mbc)
1189            struct mip6_bc *mbc;
1190    {
1191            int found;
1192            struct in6_addr *src, *dst;
....
1194            struct inpcb *inp;
....
```

```
1200            found = 0;
1201            src = &mbc->mbc_addr;
1202            dst = &mbc->mbc_phaddr;
1203
....
1205            for (inp = LIST_FIRST(&tcb); inp; inp = LIST_NEXT(inp, inp_list)) {
1206                    if ((inp->inp_vflag & INP_IPV6) == 0)
1207                            continue;
1208                    if (IN6_ARE_ADDR_EQUAL(src, &inp->in6p_laddr)
1209                        && IN6_ARE_ADDR_EQUAL(dst, &inp->in6p_faddr)) {
1210                            found++;
1211                            break;
1212                    }
1213            }
....
1224
1225            return (found);
1226    }
```
—— mip6_cncore.c

1187–1226 The `mip6_bc_need_brr()` function takes one parameter `mbc`, which points to the binding cache entry to be checked to determine whether the lifetime of the entry has to be extended.

In the loop between lines 1205 and 1213, all TCP protocol control block (`tcb`) entries are checked to find a matching entry in which the local address and foreign address are the same as the local and peer addresses in the `mbc` entry. If there is at least one entry that matches the condition, the code tries to keep the binding information.

The current implementation does not take into account other upper layer connections. If there is no TCP connection between a mobile node and a correspondent node, the cache entry between them will be removed when the lifetime expires, even if they are exchanging other types of data (e.g., UDP datagrams). If the communication continues after expiration of the related binding cache, the mobile node will perform the return routability procedure again and a new cache entry will be created.

14.8.1   Creation and Sending of Binding Refresh Request Message

A correspondent node or a home agent may send a Binding Refresh Request message before a binding cache information expires. Sending a Binding Refresh Request message is implemented in the `mip6_bc_send_brr()` function.

Listing 14-36
—— mip6_cncore.c
```
1132    static int
1133    mip6_bc_send_brr(mbc)
1134            struct mip6_bc *mbc;
1135    {
1136            struct mbuf *m;
1137            struct ip6_pktopts opt;
1138            int error;
```
—— mip6_cncore.c

1133–1134 The `mip6_bc_send_brr()` function has a pointer to the instance of the `mip6_bc{}` structure in which information is going to expire soon.

Listing 14-37

── mip6_cncore.c

```
1140            ip6_initpktopts(&opt);
1141
1142            m = mip6_create_ip6hdr(&mbc->mbc_addr, &mbc->mbc_phaddr, IPPROTO_NONE,
1143               0);
1144            if (m == NULL) {
....
1148                    return (ENOMEM);
1149            }
1150
1151            error = mip6_ip6mr_create(&opt.ip6po_mh, &mbc->mbc_addr,
1152                &mbc->mbc_phaddr);
1153            if (error) {
....
1158                    m_freem(m);
1159                    goto free_ip6pktopts;
1160            }
```
── mip6_cncore.c

1140 A Binding Refresh Request message is passed to the `ip6_output()` function as a packet option. The `in6_initpktopts()` function initializes the `ip6_pktopts{}` structure.

1142–1160 An IPv6 header is created by the `mip6_create_ip6hdr()` function with a source address set to the correspondent or home agent and a destination address set to the address of a mobile node. The `mip6_ip6mr_create()` function will fill the contents of a Binding Refresh Request message.

Listing 14-38

── mip6_cncore.c

```
1162            error = ip6_output(m, &opt, NULL, 0, NULL, NULL
....
1164                                    , NULL
....
1166                            );
1167            if (error) {
....
1171                    goto free_ip6pktopts;
1172            }
1173
1174     free_ip6pktopts:
1175            if (opt.ip6po_mh)
1176                    FREE(opt.ip6po_mh, M_IP6OPT);
1177
1178            return (error);
1179     }
```
── mip6_cncore.c

1162–1176 The `ip6_output()` function is called to send the Binding Refresh Request message. The packet option created for the message must be released before completing the function.

Listing 14-39

── mip6_cncore.c

```
2542     static int
2543     mip6_ip6mr_create(pktopt_mobility, src, dst)
2544            struct ip6_mh **pktopt_mobility;
2545            struct in6_addr *src;
2546            struct in6_addr *dst;
```

```
2547    {
2548            struct ip6_mh_binding_request *ip6mr;
2549            int ip6mr_size;
```
——— mip6_cncore.c

2542–2546 The `mip6_ip6mr_create()` function creates a Binding Refresh Request message. The function has three parameters: The `pktopt_mobility` parameter is a pointer to the address of the `pktopt_mobility{}` structure in which the created message is stored, the `src` parameter is the address of a correspondent node or a home agent, and the `dst` parameter is the home address of a mobile node.

Listing 14-40
——— mip6_cncore.c
```
2551            *pktopt_mobility = NULL;
2552
2553            ip6mr_size = sizeof(struct ip6_mh_binding_request);
2554
2555            MALLOC(ip6mr, struct ip6_mh_binding_request *,
2556                    ip6mr_size, M_IP6OPT, M_NOWAIT);
2557            if (ip6mr == NULL)
2558                    return (ENOMEM);
2559
2560            bzero(ip6mr, ip6mr_size);
2561            ip6mr->ip6mhbr_proto = IPPROTO_NONE;
2562            ip6mr->ip6mhbr_len = (ip6mr_size >> 3) - 1;
2563            ip6mr->ip6mhbr_type = IP6_MH_TYPE_BRR;
2564
2565            /* calculate checksum. */
2566            ip6mr->ip6mhbr_cksum = mip6_cksum(src, dst, ip6mr_size, IPPROTO_MH,
2567                (char *)ip6mr);
2568
2569            *pktopt_mobility = (struct ip6_mh *)ip6mr;
2570
2571            return (0);
2572    }
```
——— mip6_cncore.c

2551–2558 Memory is allocated to store the message content.

2560–2569 A Binding Refresh Request message is built. The next header (`ip6mhbr_proto`) is set to `IPPROTO_NONE`, which means there is no next header. The type (`ip6mhbr_type`) is set to `IP6_MH_TYPE_BRR`, which indicates a Binding Refresh Request message. `ip6mhbr_cksum` is filled with the return value of the `mip6_cksum()` function that computes the checksum value of a Mobility Header.

14.9   Home Registration Processing

When a node is acting as a home agent, it receives a Binding Update message for home registration from mobile nodes. The basic procedure for home registration is the same as the procedure for the binding cache creation procedure, which is done in a correspondent node. The main difference is that a home agent needs to perform the following two additional tasks:

- Set up proxy Neighbor Discovery for the mobile node
- Create an IPv6 in IPv6 tunnel to the mobile node

Home registration is implemented in the `mip6_process_hrbu()` function. The overview of the home registration process is described in Section 4.2 of Chapter 4.

Listing 14-41

── mip6_hacore.c

```
87      int
88      mip6_process_hrbu(bi)
89              struct mip6_bc *bi;
90      {
91              struct sockaddr_in6 addr_sa;
92              struct ifaddr *destifa = NULL;
93              struct ifnet *destifp = NULL;
94              struct nd_prefix *pr, *llpr = NULL;
95              struct ifnet *hifp = NULL;
96              struct in6_addr lladdr;
97              struct mip6_bc *llmbc = NULL;
98              struct mip6_bc *mbc = NULL;
99              struct mip6_bc *prim_mbc = NULL;
100             u_int32_t prlifetime = 0;
101             int busy = 0;
....
106             bi->mbc_status = IP6_MH_BAS_ACCEPTED;
107
108             /* find the interface which the destination address belongs to. */
109             bzero(&addr_sa, sizeof(addr_sa));
110             addr_sa.sin6_len = sizeof(addr_sa);
111             addr_sa.sin6_family = AF_INET6;
112             addr_sa.sin6_addr = bi->mbc_addr;
113             /* XXX ? */
114             if (in6_recoverscope(&addr_sa, &addr_sa.sin6_addr, NULL))
115                     panic("mip6_process_hrbu: recovering scope");
116             if (in6_embedscope(&addr_sa.sin6_addr, &addr_sa))
117                     panic("mip6_process_hrbu: embedding scope");
118             destifa = ifa_ifwithaddr((struct sockaddr *)&addr_sa);
119             if (!destifa) {
120                     bi->mbc_status = IP6_MH_BAS_NOT_HOME_SUBNET;
121                     bi->mbc_send_ba = 1;
122                     return (0); /* XXX is 0 OK? */
123             }
124             destifp = destifa->ifa_ifp;
```

── mip6_hacore.c

87–89 The `mip6_process_hrbu()` function has one parameter `bi`, which is a pointer to an instance of the `mip6_bc{}` structure. The `bi` parameter is a pointer to a template of a binding cache entry that will be activated as a result of this function.

106 A home agent must reply to the mobile node with a Binding Acknowledgment message when it receives a home registration request. The `mbc_status` variable is filled with an error status code if an error occurs. The default value is `IP6_MH_BAS_ACCEPTED`, which means that the registration succeeded.

109–124 The `destifa` variable will point to the network interface on which the Binding Update message was received. The destination address of the message is kept in the `mbc_addr` member variable. The instance of the `in6_ifaddr{}` structure that is related to the address is searched by calling the `ifa_ifwithaddr()` function. An error reply will be sent to the mobile node with the `IP6_MH_BAS_NOT_HOME_SUBNET` status code if the home agent failed to find the address to which the received message was sent. The `in6_ifaddr{}` structure has a pointer to the network interface structure. The pointer to `destifp` is kept for later use.

Listing 14-42

── mip6_hacore.c

```
126             /* find the home ifp of this homeaddress. */
127             for (pr = nd_prefix.lh_first; pr; pr = pr->ndpr_next) {
```

```
128                             if (pr->ndpr_ifp != destifp)
129                                     continue;
130                             if (in6_are_prefix_equal(&bi->mbc_phaddr,
131                                 &pr->ndpr_prefix.sin6_addr, pr->ndpr_plen)) {
132                                     hifp = pr->ndpr_ifp; /* home ifp. */
133                                     prlifetime = pr->ndpr_vltime;
134                             }
135                     }
136             if (hifp == NULL) {
137                     /*
138                      * the haddr0 doesn't have an online prefix.  return a
139                      * binding ack with an error NOT_HOME_SUBNET.
140                      */
141                     bi->mbc_status = IP6_MH_BAS_NOT_HOME_SUBNET;
142                     bi->mbc_send_ba = 1;
143                     return (0); /* XXX is 0 OK? */
144             }
```
—— mip6_hacore.c

127–144 When sending a Binding Acknowledgment message, a home agent needs to check
the remaining lifetime of the home prefix that is assigned to the mobile node. If the
remaining lifetime is too short, the home agent needs to notify the mobile node to update
the prefix information. The `for` loop on line 127 searches the home prefix information
of the network interface specified by the `destifp` variable and stores the valid lifetime
in the `prlifetime` variable.

    If no home-related prefix information is found, an error message with the
`IP6_MH_BAS_NOT_HOME_SUBNET` status code is sent.

Listing 14-43
—— mip6_hacore.c

```
146             /* find the link-local prefix of the home ifp. */
147             if ((bi->mbc_flags & IP6MU_LINK) != 0) {
148                     for (pr = nd_prefix.lh_first; pr; pr = pr->ndpr_next) {
149                             if (hifp != pr->ndpr_ifp) {
150                                     /* this prefix is not a home prefix. */
151                                     continue;
152                             }
153                             /* save link-local prefix. */
154                             if (IN6_IS_ADDR_LINKLOCAL(&pr->ndpr_prefix.sin6_addr)){
155                                     llpr = pr;
156                                     continue;
157                             }
158                     }
159             }
```
—— mip6_hacore.c

147–159 We also need to take care of the link-local prefix if the received Binding Update
message has the L flag (`IP6MU_LINK`) set. In the loop defined on line 148, we determine
whether there is link-local prefix information on the network interface on which the home
prefix is assigned. If such a prefix is found, the pointer to the prefix is stored to the `llpr`
variable for later use.

Listing 14-44
—— mip6_hacore.c

```
161             if (prlifetime < 4) {   /* lifetime in units of 4 sec */
     ....
166                     bi->mbc_status = IP6_MH_BAS_UNSPECIFIED;
167                     bi->mbc_send_ba = 1;
```

```
168                      bi->mbc_lifetime = 0;
169                      bi->mbc_refresh = 0;
170                      return (0); /* XXX is 0 OK? */
171              }
172              /* sanity check */
173              if (bi->mbc_lifetime < 4) {
....
179                      return (0); /* XXX is 0 OK? */
180              }
181
182              /* adjust lifetime */
183              if (bi->mbc_lifetime > prlifetime) {
184                      bi->mbc_lifetime = prlifetime;
185                      bi->mbc_status = IP6_MH_BAS_PRFX_DISCOV;
186              }
```
———————————————————————————————————— mip6_hacore.c

161–186 The lifetime field stored in the `mip6_bc{}` structure is represented in units of 4 s. If the remaining prefix lifetime is less than 4 s, the Binding Update message is rejected and an error message with the `IP6_MH_BAS_UNSPECIFIED` status code is returned.

The `mbc_lifetime` variable keeps the lifetime requested by the mobile node, which sent the Binding Update message. A Binding Update message whose lifetime is less than 4 s is also rejected.

A home agent replies with a Binding Acknowledgment message with the `IP6_MH_BAS_PRFX_DISCOV` status code if the home prefix lifetime is shorter than the lifetime of the binding information. A mobile node will perform Mobile Prefix Solicitation when receiving the status code to get the latest prefix information.

Listing 14-45
———————————————————————————————————— mip6_hacore.c
```
188              /*
189               * - L=0: defend the given address.
190               * - L=1: defend both the given non link-local unicast (home)
191               *        address and the derived link-local.
192               */
193              /*
194               * at first, check an existing binding cache entry for the
195               * link-local.
196               */
197              if ((bi->mbc_flags & IP6MU_LINK) != 0 && llpr != NULL) {
198                      mip6_create_addr(&lladdr,
199                          (const struct in6_addr *)&bi->mbc_phaddr, llpr);
200                      llmbc = mip6_bc_list_find_withphaddr(&mip6_bc_list, &lladdr);
201                      if (llmbc == NULL) {
202                              /*
203                               * create a new binding cache entry for the
204                               * link-local.
205                               */
206                              llmbc = mip6_bc_create(&lladdr, &bi->mbc_pcoa,
207                                  &bi->mbc_addr, bi->mbc_flags, bi->mbc_seqno,
208                                  bi->mbc_lifetime, hifp);
209                              if (llmbc == NULL) {
210                                      /* XXX INSUFFICIENT RESOURCE error */
211                                      return (-1);
212                              }
213
214                              /* start DAD processing. */
215                              mip6_dad_start(llmbc);
216                      } else if (MIP6_IS_BC_DAD_WAIT(llmbc)) {
217                              llmbc->mbc_pcoa = bi->mbc_pcoa;
218                              llmbc->mbc_seqno = bi->mbc_seqno;
```

```
219                         busy++;
220             } else {
221                 /*
222                  * update the existing binding cache entry for
223                  * the link-local.
224                  */
225                 llmbc->mbc_pcoa = bi->mbc_pcoa;
226                 llmbc->mbc_flags = bi->mbc_flags;
227                 llmbc->mbc_seqno = bi->mbc_seqno;
228                 llmbc->mbc_lifetime = bi->mbc_lifetime;
229                 llmbc->mbc_expire
230                         = time_second + llmbc->mbc_lifetime;
231                 /* sanity check for overflow. */
232                 if (llmbc->mbc_expire < time_second)
233                         llmbc->mbc_expire = 0x7fffffff;
234                 llmbc->mbc_state = MIP6_BC_FSM_STATE_BOUND;
235                 mip6_bc_settimer(llmbc, -1);
236                 mip6_bc_settimer(llmbc, mip6_brr_time(llmbc));
237                 /* modify encapsulation entry */
238                 /* XXX */
239                 if (mip6_tunnel_control(MIP6_TUNNEL_CHANGE, llmbc,
240                         mip6_bc_encapcheck, &llmbc->mbc_encap)) {
241                         /* XXX error */
242                 }
243             }
244             llmbc->mbc_flags |= IP6MU_CLONED;
245         }
```
———————————————————————————————— mip6_hacore.c

197–215 The existing binding cache entry of the link-local address of the mobile node is examined. If a mobile node specified the L flag, we look up the corresponding binding cache
entry on line 220. The `mip6_create_addr()` function creates a link-local address from
the interface identifier part of the IPv6 address passed as a second parameter and the
prefix passed as the third parameter.

   If there is no existing cache entry for the link-local address, a binding cache entry is
created by calling the `mip6_bc_create()` function and the DAD procedure is initiated
to make sure that the link-local address is not duplicated.

216–219 If there is an existing entry, and the entry is in DAD wait status, only the care-of
address and sequence number of the existing entry information are updated. This code
applies when a home agent receives a Binding Update message to a certain address while
performing the DAD procedure on the address.

220–244 If there is an existing entry, all information in the binding cache entry is updated with
the received information. The updated information is the care-of address, flags, sequence
number, lifetime, and registration status. Note that the overflow of expiration time of the
cache information must be checked because the expiration date is represented as a 32-bit
signed integer. The timer function also has to be scheduled to reflect the current lifetime.
This occurs on lines 235–236. Finally, the `mip6_tunnel_control()` function is called
to update the care-of address of the mobile node, which is the endpoint address of the
IPv6 in IPv6 tunnel between the home agent and the mobile node.

Listing 14-46
———————————————————————————————— mip6_hacore.c

```
247         /*
248          * next, check an existing binding cache entry for the unicast
249          * (home) address.
250          */
```

```
251                    mbc = mip6_bc_list_find_withphaddr(&mip6_bc_list, &bi->mbc_phaddr);
252                    if (mbc == NULL) {
253                            /* create a binding cache entry for the home address. */
254                            mbc = mip6_bc_create(&bi->mbc_phaddr, &bi->mbc_pcoa,
255                                &bi->mbc_addr, bi->mbc_flags, bi->mbc_seqno,
256                                bi->mbc_lifetime, hifp);
257                            if (mbc == NULL) {
258                                    /* XXX STATUS_RESOURCE */
259                                    return (-1);
260                            }
261
262                            /* mark that we should do DAD later in this function. */
263                            prim_mbc = mbc;
264
265                            /*
266                             * if the request has IP6MU_LINK flag, refer the
267                             * link-local entry.
268                             */
269                            if (bi->mbc_flags & IP6MU_LINK) {
270                                    mbc->mbc_llmbc = llmbc;
271                                    llmbc->mbc_refcnt++;
272                            }
273                    } else if (MIP6_IS_BC_DAD_WAIT(mbc)) {
274                            mbc->mbc_pcoa = bi->mbc_pcoa;
275                            mbc->mbc_seqno = bi->mbc_seqno;
276                            busy++;
277                    } else {
278                            /*
279                             * update the existing binding cache entry for the
280                             * home address.
281                             */
282                            mbc->mbc_pcoa = bi->mbc_pcoa;
283                            mbc->mbc_flags = bi->mbc_flags;
284                            mbc->mbc_seqno = bi->mbc_seqno;
285                            mbc->mbc_lifetime = bi->mbc_lifetime;
286                            mbc->mbc_expire = time_second + mbc->mbc_lifetime;
287                            /* sanity check for overflow. */
288                            if (mbc->mbc_expire < time_second)
289                                    mbc->mbc_expire = 0x7fffffff;
290                            mbc->mbc_state = MIP6_BC_FSM_STATE_BOUND;
291                            mip6_bc_settimer(mbc, -1);
292                            mip6_bc_settimer(mbc, mip6_brr_time(mbc));
293
294                            /* modify the encapsulation entry. */
295                            if (mip6_tunnel_control(MIP6_TUNNEL_CHANGE, mbc,
296                                    mip6_bc_encapcheck, &mbc->mbc_encap)) {
297                                    /* XXX UNSPECIFIED */
298                                    return (-1);
299                            }
300                    }
```
——— mip6_hacore.c

251–272 After checking the link-local address, the home address (which is always a global
address) of the mobile node is processed. If there is no existing binding cache entry for
the home address, then a new binding cache entry is created. If the mobile node specified
the L flag in the Binding Update message, a pointer to the binding cache entry of the
link-local address is kept in the llmbc variable. The llmbc variable is set to the newly
created binding cache entry and its reference counter is incremented.

273–276 If the entry exists and is in DAD status, then the care-of address and the sequence
number are updated.

277–299 If the entry exists already, the cache information is updated in a fashion similar to
updating the information for the link-local cache entry.

Listing 14-47

mip6_hacore.c

```
302              if (busy) {
....
305                      return(0);
306              }
307
308              if (prim_mbc) {
309                      /*
310                       * a new binding cache is created. start DAD
311                       * processing.
312                       */
313                      mip6_dad_start(prim_mbc);
314                      bi->mbc_send_ba = 0;
315              } else {
316                      /*
317                       * a binding cache entry is updated.  return a binding
318                       * ack.
319                       */
320                      bi->mbc_refresh = bi->mbc_lifetime *
    MIP6_REFRESH_LIFETIME_RATE / 100;
321                      if (bi->mbc_refresh < MIP6_REFRESH_MINLIFETIME)
322                              bi->mbc_refresh = bi->mbc_lifetime <
    MIP6_REFRESH_MINLIFETIME ?
323                                      bi->mbc_lifetime : MIP6_REFRESH_MINLIFETIME;
324                      bi->mbc_send_ba = 1;
325              }
326
327              return (0);
328      }
```

mip6_hacore.c

Note: Lines 320 and 322 are broken here for layout reasons. However, they are each a single line of code.

302–305 If the DAD procedure is incomplete, the `mip6_process_hrbu()` function is aborted. The remaining processing will be done after the current DAD operation has finished.

308–314 If a new home registration entry for the home address of the mobile node is created, the DAD procedure is initiated to make sure that the address is not duplicated. The `mbc_send_ba` flag is set to 0 to suppress replying to a Binding Acknowledgment message since it will be sent after the DAD procedure has completed.

316–324 If the existing binding cache information is being updated, a Binding Acknowledgment message can be sent immediately since the DAD procedure should have finished when the first registration was performed.

A refresh interval is sent to the mobile node with the Binding Acknowledgment message. The refresh time is calculated based on the lifetime. The default value is half of the lifetime since `MIP6_REFRESH_LIFETIME_RATE` is defined as 50. The refresh time must be greater than the minimum value `MIP6_REFRESH_MINLIFETIME` (= 2).

14.10   The DAD Procedure

A home agent must ensure that the addresses that a mobile node requested to register are not duplicated on the home network before replying with a Binding Acknowledgment message. The Mobile IPv6 stack utilizes the core DAD functions implemented as a part of Neighbor Discovery (see Section 5.21 of *IPv6 Core Protocols Implementation*). The DAD functions for Mobile IPv6 are implemented in the following six functions:

- `mip6_dad_start()` — start a DAD procedure for a specified address

- `mip6_dad_stop()` — cancel the running DAD procedure for a specified address

- `mip6_dad_find()` — find a pointer to the running DAD procedure for a specified address

- `mip6_dad_success()` — a callback function that is called when the DAD procedure succeeds

- `mip6_dad_duplicated()` — a wrapper function of `mip6_dad_error()`

- `mip6_dad_error()` — a callback function that is called when the DAD procedure fails

14.10.1 Starting the DAD Procedure

The `mip6_dad_start()` function is used by a home agent to initiate the DAD procedure for a given mobile node's address.

Listing 14-48

mip6_hacore.c

```
692     static int
693     mip6_dad_start(mbc)
694             struct  mip6_bc *mbc;
695     {
696             struct in6_ifaddr *ia;
697
698             if (mbc->mbc_dad != NULL)
699                     return (EEXIST);
700
701             MALLOC(ia, struct in6_ifaddr *, sizeof(*ia), M_IFADDR, M_NOWAIT);
702             if (ia == NULL)
703                     return (ENOBUFS);
704
705             bzero((caddr_t)ia, sizeof(*ia));
706             ia->ia_ifa.ifa_addr = (struct sockaddr *)&ia->ia_addr;
707             ia->ia_addr.sin6_family = AF_INET6;
708             ia->ia_addr.sin6_len = sizeof(ia->ia_addr);
709             ia->ia_ifp = mbc->mbc_ifp;
710             ia->ia6_flags |= IN6_IFF_TENTATIVE;
711             ia->ia_addr.sin6_addr = mbc->mbc_phaddr;
712             if (in6_addr2zoneid(ia->ia_ifp, &ia->ia_addr.sin6_addr,
713                                 &ia->ia_addr.sin6_scope_id)) {
714                     FREE(ia, M_IFADDR);
715                     return (EINVAL);
716             }
717             in6_embedscope(&ia->ia_addr.sin6_addr, &ia->ia_addr);
718             IFAREF(&ia->ia_ifa);
719             mbc->mbc_dad = ia;
720             nd6_dad_start((struct ifaddr *)ia, 0);
721
722             return (0);
723     }
```

mip6_hacore.c

692–694 The `mip6_dad_start()` function has one parameter that specifies a binding cache entry and includes the address for the DAD procedure of the address.

698–699 The `mbc_dad` variable points to the `in6_ifaddr{}` instance, which stores the address information that is being tested by the DAD procedure. No new DAD procedure is started if `mbc_dad` is set since it means the DAD procedure for that address has already been launched.

701–719 Memory is allocated for an `in6_ifaddr{}` structure to store the address information. The flag variable, `ia6_flags`, of the instance needs to have the `IN6_IFF_TENTATIVE` flag set to indicate that the address is being tested by the DAD procedure.

720 The `nd6_dad_start()` function is called to start the DAD operation. The `mip6_dad_success()` or the `mip6_dad_error()` function will be called when the DAD operation is completed, depending on the result of the operation.

14.10.2 Stopping the DAD Procedure

The `mip6_dad_stop()` function stops the ongoing DAD procedure executed by the `mip6_dad_start()` function.

Listing 14-49

————————————————————————————————————— mip6_hacore.c
```
725     int
726     mip6_dad_stop(mbc)
727             struct  mip6_bc *mbc;
728     {
729             struct in6_ifaddr *ia = (struct in6_ifaddr *)mbc->mbc_dad;
730
731             if (ia == NULL)
732                     return (ENOENT);
733             nd6_dad_stop((struct ifaddr *)ia);
734             FREE(ia, M_IFADDR);
735             mbc->mbc_dad = NULL;
736             return (0);
737     }
```
————————————————————————————————————— mip6_hacore.c

725–737 The `mip6_dad_stop()` function takes one parameter that points to a binding cache entry. Nothing happens if the `mbc_dad` variable of the parameter is NULL, which means no DAD procedure is running for this address; otherwise, the `nd6_dad_stop()` function is called to stop the running DAD procedure. The `mbc_dad` variable is set to NULL to indicate that no DAD procedure is being performed after stopping the DAD procedure.

Listing 14-50

————————————————————————————————————— mip6_hacore.c
```
739     struct ifaddr *
740     mip6_dad_find(taddr, ifp)
741             struct in6_addr *taddr;
742             struct ifnet *ifp;
743     {
744             struct mip6_bc *mbc;
745             struct in6_ifaddr *ia;
746
747             for (mbc = LIST_FIRST(&mip6_bc_list);
748                 mbc;
749                 mbc = LIST_NEXT(mbc, mbc_entry)) {
750                     if (!MIP6_IS_BC_DAD_WAIT(mbc))
751                             continue;
752                     if (mbc->mbc_ifp != ifp || mbc->mbc_dad == NULL)
```

```
753                              continue;
754                     ia = (struct in6_ifaddr *)mbc->mbc_dad;
755                     if (IN6_ARE_ADDR_EQUAL(&ia->ia_addr.sin6_addr, taddr))
756                             return ((struct ifaddr *)ia);
757             }
758
759             return (NULL);
760     }
```
_____ mip6_hacore.c

739–742 The mip6_dad_find() function has two parameters: the taddr parameter is the
address for which we are looking, and the ifp parameter is a pointer to the network
interface to which the address specified by the taddr variable belongs.

747–759 In the loop defined on line 747, all binding cache entries, which are performing DAD
and that belong to the same network interface as specified by ifp, are checked. If the
address stored in the mbc_dad variable and the taddr variable are the same address,
the address information is returned to the caller.

14.10.3   Finishing the DAD Procedure with Success

The mip6_dad_success() function is called when the DAD procedure succeeded and
performs remaining home registration procedures that must be done after the successful DAD
operation.

Listing 14-51
_____ mip6_hacore.c
```
762     int
763     mip6_dad_success(ifa)
764             struct ifaddr *ifa;
765     {
766             struct  mip6_bc *mbc = NULL;
767
768             for (mbc = LIST_FIRST(&mip6_bc_list);
769                 mbc;
770                 mbc = LIST_NEXT(mbc, mbc_entry)) {
771                     if (mbc->mbc_dad == ifa)
772                             break;
773             }
774             if (!mbc)
775                     return (ENOENT);
776
777             FREE(ifa, M_IFADDR);
778             mbc->mbc_dad = NULL;
779
780             /* create encapsulation entry */
781             mip6_tunnel_control(MIP6_TUNNEL_ADD, mbc, mip6_bc_encapcheck,
782                 &mbc->mbc_encap);
783
784             /* add rtable for proxy ND */
785             mip6_bc_proxy_control(&mbc->mbc_phaddr, &mbc->mbc_addr, RTM_ADD);
786
787             /* if this entry has been cloned by L=1 flag, just return. */
788             if ((mbc->mbc_flags & IP6MU_CLONED) != 0)
789                     return (0);
790
791             /* return a binding ack. */
792             if (mip6_bc_send_ba(&mbc->mbc_addr, &mbc->mbc_phaddr, &mbc->mbc_pcoa,
793                 mbc->mbc_status, mbc->mbc_seqno, mbc->mbc_lifetime,
794                 mbc->mbc_lifetime / 2 /* XXX */, NULL)) {
    ....
```

```
800              }
801
802              return (0);
803      }
```

762–764 The `mip6_dad_success()` function is called when the DAD procedure of the Neighbor Discovery mechanism has successfully completed. The parameter is a pointer to the `in6_ifaddr{}` structure that holds the address.

768–778 The code looks up the corresponding binding cache entry that has the address that has been successfully tested by the DAD procedure. Memory used to store the address information is released and the `mbc_dad` field is set to NULL to indicate that no DAD procedure is running for this cache entry.

781–785 After successful DAD, the home agent needs to set up a proxy Neighbor Discovery entry for the address of the mobile node and needs to create a tunnel to the mobile node. The `mip6_tunnel_control()` function (see Section 15.25 of Chapter 15) and the `mip6_bc_proxy_control()` function (Section 14.11) implement these mechanisms.

788–802 A Binding Acknowledgment message is sent by calling the `mip6_bc_send_ba()` function, if the binding cache entry does not have the cloned (`IP6MU_CLONED`) flag. The cloned flag indicates that the entry has been created as a side effect of another binding cache creation. An acknowledgment message will be sent when the original cache entry is processed.

14.10.4   Error Handling of the DAD Procedure

The `mip6_dad_duplicated()` function is called when the node detects the duplicated address for the mobile node's address passed by the `mip6_dad_start()` function and performs error processing.

Listing 14-52

```
805      int
806      mip6_dad_duplicated(ifa)
807              struct ifaddr *ifa;
808      {
809              return mip6_dad_error(ifa, IP6_MH_BAS_DAD_FAILED);
810      }
811
812      int
813      mip6_dad_error(ifa, err)
814              struct ifaddr *ifa;
815              int err;
816      {
817              struct mip6_bc *mbc = NULL, *llmbc = NULL;
818              struct mip6_bc *gmbc = NULL, *gmbc_next = NULL;
819              int error;
820
821              for (mbc = LIST_FIRST(&mip6_bc_list);
822                  mbc;
823                  mbc = LIST_NEXT(mbc, mbc_entry)) {
824                      if (mbc->mbc_dad == ifa)
825                              break;
826              }
```

```
827                    if (!mbc)
828                            return (ENOENT);
829
830                    FREE(ifa, M_IFADDR);
831                    mbc->mbc_dad = NULL;
```

805–810 The `mip6_dad_duplicated()` simply calls the `mip6_dad_error()` function with the status code `IP6_MH_BAS_DAD_FAILED`. This function is called from the Neighbor Discovery mechanism when the DAD procedure detects address duplication.

812–815 The `mip6_dad_error()` function has two parameters: the `ifa` is a pointer to the address information on which the DAD procedure has detected an error while performing DAD, and the `err` parameter is a status code that is used when sending back a Binding Acknowledgment message to a mobile node.

821–828 The addresses on which a home agent is performing DAD operations are stored in the `mbc_dad` member variable of the `mip6_bc{}` structure. The ENOENT error is returned if there is no matching address in the binding cache list.

830–831 There is no need for the address information stored in `ifa` any longer. `mbc_dad` is set to NULL to indicate that the DAD procedure is not running anymore.

Listing 14-53

```
833                    if ((mbc->mbc_flags & IP6MU_CLONED) != 0) {
834                            /*
835                             * DAD for a link-local address failed.  clear all
836                             * references from other binding caches.
837                             */
838                            llmbc = mbc;
839                            for (gmbc = LIST_FIRST(&mip6_bc_list);
840                                 gmbc;
841                                 gmbc = gmbc_next) {
842                                    gmbc_next = LIST_NEXT(gmbc, mbc_entry);
843                                    if (((gmbc->mbc_flags & IP6MU_LINK) != 0)
844                                        && ((gmbc->mbc_flags & IP6MU_CLONED) == 0)
845                                        && (gmbc->mbc_llmbc == llmbc)) {
846                                            gmbc_next = LIST_NEXT(gmbc, mbc_entry);
847                                            if (MIP6_IS_BC_DAD_WAIT(gmbc)) {
848                                                    mip6_dad_stop(gmbc);
849                                                    gmbc->mbc_llmbc = NULL;
850                                                    error = mip6_bc_list_remove(
851                                                        &mip6_bc_list, llmbc);
852                                                    if (error) {
....
856                                                            /* what should I do? */
857                                                    }
858
859                                            /* return a binding ack. */
860                                            mip6_bc_send_ba(&gmbc->mbc_addr,
861                                                &gmbc->mbc_phaddr, &gmbc->mbc_pcoa,
862                                                err, gmbc->mbc_seqno, 0, 0, NULL);
863
864                                            /*
865                                             * update gmbc_next, because removing
866                                             * llmbc may invalidate gmbc_next.
867                                             */
868                                            gmbc_next = LIST_NEXT(gmbc, mbc_entry);
869                                            error = mip6_bc_list_remove(
070                                                &mip6_bc_list, gmbc);
871                                            if (error) {
....
```

```
875                                             /* what should I do? */
876                                         }
877                             } else {
878                                 /*
879                                  * DAD for a lladdr failed, but
880                                  * a related BC's DAD had been
881                                  * succeeded.  does this happen?
882                                  */
883                             }
884                         }
885                     }
886                 return (0);
```
_____ mip6_hacore.c

833 Lines 833–886 are the code for binding cache entries that are cloned by other binding cache entries.

839–862 The loop checks all binding cache entries, which meet the following conditions:

- The entry has a cloned entry (IP6MU_LINK is set)

- The mbc_llmbc member variable is set to the entry that the DAD procedure failed

- The entry is not a cloned entry (IP6MU_CLONED is not set)

In these cases, the original entry must be removed as well as the cloned entry, and a Binding Acknowledgment message is sent with an error status code. The Acknowledgment message is sent by the mip6_bc_send_ba() function.

868–876 A single cloned entry may be referred to by two or more binding cache entries. The loop continues to check other cache entries that should be removed. Note that the gmbc_next variable must be updated before removing the entry from the binding cache list since removing a cloned entry (llmbc, in this case) may make the gmbc_next pointer invalid.

Listing 14-54
_____ mip6_hacore.c

```
887                 } else {
888                     /*
889                      * if this binding cache has a related link-local
890                      * binding cache entry, decrement the refcnt of the
891                      * entry.
892                      */
893                     if (mbc->mbc_llmbc != NULL) {
894                         error = mip6_bc_list_remove(&mip6_bc_list,
895                             mbc->mbc_llmbc);
896                         if (error) {
....
901                             /* what should I do? */
902                         }
903                     }
904                 }
```
_____ mip6_hacore.c

893–904 The mip6_bc_list_remove() function is called to remove the cloned entry of the entry that has a duplicated address. If the entry has a cloned entry, the mbc_llmbc holds a pointer to the cloned entry. Calling mip6_bc_list_remove() will decrement the reference count of the entry and will remove it if the reference count reaches 0.

Listing 14-55

```
                                                              ─────mip6_hacore.c
906              /* return a binding ack. */
907              mip6_bc_send_ba(&mbc->mbc_addr, &mbc->mbc_phaddr, &mbc->mbc_pcoa, err,
908                  mbc->mbc_seqno, 0, 0, NULL);
909              error = mip6_bc_list_remove(&mip6_bc_list, mbc);
910              if (error) {
  ....
914                      /* what should I do? */
915              }
916
917              return (0);
918      }
                                                              ─────mip6_hacore.c
```

907–917 After finishing the error processing of a DAD failure, a Binding Acknowledgment message is sent to the mobile node to notify that its address is duplicated. The original binding cache entry for the mobile node is removed here.

14.11   Proxy Neighbor Discovery Control

While a mobile node is away from home, the home agent of a mobile node receives all traffic sent to the home address of the mobile node at the mobile node's home address and tunnels the packets to its care-of address using an IPv6 in IPv6 tunnel. To receive packets of which the destination address is the home address of the mobile node, the home agent uses the proxy Neighbor Discovery mechanism.

Listing 14-56

```
                                                              ─────mip6_hacore.c
428      int
429      mip6_bc_proxy_control(target, local, cmd)
430              struct in6_addr *target;
431              struct in6_addr *local;
432              int cmd;
433      {
                                                              ─────mip6_hacore.c
```

428–432 The `mip6_bc_proxy_control()` function has three parameters: the `target` parameter is a pointer to the address that will be proxied by a home agent, the `local` parameter is a pointer to the address of the home agent, and the `cmd` parameter specifies the operation. In this function, either `RTM_DELETE` to stop proxying or `RTM_ADD` to start proxying can be specified.

The information is stored in the routing subsystem as a routing entry. It is used by the home agent to reply to the nodes that query for the link-layer address of the mobile node by sending Neighbor Solicitation messages. The home agent replies with Neighbor Advertisement messages, adding its link-layer address as a destination link-layer address to receive all packets sent to the mobile node.

Listing 14-57

```
                                                              ─────mip6_hacore.c
434              struct sockaddr_in6 target_sa, local_sa, mask_sa;
435              struct sockaddr_dl *sdl;
436              struct rtentry *rt, *nrt;
```

```
437                struct ifaddr *ifa;
438                struct ifnet *ifp;
439                int flags, error = 0;
440
441                /* create a sockaddr_in6 structure for my address. */
442                bzero(&local_sa, sizeof(local_sa));
443                local_sa.sin6_len = sizeof(local_sa);
444                local_sa.sin6_family = AF_INET6;
445                /* XXX */ in6_recoverscope(&local_sa, local, NULL);
446                /* XXX */ in6_embedscope(&local_sa.sin6_addr, &local_sa);
447
448                ifa = ifa_ifwithaddr((struct sockaddr *)&local_sa);
449                if (ifa == NULL)
450                        return (EINVAL);
451                ifp = ifa->ifa_ifp;
```
─── mip6_hacore.c

442–451 A sockaddr_in6{} instance, which stores the address of the home agent, is created. Lines 445–446 try to restore the scope identifier of the home address. It is impossible to decide the scope information of the address because the address information passed to this function does not have such information. If the home agent is operated with scoped addresses (such as the site-local addresses), the result will be unreliable, but this is not critical since the site-local addresses are deprecated and the home agent address that is used for proxying is usually a global address.

Listing 14-58

─── mip6_hacore.c
```
453                bzero(&target_sa, sizeof(target_sa));
454                target_sa.sin6_len = sizeof(target_sa);
455                target_sa.sin6_family = AF_INET6;
456                target_sa.sin6_addr = *target;
457                if (in6_addr2zoneid(ifp, &target_sa.sin6_addr,
458                        &target_sa.sin6_scope_id)) {
....
462                        return(EIO);
463                }
464                error = in6_embedscope(&target_sa.sin6_addr, &target_sa);
465                if (error != 0) {
466                        return(error);
467                }
```
─── mip6_hacore.c

453–467 A sockaddr_in6{} instance, which stores the address of a mobile node, is created. The scope identifier should be the same as the address of the home agent that performs proxying.

Listing 14-59

─── mip6_hacore.c
```
468                /* clear sin6_scope_id before looking up a routing table. */
469                target_sa.sin6_scope_id = 0;
470
471                switch (cmd) {
472                case RTM_DELETE:
....
474                        rt = rtalloc1((struct sockaddr *)&target_sa, 0, 0UL);
....
478                        if (rt)
479                                rt->rt_refcnt--;
480                        if (rt == NULL)
```

```
481                          return (0);
482                  if ((rt->rt_flags & RTF_HOST) == 0 ||
483                      (rt->rt_flags & RTF_ANNOUNCE) == 0) {
484                          /*
485                           * there is a rtentry, but is not a host nor
486                           * a proxy entry.
487                           */
488                          return (0);
489                  }
490                  error = rtrequest(RTM_DELETE, rt_key(rt), (struct sockaddr *)0,
491                      rt_mask(rt), 0, (struct rtentry **)0);
492                  if (error) {
 ....
497                  }
498                  rt = NULL;
499
500                  break;
```
————————————————————————————————— mip6_hacore.c

469 The sin6_scope_id field of the target_sa is cleared before using it to look up a routing entry. The sin6_scope_id field keeps scope information of the address and the KAME code utilizes this field when it handles IPv6 addresses as sockaddr_in6{} structures. The routing table is one of the exceptions that does not utilize the field because of the design of the Radix tree implementation. The current Radix tree implementation of BSD operating systems only checks the address field when looking up a routing entry. The scope information has to be embedded into the address field instead of the sin6_scope_id field.

471–500 The RTM_DELETE command deletes the existing proxy Neighbor Discovery entry from the routing table. All proxy routing entries have the RTF_HOST and the RTF_ANNOUNCE flags set. If the entry found by rtalloc1() has these flags, the rtrequest() function is called with the RTM_DELETE command to remove the entry from the routing table.

Listing 14-60
————————————————————————————————— mip6_hacore.c
```
502          case RTM_ADD:
 ....
504                  rt = rtalloc1((struct sockaddr *)&target_sa, 0, 0UL);
 ....
508                  if (rt)
509                          rt->rt_refcnt--;
510                  if (rt) {
511                          if (((rt->rt_flags & RTF_HOST) != 0) &&
512                              ((rt->rt_flags & RTF_ANNOUNCE) != 0) &&
513                              rt->rt_gateway->sa_family == AF_LINK) {
 ....
518                                  return (EEXIST);
519                          }
520                          if ((rt->rt_flags & RTF_LLINFO) != 0) {
521                                  /* nd cache exist */
522                                  rtrequest(RTM_DELETE, rt_key(rt),
523                                      (struct sockaddr *)0, rt_mask(rt), 0,
524                                      (struct rtentry **)0);
525                                  rt = NULL;
526                          } else {
527                                  /* XXX Path MTU entry? */
 ....
533                          }
534                  }
```
————————————————————————————————— mip6_hacore.c

502–525 If the command is RTM_ADD, the existing proxy Neighbor Discovery entry is removed first to avoid duplicated registration of a routing entry. A proxy entry has the RTF_LLINFO flag set indicating that the entry has link-layer address information. Such an entry is removed before adding a new proxy entry for the same address.

527 Usually, there is no routing entry for directly connected links other than link-layer address entries. If such entries exist, the code does nothing. In this case, the proxy Neighbor Discovery setup will fail. This special case is not supported.

Listing 14-61
————————————————————————————————— mip6_hacore.c

```
539                     /* sdl search */
540         {
541                     struct ifaddr *ifa_dl;
542
543                     for (ifa_dl = ifp->if_addrlist.tqh_first; ifa_dl;
544                         ifa_dl = ifa_dl->ifa_list.tqe_next) {
545                             if (ifa_dl->ifa_addr->sa_family == AF_LINK)
546                                     break;
547                     }
548
549                     if (!ifa_dl)
550                             return (EINVAL);
551
552                     sdl = (struct sockaddr_dl *)ifa_dl->ifa_addr;
553         }
```
————————————————————————————————— mip6_hacore.c

541–552 The ifa_dl variable is set to the link-layer address information of the home agent. The link-layer address is kept as one of the interface addresses in the address list of the ifnet{} structure. An error is returned if there is no address that can be used as a destination link-layer address of the proxy entry.

Listing 14-62
————————————————————————————————— mip6_hacore.c

```
556                     /* create a mask. */
557                     bzero(&mask_sa, sizeof(mask_sa));
558                     mask_sa.sin6_family = AF_INET6;
559                     mask_sa.sin6_len = sizeof(mask_sa);
560
561                     in6_prefixlen2mask(&mask_sa.sin6_addr, 128);
562                     flags = (RTF_STATIC | RTF_HOST | RTF_ANNOUNCE);
563
564                     error = rtrequest(RTM_ADD, (struct sockaddr *)&target_sa,
565                         (struct sockaddr *)sdl, (struct sockaddr *)&mask_sa, flags,
566                         &nrt);
567
568                     if (error == 0) {
569                             /* Avoid expiration */
570                             if (nrt) {
571                                     nrt->rt_rmx.rmx_expire = 0;
572                                     nrt->rt_refcnt--;
573                             } else
574                                     error = EINVAL;
575                     } else {
    ....
580                     }
```
————————————————————————————————— mip6_hacore.c

556–580 A host mask (all bits set to 1) is created and the `rtrequest()` function with the `RTM_ADD` command is called to insert the proxy Neighbor Discovery entry. The flags have the `RTF_STATIC` flag set to indicate the entry is created statically, the `RTF_HOST` flag set to indicate it is not a network route entry, and the `RTF_ANNOUNCE` flag set to indicate this is a proxy entry. After the successful call to `rtrequest()`, the expiration time of the entry is set to infinite since the lifetime of this entry is managed by the Mobile IPv6 stack.

Listing 14-63

—————————————————————————————— mip6_hacore.c

```
582                     {
583                             /* very XXX */
584                             struct sockaddr_in6 daddr_sa;
585
586                             bzero(&daddr_sa, sizeof(daddr_sa));
587                             daddr_sa.sin6_family = AF_INET6;
588                             daddr_sa.sin6_len = sizeof(daddr_sa);
589                             daddr_sa.sin6_addr = in6addr_linklocal_allnodes;
590                             if (in6_addr2zoneid(ifp, &daddr_sa.sin6_addr,
591                                 &daddr_sa.sin6_scope_id)) {
592                                     /* XXX: should not happen */
....
597                                     error = EIO; /* XXX */
598                             }
599                             if (error == 0) {
600                                     error = in6_embedscope(&daddr_sa.sin6_addr,
601                                         &daddr_sa);
602                             }
603                             if (error == 0) {
604                                     nd6_na_output(ifp, &daddr_sa.sin6_addr,
605                                         &target_sa.sin6_addr, ND_NA_FLAG_OVERRIDE,
606                                         1, (struct sockaddr *)sdl);
607                             }
608                     }
609
610                     break;
611
612             default:
....
617                     error = -1;
618                     break;
619             }
620
621             return (error);
622     }
```

—————————————————————————————— mip6_hacore.c

586–603 A Neighbor Advertisement message is sent to the all nodes link-local multicast address to inform all nodes that packets destined to the mobile node's home address should be sent to the link-layer address of the home agent. The message contains the mobile node's home address, the address of the home agent, and the home agent's link-layer address. The flag of the message has the `ND_NA_FLAG_OVERRIDE` flag set, which indicates that existing neighbor cache entries should be updated with the newly advertised information.

612–621 If the procedure succeeds, 0 is returned; otherwise an error code is returned.

14.12   Home Deregistration Procedure

When a home agent receives a message to deregister the home address of a mobile node, it removes the related binding cache entry from its binding cache list. In addition to this, a home agent needs to perform the following tasks:

- Stop proxying the home address of the mobile node
- Destroy the IPv6 in IPv6 tunnel between the home agent and mobile node

An overview of deregistration processing is described in Section 4.5 of Chapter 4.

Listing 14-64
———mip6_hacore.c

```
330     int
331     mip6_process_hurbu(bi)
332             struct mip6_bc *bi;
333     {
```
———mip6_hacore.c

330–332 The deregistration process is implemented as the `mip6_process_hurbu()` function. The function has one parameter that points to a value containing binding information to be removed. The parameter is also used to store some information needed to send a Binding Acknowledgment message by the caller.

Listing 14-65
———mip6_hacore.c

```
334             struct sockaddr_in6 addr_sa;
335             struct ifaddr *destifa = NULL;
336             struct ifnet *destifp = NULL;
337             struct mip6_bc *mbc;
338             struct nd_prefix *pr;
339             struct ifnet *hifp = NULL;
340             int error = 0;
341
342             /* find the interface which the destination address belongs to. */
343             bzero(&addr_sa, sizeof(addr_sa));
344             addr_sa.sin6_len = sizeof(addr_sa);
345             addr_sa.sin6_family = AF_INET6;
346             addr_sa.sin6_addr = bi->mbc_addr;
347             /* XXX ? */
348             if (in6_recoverscope(&addr_sa, &addr_sa.sin6_addr, NULL))
349                     panic("mip6_process_hrbu: recovering scope");
350             if (in6_embedscope(&addr_sa.sin6_addr, &addr_sa))
351                     panic("mip6_process_hrbu: embedding scope");
352             destifa = ifa_ifwithaddr((struct sockaddr *)&addr_sa);
353             if (!destifa) {
354                     bi->mbc_status = IP6_MH_BAS_NOT_HOME_SUBNET;
355                     bi->mbc_send_ba = 1;
356                     return (0); /* XXX is 0 OK? */
357             }
358             destifp = destifa->ifa_ifp;
```
———mip6_hacore.c

343–358 The `destifa` variable is set to the network interface on which the Binding Update message arrived. Usually the mbuf contains a pointer to the interface on which the packet has arrived; however, in this case the information cannot be used because a different interface may receive the message when the home agent has multiple network interfaces.

The stack needs to know which interface has the same address as the destination address of the Binding Update message. To get the interface on which the address is assigned, the `ifa_ifwithaddr()` function is used with the address of the home agent. A Binding Acknowledgment message with the status code `IP6_MH_BAS_NOT_HOME_SUBNET` is sent if there is no such interface.

Listing 14-66

```
                                                            _____ mip6_hacore.c
360              /* find the home ifp of this homeaddress. */
361              for (pr = nd_prefix.lh_first; pr; pr = pr->ndpr_next) {
362                      if (pr->ndpr_ifp != destifp)
363                              continue;
364                      if (in6_are_prefix_equal(&bi->mbc_phaddr,
365                              &pr->ndpr_prefix.sin6_addr, pr->ndpr_plen)) {
366                              hifp = pr->ndpr_ifp; /* home ifp. */
367                      }
368              }
369              if (hifp == NULL) {
370                      /*
371                       * the haddr0 doesn't have an online prefix.  return a
372                       * binding ack with an error NOT_HOME_SUBNET.
373                       */
374                      bi->mbc_status = IP6_MH_BAS_NOT_HOME_SUBNET;
375                      bi->mbc_send_ba = 1;
376                      bi->mbc_lifetime = bi->mbc_refresh = 0;
377                      return (0); /* XXX is 0 OK? */
378              }
                                                            _____ mip6_hacore.c
```

361–378 The home address must belong to the same network of the interface on which the destination address of the Binding Update message is assigned. In the `for` loop on line 361, all prefixes assigned to all interfaces are checked and the prefix that has the same prefix as the home address and that points to the same interface as `destifp` is picked. A Binding Acknowledgment message with `IP6_MH_BAS_NOT_HOME_SUBNET` is sent if there is no interface that has the same prefix as the home address of the mobile node.

Listing 14-67

```
                                                            _____ mip6_hacore.c
380              /* remove a global unicast home binding cache entry. */
381              mbc = mip6_bc_list_find_withphaddr(&mip6_bc_list, &bi->mbc_phaddr);
382              if (mbc == NULL) {
383                      /* XXX panic */
384                      return (0);
385              }
                                                            _____ mip6_hacore.c
```

381–385 The `mbc` variable is set to the binding cache entry for the home address of the mobile node that is specified as `bi->mbc_phaddr`.

Listing 14-68

```
                                                            _____ mip6_hacore.c
387              /*
388               * update the CoA of a mobile node.  this is needed to update
389               * ipsec security policy database addresses properly.
390               */
391              mbc->mbc_pcoa = bi->mbc_pcoa;
```

```
392
393              /*
394               * remove a binding cache entry and a link-local binding cache
395               * entry, if any.
396               */
397              if ((bi->mbc_flags & IP6MU_LINK) &&  (mbc->mbc_llmbc != NULL)) {
398                      /* remove a link-local binding cache entry. */
399                      error = mip6_bc_list_remove(&mip6_bc_list, mbc->mbc_llmbc);
400                      if (error) {
 ....
404                              bi->mbc_status = IP6_MH_BAS_UNSPECIFIED;
405                              bi->mbc_send_ba = 1;
406                              bi->mbc_lifetime = bi->mbc_refresh = 0;
407                              return (error);
408                      }
409              }
410              error = mip6_bc_list_remove(&mip6_bc_list, mbc);
411              if (error) {
 ....
415                      bi->mbc_status = IP6_MH_BAS_UNSPECIFIED;
416                      bi->mbc_send_ba = 1;
417                      bi->mbc_lifetime = bi->mbc_refresh = 0;
418                      return (error);
419              }
420
421              /* return BA */
422              bi->mbc_lifetime = 0; /* ID-19 10.3.2. the lifetime MUST be 0. */
423              bi->mbc_send_ba = 1;    /* Need it ? */
424
425              return (0);
426      }
```
_____mip6_hacore.c

391 The current care-of address is updated with the address stored in the Binding Update message sent by the mobile node. In the deregistration case, the care-of address should be the home address.

397–409 A binding cache entry may have a cloned entry. Before releasing the cache entry, the cloned entry specified as mbc_llmbc has to be released by calling the mip6_bc_list_remove() function. If an error occurs during the release of the cloned entry, a Binding Acknowledgment message with a status code IP6_MH_BAS_UNSPECIFIED is sent.

410–425 The binding cache entry used for home registration is removed. The error status IP6_MH_BAS_UNSPECIFIED is returned when an error occurs. After the successful removal of the binding cache entry, the home agent will return a Binding Acknowledgment message with lifetime set to 0 to the mobile node that confirms deregistration.

14.13 Sending a Binding Acknowledgment Message

A home agent replies with a Binding Acknowledgment message in response to the Binding Update message for home registration of a mobile node. A correspondent node must reply with a Binding Acknowledgment message when a Binding Update message from a mobile node has the A (Acknowledgment) flag set or when it encounters any problem during the incoming Binding Update message processing.

The mip6_bc_send_ba() function creates a Binding Acknowledgment message and sends it.

Listing 14-69

_____mip6_cncore.c

```
1036    int
1037    mip6_bc_send_ba(src, dst, dstcoa, status, seqno, lifetime, refresh, mopt)
1038           struct in6_addr *src;
1039           struct in6_addr *dst;
1040           struct in6_addr *dstcoa;
1041           u_int8_t status;
1042           u_int16_t seqno;
1043           u_int32_t lifetime;
1044           u_int32_t refresh;
1045           struct mip6_mobility_options *mopt;
1046    {
```

_____mip6_cncore.c

1037–1045 Function `mip6_bc_send_ba()` has eight parameters. The `src` and `dst` parameters are the address of a sending node and the home address of a mobile node. These addresses are used as the source and destination addresses of the Binding Acknowledgment message which will be sent. The `dstcoa` parameter is the care-of address of the mobile node. The `status` parameter is a status code of the Acknowledgment message, which must be one of the values described in Table 3-3 of Chapter 3. The `seqno` parameter is a copy of the sequence number of the corresponding Binding Update message. The `lifetime` parameter is the lifetime of the binding cache information. The value will be greater than 0 when the incoming Binding Update message is for registration, and will be 0 when the message is for deregistration. The `refresh` parameter is the refresh interval value. The `mopt` parameter is a pointer to the `mip6_mobility_options{}` structure whose contents are extracted from the Binding Update message.

Listing 14-70

_____mip6_cncore.c

```
1047           struct mbuf *m;
1048           struct ip6_pktopts opt;
1049           struct m_tag *mtag;
1050           struct ip6aux *ip6a;
1051           struct ip6_rthdr *pktopt_rthdr;
1052           int error = 0;
1053
1054           ip6_initpktopts(&opt);
1055
1056           m = mip6_create_ip6hdr(src, dst, IPPROTO_NONE, 0);
1057           if (m == NULL) {
....
1061                   return (ENOMEM);
1062           }
1063
1064           error = mip6_ip6ma_create(&opt.ip6po_mh, src, dst, dstcoa,
1065                                   status, seqno, lifetime, refresh, mopt);
1066           if (error) {
....
1070                   m_freem(m);
1071                   goto free_ip6pktopts;
1072           }
```

_____mip6_cncore.c

1056–1072 An IPv6 packet for the Binding Acknowledgment message is prepared with the `mip6_create_ip6hdr()` function. The Binding Acknowledgment message is created as an instance of the `ip6_pktopt{}` structure by the `mip6_ip6ma_create()` function. If the creation of the headers fails, an error is returned.

Listing 14-71

_____ mip6_cncore.c

```
1074                  /*
1075                   * when sending a binding ack, we use rthdr2 except when
1076                   * we are on the home link.
1077                   */
1078                  if (!IN6_ARE_ADDR_EQUAL(dst, dstcoa)) {
1079                          error = mip6_rthdr_create(&pktopt_rthdr, dstcoa, NULL);
1080                          if (error) {
....
1084                                  m_freem(m);
1085                                  goto free_ip6pktopts;
1086                          }
1087                          opt.ip6po_rthdr2 = pktopt_rthdr;
1088                  }
1089
1090                  mtag = ip6_findaux(m);
1091                  if (mtag) {
1092                          ip6a = (struct ip6aux *)(mtag + 1);
1093                          ip6a->ip6a_flags |= IP6A_NOTUSEBC;
1094                  }
```

_____ mip6_cncore.c

1078–1088 A Binding Acknowledgment message must contain a Type 2 Routing Header, except when the destination mobile node is attached to its home link. If `dst` (the home address of the mobile node) and `dstcoa` (the care-of address of the mobile node) are the same, the mobile node is at home. Otherwise, a Type 2 Routing Header is prepared with the `mip6_rthdr_create()` function.

1090–1094 When sending a Binding Acknowledgment message, binding cache entries are not used to insert a Type 2 Routing Header since the header is already inserted above.

The `IP6A_NOTUSEBC` flag indicates that there is no need to look up binding cache information when sending the packet.

Listing 14-72

_____ mip6_cncore.c

```
1096     #ifdef MIP6_HOME_AGENT
1097             /* delete proxy nd entry temporally */
1098             if ((status >= IP6_MH_BAS_ERRORBASE) &&
1099                 IN6_ARE_ADDR_EQUAL(dst, dstcoa)) {
1100                     struct mip6_bc *mbc;
1101
1102                     mbc = mip6_bc_list_find_withphaddr(&mip6_bc_list, dst);
1103                     if (mtag && mbc && mbc->mbc_flags & IP6MU_HOME &&
1104                         (mip6_bc_proxy_control(&mbc->mbc_phaddr, &mbc->mbc_addr,
     RTM_DELETE) == 0)) {
1105                             ip6a->ip6a_flags |= IP6A_TEMP_PROXYND_DEL;
1106                     }
1107             }
1108     #endif
```

_____ mip6_cncore.c

Note: Line 1104 is broken here for layout reasons. However, it is a single line of code.

1097–1107 The proxy Neighbor Discovery entry for the destination mobile node has to be disabled if the mobile node is at home and the status code is one of the error codes. If the entry is active, the Binding Acknowledgment message whose destination address is the address of the mobile node will be received by the sending node because of the proxy entry. Usually, the proxy Neighbor Discovery entry is removed before calling the `mip6_bc_send_ba()` function when no error occurs; however, when an error occurs, the entry still exists at this point. To avoid the loopback of the message, the entry has to be removed temporarily. The temporarily disabled entry will be restored after the message is sent (see Section 14.17).

Listing 14-73
_____ mip6_cncore.c

```
1112              error = ip6_output(m, &opt, NULL, 0, NULL, NULL
....
1114                                  , NULL
....
1116                                  );
1117              if (error) {
....
1121                      goto free_ip6pktopts;
1122              }
1123      free_ip6pktopts:
1124              if (opt.ip6po_rthdr2)
1125                      FREE(opt.ip6po_rthdr2, M_IP6OPT);
1126              if (opt.ip6po_mh)
1127                      FREE(opt.ip6po_mh, M_IP6OPT);
1128
1129              return (error);
1130      }
```
_____ mip6_cncore.c

1112–1130 The Binding Acknowledgment message is sent via the `ip6_output()` function. The memory allocated for the Acknowledgment message and the Routing Header has to be released here since it is not released in `ip6_output()`.

14.13.1 Creation of a Binding Acknowledgment Message

The `mip6_ip6ma_create()` function will prepare a Binding Acknowledgment message.

Listing 14-74
_____ mip6_cncore.c

```
2409      int
2410      mip6_ip6ma_create(pktopt_mobility, src, dst, dstcoa, status, seqno, lifetime,
2411          refresh, mopt)
2412              struct ip6_mh **pktopt_mobility;
2413              struct in6_addr *src;
2414              struct in6_addr *dst;
2415              struct in6_addr *dstcoa;
2416              u_int8_t status;
2417              u_int16_t seqno;
2418              u_int32_t lifetime;
2419              u_int32_t refresh;
2420              struct mip6_mobility_options *mopt;
2421      {
```
_____ mip6_cncore.c

2409–2420 The `pktopt_mobility` parameter stores a pointer to the newly created message. Other parameters are the same as the parameters of the `mip6_bc_send_ba()` function (see Listing 14-69).

Listing 14-75

mip6_cncore.c

```
2422          struct ip6_mh_binding_ack *ip6ma;
2423          struct ip6_mh_opt_refresh_advice *mopt_refresh = NULL;
2424          struct ip6_mh_opt_auth_data *mopt_auth = NULL;
2425          int need_refresh = 0;
2426          int need_auth = 0;
2427          int ip6ma_size, pad;
2428          int ba_size = 0, refresh_size = 0, auth_size = 0;
2429          u_int8_t key_bm[MIP6_KBM_LEN]; /* Stated as 'Kbm' in the spec */
2430          u_int8_t *p;
2431
2432          *pktopt_mobility = NULL;
2433
2434          ba_size = sizeof(struct ip6_mh_binding_ack);
2435          if (refresh > 3 && refresh < lifetime) {
2436                  need_refresh = 1;
2437                  ba_size += MIP6_PADLEN(ba_size, 2, 0);
2438                  refresh_size = sizeof(struct ip6_mh_opt_refresh_advice);
2439          } else {
2440                  refresh_size = 0;
2441          }
2442          if (mopt &&
2443              ((mopt->valid_options & (MOPT_NONCE_IDX | MOPT_AUTHDATA)) ==
   (MOPT_NONCE_IDX | MOPT_AUTHDATA)) &&
2444              mip6_calculate_kbm_from_index(dst, dstcoa,
2445              mopt->mopt_ho_nonce_idx, mopt->mopt_co_nonce_idx,
2446              !IS_REQUEST_TO_CACHE(lifetime, dst, dstcoa), key_bm) == 0) {
2447                  need_auth = 1;
2448                  /* Since Binding Auth Option must be the last mobility option,
2449                     an implicit alignment requirement is 8n + 2.
2450                     (6.2.7) */
2451                  if (refresh_size)
2452                          refresh_size += MIP6_PADLEN(ba_size + refresh_size,
   8, 2);
2453                  else
2454                          ba_size += MIP6_PADLEN(ba_size, 8, 2);
2455                  auth_size = IP6MOPT_AUTHDATA_SIZE;
2456          }
2457          ip6ma_size = ba_size + refresh_size + auth_size;
2458          ip6ma_size += MIP6_PADLEN(ip6ma_size, 8, 0);
```

mip6_cncore.c

Note: Lines 2443 and 2452 are broken here for layout reasons. However, they are each a single line of code.

2434–2458 The size of the Binding Acknowledgment message is calculated. Note that a Binding Acknowledgment message may have options. In this case, the Binding Refresh Advice option and the Binding Authorization Data option may exist. Each option has alignment requirements. The Binding Refresh Advice option must meet the $2n$ requirement, and the Binding Authorization Data option must meet the $8n + 2$ requirement.

When a correspondent sends a Binding Acknowledgment message, the message must have a Binding Authorization Data option to protect the contents of the message. The mechanism is the same as that for the creation of a Binding Authorization Data option for a Binding Update message. On line 2444, the key data is computed by the `mip6_calculate_kbm_from_index()` function.

Listing 14-76

mip6_cncore.c

```
2460          MALLOC(ip6ma, struct ip6_mh_binding_ack *,
2461                  ip6ma_size, M_IP6OPT, M_NOWAIT);
2462          if (ip6ma == NULL)
```

```
2463                         return (ENOMEM);
2464                 if (need_refresh) {
2465                         mopt_refresh = (struct ip6_mh_opt_refresh_advice *)
      ((u_int8_t *)ip6ma + ba_size);
2466                 }
2467                 if (need_auth)
2468                         mopt_auth = (struct ip6_mh_opt_auth_data *)
      ((u_int8_t *)ip6ma + ba_size + refresh_size);
2469
2470                 bzero(ip6ma, ip6ma_size);
2471
2472                 ip6ma->ip6mhba_proto = IPPROTO_NONE;
2473                 ip6ma->ip6mhba_len = (ip6ma_size >> 3) - 1;
2474                 ip6ma->ip6mhba_type = IP6_MH_TYPE_BACK;
2475                 ip6ma->ip6mhba_status = status;
2476                 ip6ma->ip6mhba_seqno = htons(seqno);
2477                 ip6ma->ip6mhba_lifetime =
2478                         htons((u_int16_t)(lifetime >> 2));       /* units of 4 secs */
```
——— mip6_cncore.c

Note: Lines 2465 and 2468 are broken here for layout reasons. However, they are each a single line of code.

2460–2470 Memory is allocated and pointers to the mobility options are set.

2472–2478 Each field of the Binding Acknowledgment message is filled. The `ip6mhba_proto` field must be `IPPROTO_NONE` since the message must be the last in the chain in the IPv6 packet. The length of the message is represented in units of 8 bytes and the lifetime in units of 4 s.

Listing 14-77
——— mip6_cncore.c

```
2480            /* padN */
2481            p = (u_int8_t *)ip6ma + sizeof(struct ip6_mh_binding_ack);
2482            if ((pad = ba_size - sizeof(struct ip6_mh_binding_ack)) >= 2) {
2483                    *p = IP6_MHOPT_PADN;
2484                    *(p + 1) = pad - 2;
2485            }
2486            if (refresh_size &&
2487                ((p = (u_int8_t *)ip6ma + ba_size + sizeof(struct
      ip6_mh_opt_refresh_advice)),
2488                (pad = refresh_size - sizeof(struct ip6_mh_opt_refresh_advice))
      >= 2)) {
2489                    *p = IP6_MHOPT_PADN;
2490                    *(p + 1) = pad - 2;
2491            }
2492            if (auth_size &&
2493                ((p = (u_int8_t *)ip6ma + ba_size + refresh_size
      +IP6MOPT_AUTHDATA_SIZE),
2494                (pad = auth_size - IP6MOPT_AUTHDATA_SIZE) >= 2)) {
2495                    *p = IP6_MHOPT_PADN;
2496                    *(p + 1) = pad - 2;
2497            }
2498            if (pad + (ip6ma_size - (ba_size + refresh_size + auth_size)) >= 2) {
2499                    *p = IP6_MHOPT_PADN;
2500                    *(p + 1) += ip6ma_size - (ba_size + refresh_size + auth_size)
                    - 2;
2501            }
```
——— mip6_cncore.c

Note: Lines 2487, 2488, 2493, and 2500 are broken here for layout reasons. However, they are each a single line of code.

2481–2501 All the necessary padding data is added.

Listing 14-78

_____mip6_cncore.c
```
2503            /* binding refresh advice option */
2504            if (need_refresh) {
2505                    mopt_refresh->ip6mora_type = IP6_MHOPT_BREFRESH;
2506                    mopt_refresh->ip6mora_len
2507                        = sizeof(struct ip6_mh_opt_refresh_advice) - 2;
2508                    SET_NETVAL_S(&mopt_refresh->ip6mora_interval, refresh >> 2);
2509            }
2510
2511            if (need_auth) {
2512                    /* authorization data processing. */
2513                    mopt_auth->ip6moad_type = IP6_MHOPT_BAUTH;
2514                    mopt_auth->ip6moad_len = IP6MOPT_AUTHDATA_SIZE - 2;
2515                    mip6_calculate_authenticator(key_bm, (caddr_t)(mopt_auth + 1),
2516                        dstcoa, src, (caddr_t)ip6ma, ip6ma_size,
2517                        ba_size + refresh_size + sizeof(struct
    ip6_mh_opt_auth_data),
2518                        IP6MOPT_AUTHDATA_SIZE - 2);
2519            }
```
_____mip6_cncore.c

Note: Line 2517 is broken here for layout reasons. However, it is a single line of code.

2504–2509 A Binding Refresh Advice option is created. The refresh interval is represented in units of 4 s.

2511–2519 A Binding Authorization Data option is created. The authentication data is computed by the mip6_calculate_authenticator() function with the key computed on line 2444.

Listing 14-79

_____mip6_cncore.c
```
2532
2533            /* calculate checksum. */
2534            ip6ma->ip6mhba_cksum = mip6_cksum(src, dst, ip6ma_size, IPPROTO_MH,
2535                (char *)ip6ma);
2536
2537            *pktopt_mobility = (struct ip6_mh *)ip6ma;
2538
2539            return (0);
2540    }
```
_____mip6_cncore.c

2534–2539 Finally, the checksum value of the Binding Acknowledgment message is computed by the mip6_cksum() function. The created message is set to the pktopt_mobility parameter and returned to the caller.

Listing 14-80

_____mip6_cncore.c
```
645    int
646    mip6_rthdr_create(pktopt_rthdr, coa, opt)
647            struct ip6_rthdr **pktopt_rthdr;
648            struct in6_addr *coa;
649            struct ip6_pktopts *opt;
650    {
651            struct ip6_rthdr2 *rthdr2;
652            size_t len;
653
654            /*
655             * Mobile IPv6 uses type 2 routing header for route
656             * optimization. if the packet has a type 1 routing header
```

```
657                   * already, we must add a type 2 routing header after the type
658                   * 1 routing header.
659                   */
660
661                  len = sizeof(struct ip6_rthdr2) + sizeof(struct in6_addr);
662                  MALLOC(rthdr2, struct ip6_rthdr2 *, len, M_IP6OPT, M_NOWAIT);
663                  if (rthdr2 == NULL) {
664                          return (ENOMEM);
665                  }
666                  bzero(rthdr2, len);
667
668                  /* rthdr2->ip6r2_nxt = will be filled later in ip6_output */
669                  rthdr2->ip6r2_len = 2;
670                  rthdr2->ip6r2_type = 2;
671                  rthdr2->ip6r2_segleft = 1;
672                  rthdr2->ip6r2_reserved = 0;
673                  bcopy((caddr_t)coa, (caddr_t)(rthdr2 + 1), sizeof(struct in6_addr));
674                  *pktopt_rthdr = (struct ip6_rthdr *)rthdr2;
  ....
678                  return (0);
679          }
```
── mip6_cncore.c

645–649 The `mip6_rthdr_create()` function prepares a Type 2 Routing Header. The `pktopt_rthdr` parameter stores the created Routing Header, while the `coa` parameter is the care-of address of a mobile node, which is set in the Routing Header. The `opt` parameter is not used currently.

661–678 The length of the Type 2 Routing Header is fixed to 2. Memory to store the `ip6_rthdr2{}` structure and the `in6_addr{}` structure is allocated. The type field, `ip6r2_type`, is 2. The segment left field, `ip6r2_segleft`, is fixed to 1 since the header only has one address. The care-of address is copied just after the header.

14.14   Nonce and Nodekey Management

A correspondent node must maintain the nonce and nodekey arrays to perform the return routability procedure. Nonce and nodekey have associated lifetimes. The lifetime must be less than 240 s (`MIP6_MAX_NONCE_LIFE`) as specified in [RFC3775]. The KAME implementation keeps the latest 10 nonces/nodekeys (`MIP6_NONCE_HISTORY`) as the `mip6_nonce` and `mip6_nodekey` global variables. Figure 14-5 shows the structure of the nonce array.

Listing 14-81
── mip6_cncore.c
```
160      #define NONCE_UPDATE_PERIOD     (MIP6_MAX_NONCE_LIFE / MIP6_NONCE_HISTORY)
161      mip6_nonce_t mip6_nonce[MIP6_NONCE_HISTORY];
162      mip6_nodekey_t mip6_nodekey[MIP6_NONCE_HISTORY];          /* this is described
  as 'Kcn' in the spec */
163      u_int16_t nonce_index;          /* the idx value pointed by nonce_head */
164      mip6_nonce_t *nonce_head;       /* Current position of nonce on the array
  mip6_nonce */
```
── mip6_cncore.c
Note: Lines 162 and 164 are broken here for layout reasons. However, they are each a single line of code.

160–164 The `NONCE_UPDATE_PERIOD` macro represents the interval of nonce creation. The `mip6_nonce` and `mip6_nodekey` arrays hold this information. The `nonce_index` variable indicates the latest index of the nonce array. The `nonce_head` variable is a pointer to the latest nonce entry in the array.

FIGURE 14-5

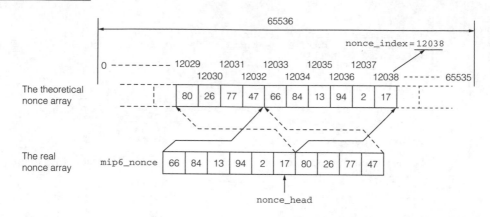

Nonce array management.

The nonce management mechanism consists of the following functions:

- `mip6_update_nonce_nodekey()` — a timer function to update the array
- `mip6_create_nonce()` — generates a new nonce
- `mip6_create_nodekey()` — generates a new nodekey
- `mip6_get_nonce()` — retrieves a nonce from the array
- `mip6_get_nodekey()` — retrieves a nodekey from the array

14.14.1   Update Nonce and Nodekeys

The `mip6_update_nonce_nodekey()` function is called periodically to update the nonce and nodekey arrays.

Listing 14-82

```
                                                            mip6_cncore.c
1343    static void
1344    mip6_update_nonce_nodekey(ignored_arg)
1345            void    *ignored_arg;
1346    {
1347            int s;
1348
....
1352            s = splnet();
....
1355            callout_reset(&mip6_nonce_upd_ch, hz * NONCE_UPDATE_PERIOD,
1356                    mip6_update_nonce_nodekey, NULL);
....
1364            nonce_index++;
1365            if (++nonce_head >= mip6_nonce + MIP6_NONCE_HISTORY)
1366                    nonce_head = mip6_nonce;
1367
1368            mip6_create_nonce(nonce_head);
1369            mip6_create_nodekey(mip6_nodekey + (nonce_head - mip6_nonce));
1370
```

```
1371             splx(s);
1372     }
```
_____ mip6_cncore.c

1343–1372 The interval is set to NONCE_UPDATE_PERIOD, which is 24s in the KAME implementation. The nonce_index variable is an unsigned 16-bit integer and is incremented
by 1 every time a new nonce is generated. The nonce_head variable points to the
latest entry of the nonce array. The function calls the mip6_create_nonce() and
mip6_create_nodekey() functions before returning.

14.14.2 Generate Nonce and Nodekeys

The mip6_create_nonce() and mip6_create_nodekey() functions are called to
generate a new nonce and a nodekey.

Listing 14-83
_____ mip6_cncore.c

```
1322     static void
1323     mip6_create_nonce(nonce)
1324             mip6_nonce_t *nonce;
1325     {
1326             int i;
1327
1328             for (i = 0; i < MIP6_NONCE_SIZE / sizeof(u_long); i++)
1329                     ((u_long *)nonce)[i] = random();
1330     }
1331
1332     static void
1333     mip6_create_nodekey(nodekey)
1334             mip6_nodekey_t *nodekey;
1335     {
1336             int i;
1337
1338             for (i = 0; i < MIP6_NODEKEY_SIZE / sizeof(u_long); i++)
1339                     ((u_long *)nodekey)[i] = random();
1340     }
```
_____ mip6_cncore.c

1322–1330 The mip6_create_nonce() function takes one parameter to store the result.
In this function, the nonce is filled with a random value generated by the random()
function.

1332–1340 The mip6_create_nodekey() function is almost the same as the
mip6_create_nonce() function. It puts a random value into the pointer specified
in the parameter nodekey.

14.14.3 Retrieve Nonce and Nodekeys

The mip6_get_nonce() function returns the nonce value that is related to the specified index
value.

Listing 14-84
_____ mip6_cncore.c

```
1374     int
1375     mip6_get_nonce(index, nonce)
1376             u_int16_t index;          /* nonce index */
1377             mip6_nonce_t *nonce;
```

```
1378     {
1379             int32_t offset;
1380
1381             offset = index - nonce_index;
1382             if (offset > 0) {
1383                     /* nonce_index was wrapped. */
1384                     offset = offset - 0xffff;
1385             }
1386
1387             if (offset <= -MIP6_NONCE_HISTORY) {
1388                     /* too old index. */
1389                     return (-1);
1390             }
1391
1392             if (nonce_head + offset < mip6_nonce)
1393                     offset = nonce_head - mip6_nonce - offset;
1394
1395             bcopy(nonce_head + offset, nonce, sizeof(mip6_nonce_t));
1396             return (0);
1397     }
```
——————————————————————————————————— mip6_cncore.c

1374–1377 The mip6_get_nonce() function has two parameters: the index parameter is the index to the nonce being requested and the nonce parameter is for the returned nonce value.

1381–1385 The KAME implementation keeps only 10 recent values. The offset from the head of the nonce array is calculated from the index parameter.

1387–1390 If the specified index is older than the last 10 nonce values, the nonce is invalid and not returned to the caller.

1392–1396 The nonce value specified by index is copied to the nonce variable.

Listing 14-85

——————————————————————————————————— mip6_cncore.c
```
1399     int
1400     mip6_get_nodekey(index, nodekey)
1401             u_int16_t index;            /* nonce index */
1402             mip6_nodekey_t *nodekey;
1403     {
1404             int32_t offset;
1405             mip6_nodekey_t *nodekey_head;
1406
1407             offset = index - nonce_index;
1408             if (offset > 0) {
1409                     /* nonce_index was wrapped. */
1410                     offset = offset - 0xffff;
1411             }
1412
1413             if (offset <= -MIP6_NONCE_HISTORY) {
1414                     /* too old index. */
1415                     return (-1);
1416             }
1417
1418             if (nonce_head + offset < mip6_nonce)
1419                     offset = nonce_head - mip6_nonce - offset;
1420
1421             nodekey_head = mip6_nodekey + (nonce_head - mip6_nonce);
1422             bcopy(nodekey_head + offset, nodekey, sizeof(mip6_nodekey_t));
1423
1424             return (0);
1425     }
```
——————————————————————————————————— mip6_cncore.c

1399–1425 The code is almost the same as the `mip6_get_nonce()` function. The only difference is that the array referred to in this function is the nodekey array.

14.15 Receiving a Home Address Option

When a correspondent node communicates with a mobile node using route optimization, the correspondent node receives packets with a Home Address option. The option indicates the real source address of the mobile node; however, the correspondent node cannot believe it without verifying the validity of the address.

14.15.1 Destination Option Processing

The Home Address option is defined as a destination option. The option is processed in the `dest6_input()` function defined in `dest6.c`, which is called as a part of extension header processing.

Listing 14-86
```
                                                                    dest6.c
127             /* search header for all options. */
128             for (optlen = 0; dstoptlen > 0; dstoptlen -= optlen, opt += optlen)
 ....
142     #ifdef MIP6
143                 case IP6OPT_HOME_ADDRESS:
144                     /* HAO must appear only once */
145                     n = ip6_addaux(m);
146                     if (!n) {
147                         /* not enough core */
148                         goto bad;
149                     }
150                     ip6a = (struct ip6aux *) (n + 1);
151                     if ((ip6a->ip6a_flags & IP6A_HASEEN) != 0) {
152                         /* XXX icmp6 paramprob? */
153                         goto bad;
154                     }
                                                                    dest6.c
```

143–154 The loop defined on line 128 processes each destination option included in an incoming Destination Options header. The Home Address option can appear only once in an IPv6 packet. The `IP6A_HASEEN` flag is set in an auxiliary mbuf when the `dest6_input()` function processes a Home Address option. If the mbuf has the `IP6A_HASEEN` flag at this point, it means there is one more Home Address option in the packet.

Listing 14-87
```
                                                                    dest6.c
156                     haopt = (struct ip6_opt_home_address *)opt;
157                     optlen = haopt->ip6oh_len + 2;
158
159                     if (optlen != sizeof(*haopt)) {
 ....
161                         goto bad;
162                     }
163
164                     /* XXX check header ordering */
165
166                     bcopy(haopt->ip6oh_addr, &home,
```

```
167                                sizeof(struct in6_addr));
168
169                         bcopy(&home, &ip6a->ip6a_coa, sizeof(ip6a->ip6a_coa));
170                         ip6a->ip6a_flags |= IP6A_HASEEN;
```
                                                                         ____ dest6.c

156–170 The length of the Home Address option must be the size of the
`ip6_opt_home_address{}` structure. If the length is correct, the address information
is copied to the home variable. The address is also copied to the auxiliary mbuf of the
input packet as the `ip6a_coa` variable. At this point, the source address of the IPv6
packet is the care-of address of a mobile node, and the address in the Home Address
option is the home address of the mobile node. These addresses will be swapped
when the validity of the addresses is verified.

Listing 14-88

                                                                         ____ dest6.c

```
174                              /* check whether this HAO is 'verified'. */
175                              if ((mbc = mip6_bc_list_find_withphaddr(
176                                      &mip6_bc_list, &home)) != NULL) {
177                                      /*
178                                       * we have a corresponding binding
179                                       * cache entry for the home address
180                                       * includes in this HAO.
181                                       */
182                                      if (IN6_ARE_ADDR_EQUAL(&mbc->mbc_pcoa,
183                                          &ip6->ip6_src))
184                                              verified = 1;
185                              }
186                              /*
187                               * we have neither a corresponding binding
188                               * cache nor ESP header. we have no clue to
189                               * believe this HAO is a correct one.
190                               */
191                              /*
192                               * Currently, no valid sub-options are
193                               * defined for use in a Home Address option.
194                               */
195
196                              break;
197      #endif /* MIP6 */
198              default:                        /* unknown option */
....
204                              break;
205              }
206          }
```
                                                                         ____ dest6.c

175–185 The pair of care-of and home addresses of the mobile node must be registered as a
binding cache entry. The existence of a cache for the same pair of addresses indicates that
the mobile node and the correspondent node have already performed the return routability
procedure, or the home registration procedure if the node is acting as a home agent of
the mobile node, and previously successfully exchanged a Binding Update message. In
this case, the home address stored in the Home Address option is valid.

187–190 Otherwise, the address has to be verified in other ways. If the packet is a Binding
Update message for initial registration, there is no binding cache entry related to the
home address. These packets are verified as follows:

- If the packet is for home registration, the packet must be secured by ESP. If the Home Address option contains an incorrect home address, the ESP processing will fail.

- If the packet is for registration to a correspondent node, the packet must contain a Binding Authorization Data option. In this case, the `mip6_ip6mu_input()` function will verify the validity of the option (see Section 14.5).

Listing 14-89
dest6.c

```
207
208     #ifdef MIP6
209             /* if haopt is non-NULL, we are sure we have seen fresh HA option */
210             if (verified)
211                     if (dest6_swap_hao(ip6, ip6a, haopt) < 0)
212                             goto bad;
213     #endif /* MIP6 */
214
215             *offp = off;
216             return (dstopts->ip6d_nxt);
217
218       bad:
219             m_freem(m);
220             return (IPPROTO_DONE);
221     }
```
dest6.c

210–212 The source address (which is the care-of address of the mobile node) and the address stored in the Home Address option (which is the home address of the mobile node) are swapped.

  Note that if there is no existing binding cache entry for the incoming Binding Update message, the address swapping is done in the `ip6_input()` function (Listing 14-91). The detailed discussion will be done in Listings 14-92–14-96.

14.15.2 Swap Home Address and Source Address

The `dest6_swap_hao()` function swaps the source address of a received IPv6 header and the address stored in the Home Address option.

Listing 14-90
dest6.c

```
224     static int
225     dest6_swap_hao(ip6, ip6a, haopt)
226             struct ip6_hdr *ip6;
227             struct ip6aux *ip6a;
228             struct ip6_opt_home_address *haopt;
229     {
230
231             if ((ip6a->ip6a_flags & (IP6A_HASEEN | IP6A_SWAP)) != IP6A_HASEEN)
232                     return (EINVAL);
233
....
238             bcopy(&ip6->ip6_src, &ip6a->ip6a_coa, sizeof(ip6a->ip6a_coa));
239             bcopy(haopt->ip6oh_addr, &ip6->ip6_src, sizeof(ip6->ip6_src));
240             bcopy(&ip6a->ip6a_coa, haopt->ip6oh_addr, sizeof(haopt->ip6oh_addr));
....
246             ip6a->ip6a_flags |= IP6A_SWAP;
```

```
247
248              return (0);
249      }
```
                                                                              dest6.c

224–228 The `dest6_swap_hao()` function has three parameters: the `ip6` parameter is a
pointer to the incoming IPv6 packet, the `ip6a` parameter is a pointer to the auxiliary
mbuf of the incoming packet, and the `haopt` parameter is a pointer to the Home Address
option.

231–232 If the addresses are already swapped, that is, the `IP6A_SWAP` flag is set, an error is
returned.

238–246 The care-of address and the home address are swapped. The `IP6A_SWAP` flag is set
to indicate that the addresses are swapped.

To handle the case that the Home Address option and the source address of the incoming
IPv6 header are not swapped in the destination option processing, the `ip6_input()` function
has to deal with swapping the addresses. The following code is quoted from `ip6_input()`.

Listing 14-91

                                                                              ip6_input.c
```
1092              while (nxt != IPPROTO_DONE) {
....
1136    #ifdef MIP6
1137                    if (dest6_mip6_hao(m, off, nxt) < 0)
1138                        goto bad;
....
1154    #endif /* MIP6 */
1155                    nxt = (*inet6sw[ip6_protox[nxt]].pr_input)(&m, &off, nxt);
1156              }
```
                                                                              ip6_input.c

1092–1156 The loop processes all extension headers inserted in an IPv6 packet. The
`dest6_mip6_hao()` function checks the next extension header to be processed and
swaps the home and care-of addresses if the next header is an ESP, an AH, or a Binding Update message. If the addresses are correct, the ESP or the AH processing succeeds;
otherwise, an error will occur while processing the ESP or AH. If the packet does not have
either ESP or AH but has a Binding Update message, the validation will be performed
during the Binding Update message input processing (see Section 14.5). Any bogus Binding Update messages in which addresses are forged can be safely dropped.

14.15.3   Swap Home Address and Source Address in Unverified Packet

The `dest6_mip6_hao()` function is called before processing every header chained in a
received IPv6 packet and swaps the source address of the IPv6 packet and the address in
the Home Address option using the `dest6_swap_hao()` function when necessary.

Listing 14-92

                                                                              dest6.c
```
282      int
283      dest6_mip6_hao(m, mhoff, nxt)
284          struct mbuf *m;
```

```
285                     int mhoff, nxt;
286         {
287                     struct ip6_hdr *ip6;
288                     struct ip6aux *ip6a;
289                     struct ip6_opt ip6o;
290                     struct m_tag *n;
291                     struct in6_addr home;
292                     struct ip6_opt_home_address haopt;
293                     struct ip6_mh mh;
294                     int newoff, off, proto, swap;
295
296                     /* XXX should care about destopt1 and destopt2.  in destopt2,
297                        hao and src must be swapped. */
298                     if ((nxt == IPPROTO_HOPOPTS) || (nxt == IPPROTO_DSTOPTS)) {
299                             return (0);
300                     }
301                     n = ip6_findaux(m);
302                     if (!n)
303                             return (0);
304                     ip6a = (struct ip6aux *) (n + 1);
305
306                     if ((ip6a->ip6a_flags & (IP6A_HASEEN | IP6A_SWAP)) != IP6A_HASEEN)
307                             return (0);
```
——— dest6.c

282–285 The `dest6_mip6_hao()` function has three parameters: the `m` parameter is a pointer
to the mbuf that contains the incoming packet, the `mhoff` parameter is the offset to the
header to be processed, and the `nxt` parameter is the value of protocol number of the
header to be processed.

298–306 There is no need to swap addresses before processing a Hop-by-Hop Options Header
or a Destination Options Header, and there is nothing to do if the addresses have already
been swapped.

Listing 14-93
——— dest6.c

```
309                     ip6 = mtod(m, struct ip6_hdr *);
310                     /* find home address */
311                     off = 0;
312                     proto = IPPROTO_IPV6;
313                     while (1) {
314                             int nxt;
315                             newoff = ip6_nexthdr(m, off, proto, &nxt);
316                             if (newoff < 0 || newoff < off)
317                                     return (0);       /* XXX */
318                             off = newoff;
319                             proto = nxt;
320                             if (proto == IPPROTO_DSTOPTS)
321                                     break;
322                     }
323                     ip6o.ip6o_type = IP6OPT_PADN;
324                     ip6o.ip6o_len = 0;
325                     while (1) {
326                             newoff = dest6_nextopt(m, off, &ip6o);
327                             if (newoff < 0)
328                                     return (0);       /* XXX */
329                             off = newoff;
330                             if (ip6o.ip6o_type == IP6OPT_HOME_ADDRESS)
331                                     break;
332                     }
333                     m_copydata(m, off, sizeof(struct ip6_opt_home_address),
334                         (caddr_t)&haopt);
```
——— dest6.c

312–333 A Home Address option is searched for in the incoming Destination Options
header. If there is a Home Address option, the contents are copied to the `haopt`
variable.

Note that it is impossible to keep the pointer to the Home Address option in the
auxiliary area during the destination option processing at the `dest6_input()` function for later use because such a pointer may become invalid when some part of the
packet is relocated by the `m_pullup()` function during processing other extension
headers.

Listing 14-94

```
                                                                        dest6.c
336             swap = 0;
337             if (nxt == IPPROTO_AH || nxt == IPPROTO_ESP)
338                     swap = 1;
339             if (nxt == IPPROTO_MH) {
340                     m_copydata(m, mhoff, sizeof(mh), (caddr_t)&mh);
341                     if (mh.ip6mh_type == IP6_MH_TYPE_BU)
342                             swap = 1;
343                     else if (mh.ip6mh_type == IP6_MH_TYPE_HOTI ||
344                              mh.ip6mh_type == IP6_MH_TYPE_COTI)
345                             return (-1);
346                     else if (mh.ip6mh_type > IP6_MH_TYPE_MAX)
347                             swap = 1;        /* must be sent BE with
    UNRECOGNIZED_TYPE */
348             }
                                                                        dest6.c
```

Note: Line 347 is broken here for layout reasons. However, it is a single line of code.

337–348 The addresses are swapped if the next header is either an ESP or an AH. The addresses
must also be swapped when the next header is an MH. There are three types of MH
messages, which a correspondent node may receive: the Binding Update, Home Test Init,
and Care-of Test Init messages. The addresses will not be swapped in the Home Test
Init and the Care-of Test Init cases, and the code explicitly excludes these cases on lines
343–345. However, these conditions should not happen since these types of message must
not contain a Home Address option.

Listing 14-95

```
                                                                        dest6.c
350             home = *(struct in6_addr *)haopt.ip6oh_addr;
351             /*
352              * reject invalid home-addresses
353              */
354             if (IN6_IS_ADDR_MULTICAST(&home) ||
355                 IN6_IS_ADDR_LINKLOCAL(&home) ||
356                 IN6_IS_ADDR_V4MAPPED(&home) ||
357                 IN6_IS_ADDR_UNSPECIFIED(&home) ||
358                 IN6_IS_ADDR_LOOPBACK(&home)) {
....
360                     if (!(nxt == IPPROTO_MH && mh.ip6mh_type == IP6_MH_TYPE_BU)) {
361                             /* BE is sent only when the received packet is
362                                not BU */
363                             (void)mobility6_send_be(&ip6->ip6_dst, &ip6->ip6_src,
364                                 IP6_MH_BES_UNKNOWN_HAO, &home);
365                     }
366                     return (-1);
367             }
                                                                        dest6.c
```

354–366 There are some kinds of addresses that cannot be used as a home address. These prohibited addresses are explicitly excluded. A Binding Error message has to be sent when an invalid type of home address is received, except in one case. An error message is not sent if the incoming packet is a Binding Update message.[1]

Listing 14-96

```
                                                          dest6.c
369            if (swap) {
370                    int error;
371                    error = dest6_swap_hao(ip6, ip6a, &haopt);
372                    if (error)
373                            return (error);
374                    m_copyback(m, off, sizeof(struct ip6_opt_home_address),
375                        (caddr_t)&haopt);                /* XXX */
376                    return (0);
377            }
378
379            /* reject */
....
381            mobility6_send_be(&ip6->ip6_dst, &ip6->ip6_src,
382                IP6_MH_BES_UNKNOWN_HAO, &home);
383
384            return (-1);
385    }
386    #endif /* MIP6 */
                                                          dest6.c
```

369–384 The addresses are swapped. If the node receives a packet that has a Home Address option while it does not have a binding cache entry related to the home address included in the option, the code on lines 381–384 generates a Binding Error message to notify the sender that the received packet has an unverified Home Address option.

Listing 14-97

```
                                                          dest6.c
251    static int
252    dest6_nextopt(m, off, ip6o)
253            struct mbuf *m;
254            int off;
255            struct ip6_opt *ip6o;
256    {
257            u_int8_t type;
258
259            if (ip6o->ip6o_type != IP6OPT_PAD1)
260                    off += 2 + ip6o->ip6o_len;
261            else
262                    off += 1;
263            if (m->m_pkthdr.len < off + 1)
264                    return -1;
265            m_copydata(m, off, sizeof(type), (caddr_t)&type);
266
267            switch (type) {
268            case IP6OPT_PAD1:
269                    ip6o->ip6o_type = type;
270                    ip6o->ip6o_len = 0;
271                    return off;
272            default:
```

---

1. The exception seems to have no meaning, perhaps due to misreading of the RFC.

```
273                        if (m->m_pkthdr.len < off + 2)
274                                return -1;
275                        m_copydata(m, off, sizeof(ip6o), (caddr_t)ip6o);
276                        if (m->m_pkthdr.len < off + 2 + ip6o->ip6o_len)
277                                return -1;
278                        return off;
279                }
280        }
```
—— dest6.c

251–280 The `dest6_nextopt()` function locates the next option stored in a Destination Options Header beginning from the offset specified by the second parameter `off` and returns the offset of the next option. This function is used by the `dest6_mip6_hao()` function to find a Home Address option.

14.16   Sending Packets to Mobile Nodes via Tunnel

When a node is acting as a home agent of a mobile node, it must intercept all packets sent to the home address of the mobile node and tunnel them to the care-of address of the mobile node using an IPv6 in IPv6 tunnel. The home agent will receive packets by using the proxy Neighbor Discovery mechanism. Received packets are sent to the `ip6_input()` function and are passed to the `ip6_forward()` function since the destination address of these packets is not the address of the home agent. In the `ip6_forward()` function, the packets are tunneled to the mobile node based on the binding cache information.

Listing 14-98
—— ip6_forward.c
```
137     void
138     ip6_forward(m, srcrt)
139             struct mbuf *m;
140             int srcrt;
141     {
....
403     #if defined(MIP6) && defined(MIP6_HOME_AGENT)
404             {
405                     /*
406                      * intercept and tunnel packets for home addresses
407                      * which we are acting a home agent for.
408                      */
409                     struct mip6_bc *mbc;
410
411                     mbc = mip6_bc_list_find_withphaddr(&mip6_bc_list,
412                         &ip6->ip6_dst);
413                     if (mbc &&
414                         (mbc->mbc_flags & IP6MU_HOME) &&
415                         (mbc->mbc_encap != NULL)) {
```
—— ip6_forward.c

411–415 When forwarding a packet, a home agent checks the binding cache entries to determine whether it has a matching cache entry for the forwarded packet. The following two conditions are checked before forwarding:

(1) An entry exists and it is for home registration; that is, the `IP6MU_HOME` flag is set.

(2) The IPv6 in IPv6 tunnel of the entry is active; that is, the `mbc_encap` variable is not NULL.

If these two conditions are met, then the home agent tries to forward the packet to the
mobile node associated with the binding cache entry using the tunnel.

Listing 14-99

```
                                                                        ip6_forward.c
416                            if (IN6_IS_ADDR_LINKLOCAL(&mbc->mbc_phaddr)
417                                || IN6_IS_ADDR_SITELOCAL(&mbc->mbc_phaddr)
418                                )
419                            {
420                                    ip6stat.ip6s_cantforward++;
421                                    if (mcopy) {
422                                            icmp6_error(mcopy, ICMP6_DST_UNREACH,
423                                                    ICMP6_DST_UNREACH_ADDR, 0);
424                                    }
425                                    m_freem(m);
426                                    return;
427                            }
428
429                            if (m->m_pkthdr.len > IPV6_MMTU) {
430                                    u_long mtu = IPV6_MMTU;
....
432                                    if (mcopy) {
433                                            icmp6_error(mcopy,
434                                                    ICMP6_PACKET_TOO_BIG, 0, mtu);
435                                    }
436                                    m_freem(m);
437                                    return;
438                            }
                                                                        ip6_forward.c
```

416–438 Some validation checks are done. If the home address of the mobile node is not
in the proper scope, the home agent sends an ICMPv6 error message with a type of
ICMP6_DST_UNREACH. This should not happen since the home agent checks the validity
of the home address when accepting the home registration message for the home address.

The MTU size of the tunnel interface between a mobile node and a home agent is
hard coded to the minimum MTU size. If the size of a forwarded packet exceeds the minimum MTU value, the home agent sends an ICMPv6 error with ICMP6_PACKET_TOO_BIG
to perform the Path MTU Discovery procedure (see Section 4.3 of *IPv6 Core Protocols
Implementation*).

Listing 14-100

```
                                                                        ip6_forward.c
440                            /*
441                             * if we have a binding cache entry for the
442                             * ip6_dst, we are acting as a home agent for
443                             * that node.  before sending a packet as a
444                             * tunneled packet, we must make sure that
445                             * encaptab is ready.  if dad is enabled and
446                             * not completed yet, encaptab will be NULL.
447                             */
448                            if (mip6_tunnel_output(&m, mbc) != 0) {
....
450                            }
451                            if (mcopy)
452                                    m_freem(mcopy);
453                            return;
454                    }
455            mbc = mip6_bc_list_find_withphaddr(&mip6_bc_list,
456                &ip6->ip6_src);
```

```
457                         if (mbc &&
458                             (mbc->mbc_flags & IP6MU_HOME) &&
459                             (mbc->mbc_encap != NULL)) {
460                                 tunnel_out = 1;
461                         }
462                 }
463     #endif /* MIP6 && MIP6_HOME_AGENT */
```
─── ip6_forward.c

448–453 The `mip6_tunnel_output()` function is called if the home address is valid.

455–461 When the home agent forwards a packet, the `tunnel_out` variable is set to true. This variable will be examined when sending an ICMPv6 redirect message later.

Listing 14-101
─── ip6_forward.c
```
659                 if (rt->rt_ifp == m->m_pkthdr.rcvif && !srcrt && ip6_sendredirects &&
660     #ifdef IPSEC
661                     !ipsecrt &&
662     #endif
663     #ifdef MIP6
664                     !tunnel_out &&
665     #endif
666                     (rt->rt_flags & (RTF_DYNAMIC|RTF_MODIFIED)) == 0) {
```
    (Redirect processing)
─── ip6_forward.c

659–666 During forwarding, a packet may be resent to the same network interface. In this case, a router usually sends an ICMPv6 Redirect message to notify the sender that there is a better route. In the case of Mobile IPv6, this condition may happen even during normal operation. For example, a home agent will resend a packet to the same interface if the home agent has only one network interface that is attached to the home network. Because of this, the home agent does not send a redirect message if `tunnel_out` is set to true.

14.16.1   Sending Packets in IPv6 in IPv6 Format

The `mip6_tunnel_output()` function encapsulates the packet addressed to a mobile node and passes it to the `ip6_output()` function.

Listing 14-102
─── mip6_hacore.c
```
926     int
927     mip6_tunnel_output(mp, mbc)
928             struct mbuf **mp;     /* the original ipv6 packet */
929             struct mip6_bc *mbc;  /* the bc entry for the dst of the pkt */
930     {
931             const struct encaptab *ep = mbc->mbc_encap;
932             struct mbuf *m = *mp;
933             struct in6_addr *encap_src = &mbc->mbc_addr;
934             struct in6_addr *encap_dst = &mbc->mbc_pcoa;
935             struct ip6_hdr *ip6;
936             int len;
937
938             if (ep->af != AF_INET6) {
    ....
942                     return (EFAULT);
```

```
943                     }
944
945                     /* Recursion problems? */
946
947                     if (IN6_IS_ADDR_UNSPECIFIED(encap_src)) {
....
951                             return (EFAULT);
952                     }
```
                                                                    ―――――― mip6_hacore.c

926–929 The `mip6_tunnel_output()` function has two parameters: the `mp` parameter is a
double pointer to the mbuf in which the tunneled packet is stored, and the `mbc` parameter
is a pointer to the related binding cache entry.

938–952 The tunneling function only supports IPv6 in IPv6 tunnels. If the tunnel, represented
by the `encaptab{}` structure, does not support the IPv6 protocol family, forwarding
cannot be done. If the tunnel source address is not specified, the packet cannot be sent
through the tunnel.

Listing 14-103
                                                                    ―――――― mip6_hacore.c

```
954                     len = m->m_pkthdr.len; /* payload length */
955
956                     if (m->m_len < sizeof(*ip6)) {
957                             m = m_pullup(m, sizeof(*ip6));
958                             if (!m) {
....
962                                     return (ENOBUFS);
963                             }
964                     }
965                     ip6 = mtod(m, struct ip6_hdr *);
```
                                                                    ―――――― mip6_hacore.c

956–965 Before accessing the internal member variables of the IPv6 header to be tunneled,
the contiguity of the header memory is verified. If the header is not contiguous, the
`m_pullup()` function is called to rearrange it. An error is returned if getting the header
into contiguous memory fails.

Listing 14-104
                                                                    ―――――― mip6_hacore.c

```
967                     /* prepend new, outer ipv6 header */
968                     M_PREPEND(m, sizeof(struct ip6_hdr), M_DONTWAIT);
969                     if (m && m->m_len < sizeof(struct ip6_hdr))
970                             m = m_pullup(m, sizeof(struct ip6_hdr));
971                     if (m == NULL) {
....
975                             return (ENOBUFS);
976                     }
```
                                                                    ―――――― mip6_hacore.c

967–976 The `M_PREPEND()` macro is called to prepare an extra IPv6 header for tunneling.

Listing 14-105
                                                                    ―――――― mip6_hacore.c

```
978                     /* fill the outer header */
979                     ip6 = mtod(m, struct ip6_hdr *);
980                     ip6->ip6_flow = 0;
```

```
981              ip6->ip6_vfc &= ~IPV6_VERSION_MASK;
982              ip6->ip6_vfc |= IPV6_VERSION;
 ....
986              ip6->ip6_nxt = IPPROTO_IPV6;
987              ip6->ip6_hlim = ip6_defhlim;
988              ip6->ip6_src = *encap_src;
989
990              /* bidirectional configured tunnel mode */
991              if (!IN6_IS_ADDR_UNSPECIFIED(encap_dst))
992                      ip6->ip6_dst = *encap_dst;
993              else {
 ....
997                      m_freem(m);
998                      return (ENETUNREACH);
999              }
 ....
1004             /*
1005              * force fragmentation to minimum MTU, to avoid path MTU discovery.
1006              * it is too painful to ask for resend of inner packet, to achieve
1007              * path MTU discovery for encapsulated packets.
1008              */
1009             return (ip6_output(m, 0, 0, IPV6_MINMTU, 0, NULL
 ....
1011                                    , NULL
 ....
1013                                           ));
 ....
1021     }
```
_____mip6_hacore.c

978–1021 The fields of the outer IPv6 header are filled and the IPv6 in IPv6 packet is sent
via the ip6_output() function. The source and the destination addresses of the outer
header are taken from the tunnel information stored in the encaptab{} structure of the
binding cache entry.

14.17   Recovery of Temporarily Disabled Proxy Entry

When a home agent sends a Binding Acknowledgment message with an error status to a
mobile node that is attached to its home network, the home agent needs to disable the proxy
Neighbor Discovery entry for the mobile node temporarily; otherwise, the Binding Acknowledgment message sent to the mobile node will be caught by the home agent because of the
proxy Neighbor Discovery mechanism. The following two functions restore the proxy Neighbor
Discovery entry of a particular binding cache entry that is temporarily disabled.

(1) mip6_restore_proxynd_entry() — called from the nd6_output() function when
     the neighbor discovery entry for the home address of a mobile node is removed to restore
     the proxy entry.

(2) mip6_temp_deleted_proxy() — finds a binding cache entry related to the IPv6 packet
     passed as its argument.

Listing 14-106
_____mip6_hacore.c

```
624     struct mip6_bc *
625     mip6_restore_proxynd_entry(m)
626             struct mbuf *m;
```

```
627        {
628                struct mip6_bc *mbc;
629
630                mbc = mip6_temp_deleted_proxy(m);
631                if (mbc)
632                        mip6_bc_proxy_control(&mbc->mbc_phaddr, &mbc->mbc_addr,
      RTM_ADD);
633
634                return (mbc);
635        }
```
——— mip6_hacore.c

Note: Line 632 is broken here for layout reasons. However, it is a single line of code.

624–635 The mip6_restore_proxynd_entry() function recovers the proxy Neighbor Discovery entry that is temporarily disabled in the mip6_bc_send_ba() function. The function takes one parameter, which is a pointer to the mbuf that contains the IPv6 packet sent while proxy Neighbor Discovery is disabled. We use the mip6_bc_proxy_control() function to restore the proxy Neighbor Discovery entry. The binding cache entry which the mip6_bc_proxy_control() function requires can be obtained by calling the mip6_temp_deleted_proxy() function.

Listing 14-107
——— mip6_hacore.c

```
637        struct mip6_bc *
638        mip6_temp_deleted_proxy(m)
639                struct mbuf *m;
640        {
641                struct ip6_hdr *ip6;
642                struct m_tag *mtag;
643                struct mip6_bc *mbc = NULL;
644                struct ip6aux *ip6a;
645
646                ip6 = mtod(m, struct ip6_hdr *);
647
648                mtag = ip6_findaux(m);
649                if (!mtag)
650                        return (NULL);
651                ip6a = (struct ip6aux *) (mtag + 1);
652
653                if (ip6a->ip6a_flags & IP6A_TEMP_PROXYND_DEL) {
654                        mbc = mip6_bc_list_find_withphaddr(&mip6_bc_list,
655                            &ip6->ip6_dst);
656                        ip6a->ip6a_flags &= ~IP6A_TEMP_PROXYND_DEL;
657                }
658
659                return (mbc);
660        }
```
——— mip6_hacore.c

637–660 The mip6_temp_deleted_proxy() function takes one parameter, which is a pointer to the IPv6 packet. This function checks the auxiliary mbuf specified as the parameter m and checks to see if the IP6A_TEMP_PROXYND_DEL flag is set. The flag indicates that the proxy entry was temporarily disabled and needs to be restored. If the flag is set, the mip6_bc_list_find_withphaddr() function is called to get the binding cache entry related to the packet, and the binding cache entry is returned.

14.18   Receiving ICMPv6 Error Messages

A correspondent node may receive an ICMPv6 Destination Unreachable message when a mobile node moves. Usually, a mobile node sends a Binding Update message when it moves; however, the message may be lost and the correspondent node may not know the new care-of address of the mobile node. In this case, the correspondent node sends packets to the old care-of address of the mobile node and will receive an ICMPv6 error message. ICMPv6 messages are not reliable and the node should not rely on them. As the binding information to the destination mobile node has a lifetime even if the correspondent node does not receive either a Binding Update message or an ICMPv6 error message, the information will be deleted shortly. ICMPv6 is used as a supportive mechanism to reduce the detection time.

ICMPv6 messages are processed in the `mip6_icmp6_input()` function, which is called from the `icmp6_input()` function (for more information on basic ICMPv6 message processing, see Chapter 4 of *IPv6 Core Protocols Implementation*).

Listing 14-108

```
                                                              mip6_icmp6.c
129     int
130     mip6_icmp6_input(m, off, icmp6len)
131             struct mbuf *m;
132             int off;
133             int icmp6len;
134     {
135             struct ip6_hdr *ip6;
136             struct icmp6_hdr *icmp6;
137             struct mip6_bc *mbc;
138             struct in6_addr laddr, paddr;
                                                              mip6_icmp6.c
```

129–133 The `mip6_icmp6_input()` function has three parameters: the m parameter is a pointer to the mbuf that contains the incoming ICMPv6 packet, the `off` parameter is an offset from the head of the packet to the head of the ICMPv6 header, and the `icmp6len` parameter is the length of the ICMPv6 message.

Listing 14-109

```
                                                              mip6_icmp6.c
147             /* header pullup/down is already done in icmp6_input(). */
148             ip6 = mtod(m, struct ip6_hdr *);
149             icmp6 = (struct icmp6_hdr *)((caddr_t)ip6 + off);
150
151             switch (icmp6->icmp6_type) {
152             case ICMP6_DST_UNREACH:
153                     /*
154                      * the contacting mobile node might move to somewhere.
155                      * in current code, we remove the corresponding
156                      * binding cache entry immediately.  should we be more
157                      * patient?
158                      */
159                     IP6_EXTHDR_CHECK(m, off, icmp6len, EINVAL);
160                     mip6_icmp6_find_addr(m, off, icmp6len, &laddr, &paddr);
161                     mbc = mip6_bc_list_find_withphaddr(&mip6_bc_list, &paddr);
162                     if (mbc && (mbc->mbc_flags & IP6MU_HOME) == 0) {
....
166                             mip6_bc_list_remove(&mip6_bc_list, mbc);
167                     }
168             break;
```

```
169
```
... [code for a mobile node]
```
269              }
270
271              return (0);
272      }
```

152–168 If the type value is `ICMP6_DST_UNREACH`, the `mip6_icmp6_find_addr()` function is called to extract the source and destination addresses of the original packet that caused the error. If a correspondent node has a binding cache entry for the address pair extracted by function `mip6_icmp6_find_addr()`, the `mip6_bc_list_remove()` is called to remove the entry since the entry is not valid anymore.

If the node is acting as a home agent, the node may have home registration entries. Home registration entries are not removed even if ICMPv6 error messages are received. A mobile node will continue to send a Binding Update message for its home registration entry until the registration succeeds. The home agent has to keep the entry.

14.18.1 Find Source and Destination Addresses of the Original Packet

The `mip6_icmp6_find_addr()` function extracts the source and destination addresses of the original packet stored in the ICMPv6 payload.

Listing 14-110

```
274      static void
275      mip6_icmp6_find_addr(m, icmp6off, icmp6len, local, peer)
276              struct mbuf *m;
277              int icmp6off;
278              int icmp6len; /* Total icmp6 payload length */
279              struct in6_addr *local; /* Local home address */
280              struct in6_addr *peer; /* Peer home address */
281      {
282              caddr_t icmp6;
283              struct ip6_opt_home_address *haddr_opt; /* Home Address option */
284              struct ip6_hdr *ip6;                    /* IPv6 header */
285              struct ip6_ext *ehdr;                   /* Extension header */
286              struct sockaddr_in6 local_sa, peer_sa;
287              struct ip6_rthdr2 *rh;                  /* Routing header */
288              u_int8_t *eopt, nxt, optlen;
289              int off, elen, eoff;
290              int rlen, addr_off;
```

274–280 The `mip6_icmp6_find_addr()` function has five parameters: the `m`, `icmp6off`, and `icmp6len` parameters are the same as those of the `mip6_icmp6_input()` function. The `local` and `peer` parameters are the pointers to the memory, which are used to store the source and destination addresses extracted from the ICMPv6 message.

Listing 14-111

```
292              icmp6 = mtod(m, caddr_t) + icmp6off;
293              off = sizeof(struct icmp6_hdr);
294              ip6 = (struct ip6_hdr *)(icmp6 + off);
```

```
295              nxt = ip6->ip6_nxt;
296              off += sizeof(struct ip6_hdr);
297
298              *local = ip6->ip6_src;
299              *peer = ip6->ip6_dst;
```
_____ mip6_icmp6.c

292–296 An ICMPv6 message contains the original packet that caused the error immediately after the ICMPv6 header. The `ip6` variable points to the address of the original packet.

298–299 `local` and `peer` are initialized to the source and the destination addresses of the original IPv6 packet. These values may be updated later in this function if the original packet contains a Home Address option or a Routing Header.

Listing 14-112

_____ mip6_icmp6.c
```
301              /* Search original IPv6 header extensions for Routing Header type 0
302                 and for home address option (if I'm a mobile node). */
303              while ((off + 2) < icmp6len) {
304                      if (nxt == IPPROTO_DSTOPTS) {
305                              ehdr = (struct ip6_ext *)(icmp6 + off);
306                              elen = (ehdr->ip6e_len + 1) << 3;
307                              eoff = 2;
308                              eopt = icmp6 + off + eoff;
309                              while ((eoff + 2) < elen) {
310                                      if (*eopt == IP6OPT_PAD1) {
311                                              eoff += 1;
312                                              eopt += 1;
313                                              continue;
314                                      }
315                                      if (*eopt == IP6OPT_HOME_ADDRESS) {
316                                              optlen = *(eopt + 1) + 2;
317                                              if ((off + eoff + optlen) > icmp6len)
318                                                      break;
319
320                                              haddr_opt = (struct
   ip6_opt_home_address *)eopt;
321                                              bcopy((caddr_t)haddr_opt->ip6oh_addr,
322                                                  local, sizeof(struct in6_addr));
323                                              eoff += optlen;
324                                              eopt += optlen;
325                                              continue;
326                                      }
327                                      eoff += *(eopt + 1) + 2;
328                                      eopt += *(eopt + 1) + 2;
329                              }
330                              nxt = ehdr->ip6e_nxt;
331                              off += (ehdr->ip6e_len + 1) << 3;
332                              continue;
333                      }
```
_____ mip6_icmp6.c
Note: Line 320 is broken here for layout reasons. However, it is a single line of code.

303 Extension headers of the original IPv6 packet, which are contained in the incoming ICMPv6 message, are examined.

304–328 If the original packet has a Destination Options Header, we need to check to see if there is a Home Address option. If a Home Address option exists, the home address is copied to the `local` variable as the source address of the original packet. All other options are skipped. The PAD1 option is skipped on lines 310–314, and other options are skipped on lines 327–328.

330–332 The `nxt` variable is set to the protocol number of the next extension header of the Destination Options Header, and the `off` variable is set to the offset of the next extension header.

Listing 14-113
_____ mip6_icmp6.c
```
334                        if (nxt == IPPROTO_ROUTING) {
335                                rh = (struct ip6_rthdr2 *)(icmp6 + off);
336                                rlen = (rh->ip6r2_len + 1) << 3;
337                                if ((off + rlen) > icmp6len) break;
338                                if ((rh->ip6r2_type != 2) || (rh->ip6r2_len % 2)) {
339                                        nxt = rh->ip6r2_nxt;
340                                        off += (rh->ip6r2_len + 1) << 3;
341                                        continue;
342                                }
343
344                                addr_off = 8 + (((rh->ip6r2_len / 2) - 1) << 3);
345                                bcopy((caddr_t)(icmp6 + off + addr_off), peer,
346                                    sizeof(struct in6_addr));
347
348                                nxt = rh->ip6r2_nxt;
349                                off += (rh->ip6r2_len + 1) << 3;
350                                continue;
351                        }
```
_____ mip6_icmp6.c

334–351 If the original packet has a Type 2 Routing Header, the address contained in the routing header is copied to the `peer` variable as the destination address of the original packet. Note that we do not deal with Type 0 Routing Headers, since the Type 0 Routing Header has nothing to do with Mobile IPv6.

The `nxt` and `off` variables are adjusted to the protocol value and the offset of the next extension header, respectively.

Listing 14-114
_____ mip6_icmp6.c
```
352                        if (nxt == IPPROTO_HOPOPTS) {
353                                ehdr = (struct ip6_ext *)(icmp6 + off);
354                                nxt = ehdr->ip6e_nxt;
355                                off += (ehdr->ip6e_len + 1) << 3;
356                                continue;
357                        }
358
359                        /* Only look at the unfragmentable part.  Other headers
360                           may be present but they are of no interest. */
361                        break;
362                }
```
_____ mip6_icmp6.c

352–357 Hop-by-Hop Options Headers are ignored since they contain no data related to Mobile IPv6.

359–361 Only the Destination Options Header and (Type 2) Routing Header are checked. The rest of the packet contains no data related to the source and destination addresses of the original packet.

Listing 14-115

                                                                          ─────── mip6_icmp6.c

```
364            local_sa.sin6_len = sizeof(local_sa);
365            local_sa.sin6_family = AF_INET6;
366            local_sa.sin6_addr = *local;
367            /* XXX */
368            in6_addr2zoneid(m->m_pkthdr.rcvif, &local_sa.sin6_addr,
369                &local_sa.sin6_scope_id);
370            in6_embedscope(local, &local_sa);
371
372            peer_sa.sin6_len = sizeof(peer_sa);
373            peer_sa.sin6_family = AF_INET6;
374            peer_sa.sin6_addr = *peer;
375            /* XXX */
376            in6_addr2zoneid(m->m_pkthdr.rcvif, &peer_sa.sin6_addr,
377                &peer_sa.sin6_scope_id);
378            in6_embedscope(peer, &peer_sa);
379    }
```
                                                                          ─────── mip6_icmp6.c

364–378 The scope identifier of the address has to be recovered if the source or destination address is a scoped address. The scope identifier is recovered from the information about the interface on which the ICMPv6 packet arrived. Unfortunately, it is not always possible to know the original scope identifier since an error packet may come from a different interface than the one on which the original packet arrived. The above mechanism is the best effort to recover the information.

14.19   Home Agent List Management

The home agent list is managed by a user space daemon program called **had** on the home agent side. The **had** program listens for Router Advertisement messages (Section 10.20 of Chapter 10) and constructs the list of home agents on its home network. The program also responds to Dynamic Home Agent Address Discovery request messages (Section 10.24 of Chapter 10) and Mobile Prefix Solicitation messages (Section 10.26 of Chapter 10) sent from a mobile node and replies with Dynamic Home Agent Address Discovery response messages (Section 10.25 of Chapter 10) and Mobile Prefix Advertisement messages (Section 10.27 of Chapter 10), respectively.

In this section, we discuss the implementation of the **had** program briefly.

14.19.1   Home Agent Information Structures

The structures to keep home agent information are defined in `halist.h`.

Listing 14-116

                                                                          ─────── halist.h

```
121    struct hagent_gaddr {
122            struct hagent_gaddr     *hagent_next_gaddr, *hagent_prev_gaddr;
123            struct hagent_gaddr     *hagent_next_expire, *hagent_prev_expire;
124            struct in6_addr         hagent_gaddr;
125            u_int8_t                hagent_prefixlen;
126            struct hagent_flags {
127                    u_char          onlink : 1;
128                    u_char          autonomous : 1;
129                    u_char          router : 1;
130            } hagent_flags;
131            u_int32_t               hagent_vltime;
132            u_int32_t               hagent_pltime;
```

```
133                 long                         hagent_expire;
134                 long                         hagent_preferred;
135       };
```
halist.h

121–135 The `hagent_gaddr{}` structure represents the global address of a home agent. The `hagent_next_gaddr` and the `hagent_prev_gaddr` variables point to a list of `hagent_gaddr{}` structures in one `hagent_entry{}` structure described later. The `hagent_next_expire` and the `hagent_prev_expire` variables point to a list of `hagent_gaddr{}` structures ordered by lifetime. The list constructed by `hagent_next_expire` and `hagent_prev_expire` includes all `hagent_gaddr{}` instances kept in a node. The `hagent_gaddr` and the `hagent_prefixlen` variables are address and prefix length. The `hagent_flags` variable is a copy of the flags, which are received in Router Advertisement messages. The `hagent_vltime` and `hagent_expire` variables are the valid lifetime and the time when the valid lifetime expires. The `hagent_pltime` and the `hagent_preferred` variables are the preferred lifetime and the time when the preferred lifetime expires, respectively.

Listing 14-117

halist.h

```
138       struct hagent_entry {
139                 struct hagent_entry        *hagent_next_expire, *hagent_prev_expire,
140                                            *hagent_next_pref, *hagent_prev_pref;
141                 struct in6_addr            hagent_addr;
142                 u_int16_t                  hagent_pref;
143                 u_int16_t                  hagent_lifetime;
144                 long                       hagent_expire;
145                 struct hagent_gaddr        hagent_galist;
146       };
```
halist.h

138–146 The `hagent_entry{}` structure represents a home agent. The `hagent_next_expire` and the `hagent_prev_expire` variables construct a list of `hagent_entry{}` structures ordered by lifetime. The `hagent_next_pref` and `hagent_prev_pref` variables construct another list of `hagent_entry{}` structures ordered by the preference value. The `hagent_addr` variable is a link-local address of a home agent. The `hagent_pref` and the `hagent_lifetime` variables are the preference value and the lifetime, respectively. The `hagent_expire` variable is the time when the lifetime expires. The `hagent_galist` variable is a pointer to the `hagent_gaddr{}` structure that contains a list of global addresses of this home agent.

Listing 14-118

halist.h

```
151       struct hagent_ifinfo {
152                 struct hagent_entry        halist_pref;
153                 int                        ifindex;
154                 char                       ifname[IF_NAMESIZE];
155                 struct ifaddrs             *linklocal;
156                 struct hagent_ifa_pair     *haif_gavec;
157                 int                        gavec_used;
158                 int                        gavec_size;
159       };
160
161       #define GAVEC_INIT_SIZE            (16)
```
halist.h

151–159 The `hagent_ifinfo{}` structure represents a network interface on a home agent. The `halist_pref` variable is a pointer to the list of `hagent_entry{}` structures that contain home agent information on the interface. The `ifindex` and the `ifname` variables are the interface index and name, respectively. The `linklocal` variable is a pointer to the `ifaddrs{}` structure that contains a link-local address for the interface. The `haif_gavec` variable is a pointer to the array of `hagent_ifa_pair{}` structures, which contains a pair of a global address and an anycast address of the interface. The `gavec_used` and the `gavec_size` variables are the number of entries used in the `haif_gavec{}` array and the number of allocated entries of the `haif_gavec{}` array.

Listing 14-119

```
                                                                    halist.h
164     struct hagent_ifa_pair {
165             struct ifaddrs          *global;
166             struct ifaddrs          *anycast;
167     };
                                                                    halist.h
```

164–167 The `hagent_ifa_pair{}` structure keeps a set of a global address and an anycast address.

Figures 14-6 and 14-7 show the relationship of the above structures.

FIGURE 14-6

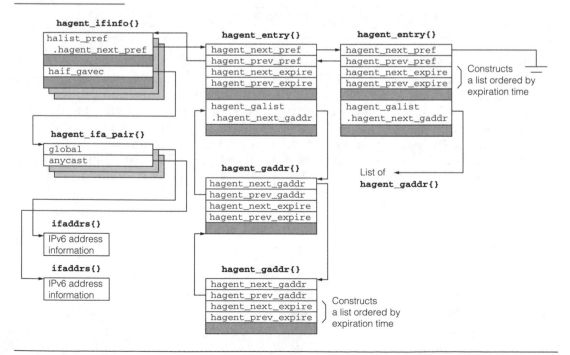

The relationship of the `hagent_ifinfo{}` and related structures.

FIGURE 14-7

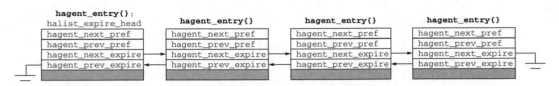

hagent_next_pref and hagent_prev_pref constructs a list of hagent_entry{}
which is ordered by the preference value per hagent_ifinfo{}

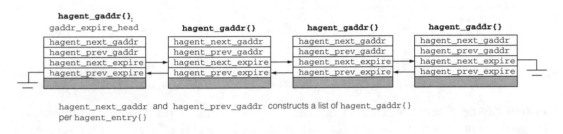

hagent_next_gaddr and hagent_prev_gaddr constructs a list of hagent_gaddr{}
per hagent_entry{}

The global list of the hagent_entry{} and the hagent_gaddr{} structures.

14.19.2 Updating Home Agent Information

When a home agent receives a Router Advertisement message, the **had** program updates its internal information by extracting data related to the home agent management from the message.

Listing 14-120
_____ haadisc.c

```
607     static void
608     ra_input(len, ra, pinfo, from)
609         int len;
610         struct nd_router_advert *ra;
611         struct in6_pktinfo *pinfo;
612         struct sockaddr_in6 *from;
613     {
```
_____ haadisc.c

607–612 The ra_input() function is called when the **had** program receives a Router Advertisement message. The len parameter is the length of the message; the ra parameter is a pointer of the nd_router_advert{} structure that points to the head of the message; the pinfo parameter is a pointer to the in6_pktinfo{} structure that contains an interface index and hoplimit information; and the from parameter is the source address of the message.

Listing 14-121
_____ haadisc.c

```
623         haif = haif_find(pinfo->ipi6_ifindex);
624         if (haif == NULL) {
```

```
....
630               goto done;
631           }
```
_____ haadisc.c

623–631 The `haif_find()` function finds an instance of the `hagent_ifinfo{}` structure with the specified interface index value from the list of `hagent_ifinfo{}`. If there is no `hagent_ifinfo{}` structure that has the specified index value, the packet is dropped.

Listing 14-122
_____ haadisc.c
```
634           bzero(&ndopts, sizeof(union nd_opts));
635           nd6_option_init(ra + 1, len - sizeof(struct nd_router_advert), &ndopts);
636           if (nd6_options(&ndopts) < 0) {
....
638               goto done;
639           }
```
_____ haadisc.c

635–639 The `nd6_option_init()` and the `nd6_options()` functions do the same tasks as the functions that have the same names in the kernel. The contents of the options of the incoming Router Advertisement message are extracted into the `ndopts` variable.

Listing 14-123
_____ haadisc.c
```
641           /* Is this RA from some home agent or not? */
642           if (0 == (ra->nd_ra_flags_reserved & ND_RA_FLAG_HOME_AGENT)) {
643               /* IMPLID:MIP6HA#7 */
644               /*
645                * delete home agent list entry if it exists,
646                * because this router is not a home agent.
647                */
648               hal_delete(haif, &(from->sin6_addr));
649               goto done;
650           }
```
_____ haadisc.c

641–650 Every home agent must set the home agent flag, the ND_RA_FLAG_HOME_AGENT flag, when sending a Router Advertisement message. If the received message does not have the flag, and there is an entry for the router, the home agent removes the entry from the list by calling the `hal_delete()` function since the node is not acting as a home agent anymore. The `hal_delete()` function removes the `hagent_entry{}` instance specified by its second parameter from the `hagent_ifinfo{}` structure specified by the first parameter.

Listing 14-124
_____ haadisc.c
```
652           /* determine HA lifetime and preference */
653           /* IMPLID:MIP6HA#9 */
654           ha_lifetime = ntohs(ra->nd_ra_router_lifetime);
655           if (ndopts.nd_opts_hai) {
656               ha_lifetime = ntohs(ndopts.nd_opts_hai->nd_opt_hai_lifetime);
657               /* IMPLID:MIP6HA#10 */
658               ha_pref = ntohs(ndopts.nd_opts_hai->nd_opt_hai_preference);
```

```
659             }
660
661             /* update and get home agent list entry */
662             halp = hal_update(pinfo->ipi6_ifindex, &from->sin6_addr, ha_lifetime,
    ha_pref);
663
664             if (!halp) {
665                 /*
666                  * no home agent list entry (deleted or cannot create)
667                  */
668                 goto done;
669             }
```
                                                                            ————haadisc.c

Note: Line 662 is broken here for layout reasons. However, it is a single line of code.

654–659 The lifetime of a home agent is considered the same as the lifetime of a router, nd_ra_router_lifetime, unless the lifetime as a home agent is explicitly specified by the Home Agent Information option, nd_opt_hai_lifetime. The preference is set to 0 unless explicitly specified by the Home Agent Information option, nd_opt_hai_preference.

662–669 The hal_update() function updates the home agent information of the specified address as its second parameter.

Listing 14-125
                                                                            ————haadisc.c

```
671             /* proceee prefix information option in RA
672              * in order to accumulate home agent global address
673              * information in home agent list
674              */
675             if (ndopts.nd_opts_pi) {
676                 /*
677                  * parse prefix information option and
678                  * get global address(es) in it.
679                  */
680                 struct nd_opt_hdr *pt;
681                 struct nd_opt_prefix_info *pi;
682                 struct hagent_gaddr *lastp;
683                 /* temporary global address list */
684                 struct hagent_gaddr newgaddrs;
```
        .... (validation of prefix information option)
```
737                     if ((pi->nd_opt_pi_flags_reserved & ND_OPT_PI_FLAG_ROUTER) != 0) {
738                         /* IMPLID:MIP6HA#14 */
739                         lastp = hal_gaddr_add(halp, lastp, pi);
740                     }
741                 }
742                 /* replace home agent global address list to new one */
743                 if (newgaddrs.hagent_next_gaddr == NULL) goto done;
744                 hal_gaddr_last(halp, newgaddrs.hagent_next_gaddr);
        ....
750             }
751     done:
752     }
```
                                                                            ————haadisc.c

675–741 A Prefix Information option may include a global address for the home agent, in which case the option has the ND_OPT_PI_FLAG_ROUTER flag. The hal_gaddr_add() function collects the global addresses from the incoming Router Advertisement message and makes a list of addresses in the newgaddrs variable, which is an instance of the hagent_gaddr{} structure. If the incoming message contains global addresses,

the global address list of the home agent who sent the message is updated by the
`hal_gaddr_last()` function.

14.19.3 Updating the `hagent_entry{}` Structure

The `hagent_entry{}` structure is updated when a home agent receives a Router Advertisement message. The update code is implemented as the `hal_update()` function.

Listing 14-126

```
                                                                      halist.c
131     struct hagent_entry *
132     hal_update(ifindex, ha_addr, ha_lifetime, ha_pref)
133         int ifindex;
134         struct in6_addr *ha_addr;
135         u_int16_t ha_lifetime;
136         u_int16_t ha_pref;
137     {
                                                                      halist.c
```

131–136 The `hal_update()` function updates the `hagent_entry{}` structure identified by
the interface index specified by the first parameter and the address specified by the second
parameter. The `ha_lifetime` and `ha_pref` parameters are the new values of the lifetime
and preference of the home agent.

Listing 14-127

```
                                                                      halist.c
148         /* lookup home agent i/f info from ifindex */
149         haif = haif_find(ifindex);
150
151         if (!haif) {
....
153             goto err;
154         }
                                                                      halist.c
```

149–154 If a home agent does not have a `hagent_ifinfo{}` instance whose interface index
is the same as the specified value as the `ifindex` parameter, it ignores the message.

Listing 14-128

```
                                                                      halist.c
156         /* lookup home agent entry from home agent list of specified i/f */
157         halp = hal_find(haif, ha_addr);
158
159         /* if HA entry exists, remove it from list first */
160         if (halp) {
....
164             /* remove from preference list */
165             if (halp->hagent_next_pref) {
166                 halp->hagent_next_pref->hagent_prev_pref
167                     = halp->hagent_prev_pref;
168             }
169             if (halp->hagent_prev_pref) {
170                 halp->hagent_prev_pref->hagent_next_pref
171                     = halp->hagent_next_pref;
172             }
173             halp->hagent_next_pref = halp->hagent_prev_pref = NULL;
174
175             /* remove from expire list */
```

```
176                 if (halp->hagent_next_expire) {
177                     halp->hagent_next_expire->hagent_prev_expire
178                         = halp->hagent_prev_expire;
179                 }
180                 if (halp->hagent_prev_expire) {
181                     halp->hagent_prev_expire->hagent_next_expire
182                         = halp->hagent_next_expire;
183                 }
184                 halp->hagent_next_expire = halp->hagent_prev_expire = NULL;
185             }
```
——— halist.c

157–185 If a home agent has a `hagent_entry{}` instance that matches the specified parameters, it first removes the entry from the list. The `hagent_entry{}` structure has two
kinds of lists: one is a list ordered by the preference value and the other is ordered by
lifetime. The entry is removed from both lists and the pointers, which construct the lists
are initialized with a NULL pointer.

Listing 14-129
——— halist.c

```
187         if (ha_lifetime > 0) {
188             /* create list entry if not already exist */
189             if (! halp) {
190
191                 /* IMPLID:MIP6HA#13 */
192                 halp = malloc(sizeof (struct hagent_entry));
193                 if (halp) {
194                     bzero(halp, sizeof (struct hagent_entry));
195                     bcopy(ha_addr, &halp->hagent_addr, sizeof (struct in6_addr));
....
197                 }
198                 else {
....
200                     goto err;
201                 }
202             }
```
——— halist.c

187–202 If `ha_lifetime` is a positive value and there is no existing `hagent_entry{}`
instance to update, the home agent creates a new `hagent_entry{}` instance.

Listing 14-130
——— halist.c

```
204             /* IMPLID:MIP6HA#12 */
205             /* update parameters */
206             halp->hagent_pref = ha_pref;
207             halp->hagent_lifetime = ha_lifetime;
208             halp->hagent_expire = now + ha_lifetime;
```
——— halist.c

206–208 Each element of the `hagent_entry{}` structure is updated with the specified lifetime
and preference values as parameters.

Listing 14-131
——— halist.c

```
210             /* insert entry to preference list */
211             for (prevp = curp = haif->halist_pref.hagent_next_pref;
212                 curp; curp = curp->hagent_next_pref) {
```

```
213                     if (halp->hagent_pref > curp->hagent_pref) {
214                         halp->hagent_prev_pref = curp->hagent_prev_pref;
215                         halp->hagent_next_pref = curp;
216                         if (curp->hagent_prev_pref) {
217                             curp->hagent_prev_pref->hagent_next_pref = halp;
218                         }
219                         curp->hagent_prev_pref = halp;
220
221                         break;
222                     }
223                     prevp = curp;
224                 }
225                 if (! curp) {
226                     if (prevp) {
227                         /* append tail */
228                         prevp->hagent_next_pref = halp;
229                         halp->hagent_prev_pref = prevp;
230                     }
231                     else {
232                         /* insert head */
233                         haif->halist_pref.hagent_next_pref = halp;
234                         halp->hagent_prev_pref = &haif->halist_pref;
235                     }
236                 }
```
———————————————————————————————————— halist.c

211–223 The updated entry or the newly created entry is inserted into the list whose entries
are ordered by the preference value. Each pointer to the entry in the list is set to the `curp`
variable and compared to the updated/created entry, while `prevp` points to the previous
entry of the entry specified by the `curp`.

225–235 If `curp` is a NULL pointer and `prevp` has a valid pointer to the entry in the list, the
updated/created entry will be inserted at the tail of the list. If `curp` and `prevp` are both
NULL pointers, then it means the list is empty. The updated/new entry is inserted at the
head of the list.

Listing 14-132

———————————————————————————————————— halist.c
```
238                 /* insert entry to expire list */
239                 for (prevp = curp = halist_expire_head.hagent_next_expire;
240                     curp; curp = curp->hagent_next_expire) {
241                     if (curp->hagent_expire > halp->hagent_expire) {
242                         halp->hagent_prev_expire = curp->hagent_prev_expire;
243                         halp->hagent_next_expire = curp;
244                         if (curp->hagent_prev_expire) {
245                             curp->hagent_prev_expire->hagent_next_expire = halp;
246                         }
247                         curp->hagent_prev_expire = halp;
248
249                         break;
250                     }
251                     prevp = curp;
252                 }
253                 if (! curp) {
254                     if (prevp) {
255                         /* append tail */
256                         prevp->hagent_next_expire = halp;
257                         halp->hagent_prev_expire = prevp;
258                     }
259                     else {
260                         /* insert head */
261                         halist_expire_head.hagent_next_expire = halp;
262                         halp->hagent_prev_expire = &halist_expire_head;
```

```
263                     }
264                 }
265
266             }
```
_____ halist.c

238–266 The entry is also inserted into the other list ordered by the lifetime value.

Listing 14-133
_____ halist.c

```
267         else if (halp) { /* must be deleted */
268             /* IMPLID:MIP6HA#11 */
269             /* clear global address list */
270             hal_gaddr_clean(halp);
271             free(halp);
272             halp = NULL;
    ....
275         }
276
277    done:
    ....
284         return halp;
285    err:
    ....
288         halp = NULL;
289         goto done;
290     }
```
_____ halist.c

267–272 If the specified lifetime is 0, the entry will be removed. The `hal_gaddr_clean()` function removes all `hagent_gaddr{}` instances pointed to by the `hagent_galist` variable, which contains the list of global addresses of the home agent.

14.19.4   Sending a Dynamic Home Agent Address Discovery Reply Message

When the **had** program receives a Dynamic Home Agent Address Discovery request message from a mobile node, it replies with a Dynamic Home Agent Address Discovery reply message with home agent addresses of the home network of the mobile node as described in Chapter 6. Receiving and Replying to these messages is implemented in the `haad_request_input()` and the `haad_reply_output()` functions.

Listing 14-134
_____ haadisc.c

```
900    static void
901    haad_request_input(len, haad_req, pi, src, type)
902         int len;
903         struct mip6_dhaad_req *haad_req;
904         struct in6_pktinfo *pi;
905         struct sockaddr_in6 *src;
906         int type;
907    {
908         u_int16_t msgid;
909         struct hagent_ifinfo *haif;
910         int ifga_index = -1;
```
_____ haadisc.c

900–906 The `haad_request_input()` function is called when a mobile node receives a Dynamic Home Agent Address Discovery request message. The `len` parameter is the length of the message; the `haad_req` parameter is a pointer to the head of the incoming message; the `pi` parameter is a pointer to the `in6_pktinfo{}` structure that contains the destination address of the message; and the `type` parameter is the type number of the ICMPv6 message, which should be `MIP6_HA_DISCOVERY_REQUEST` in this case.

Listing 14-135

```
                                                                    haadisc.c
912        msgid = haad_req->mip6_dhreq_id;
913
914        /* determine home link by global address */
915        haif = haif_findwithanycast(&pi->ipi6_addr, &ifga_index);
916
917        if (! haif) {
....
919            goto err;
920        }
921
922        /* send home agent address discovery response message */
923        haad_reply_output(msgid, src,
....
927                        &(pi->ipi6_addr),            /* anycast addr. */
....
929                        haif, type, ifga_index);
930    err:
931    }
                                                                    haadisc.c
```

912 The identifier, `mip6_dhreq_id`, of the message that is included in the received Dynamic Home Agent Address Discovery request message must be copied to the reply message.

915–920 The destination address of the request message must be one of the addresses of the interface that is acting as the home network. If there is no `hagent_ifinfo{}` instance, which has the address used as the destination address of the received Dynamic Home Agent Address Discovery request message, the home agent ignores the message. The `ifga_index` variable is passed to the `haif_findwithanycast()` function and updated in the function. The index will point to the entry in the `haif_gavec{}` array of `hagent_ifinfo{}` structures, which has the address specified by the first parameter of the `haif_findwithanycast()` function.

923–929 The `haad_reply_output()` function sends a Dynamic Home Agent Address Discovery reply message.

Listing 14-136

```
                                                                    haadisc.c
936    static void
937    haad_reply_output(msgid, coaddr, reqaddr, haif, type, ifga_index)
938        u_int16_t msgid;
939        struct sockaddr_in6 *coaddr;
940        struct in6_addr *reqaddr;
941        struct hagent_ifinfo *haif;
942        int type, ifga_index;
943    {
944        struct cmsghdr *cm;
945        struct in6_pktinfo *pi;
946        struct mip6_dhaad_rep *hap;
947        struct in6_addr *hagent_addr;
```

```
948                struct in6_addr src = in6addr_any;
949                int len, nhaa, count;
950                u_int8_t buf[IPV6_MMTU];
```
——— haadisc.c

936–942 The `haad_reply_output()` function has six parameters: the `msgid` parameter is an identifier that is included in the Dynamic Home Agent Address Discovery request message; the `coaddr` and `reqaddr` parameters are the care-of address of a mobile node and the destination address of the incoming request message, respectively; the `haif` parameter is a pointer to the `hagent_ifinfo{}` structure that indicates the incoming interface; the `type` parameter is the type number of the incoming ICMPv6 message, which is `MIP6_HA_DISCOVERY_REQUEST` in this case; and the `ifga_index` parameter is an index number to the `haif_gavec{}` array, which contains the pair of global and anycast addresses of the home agent.

Listing 14-137
——— haadisc.c

```
959                if (haif->haif_gavec[ifga_index].global != NULL)
960                    src = ((struct sockaddr_in6 *)(haif->
   haif_gavec[ifga_index].global->ifa_addr))->sin6_addr;
961
962                /* create ICMPv6 message */
963                hap = (struct mip6_dhaad_rep *)buf;
964                bzero(hap, sizeof (struct mip6_dhaad_rep));
965                hap->mip6_dhrep_type = MIP6_HA_DISCOVERY_REPLY;
966                hap->mip6_dhrep_code = 0;
967                hap->mip6_dhrep_cksum = 0;
968                hap->mip6_dhrep_id = msgid;
969                len = sizeof (struct mip6_dhaad_rep);
```
——— haadisc.c

Note: Line 960 is broken here for layout reasons. However, it is a single line of code.

959–960 The global address that corresponds to the anycast address used as the destination address of the incoming request message is used as a source address of the reply message.

963–969 All ICMPv6 message fields are filled. The identifier, `mip6_dhrep_id`, must be copied from the identifier in the request message.

Listing 14-138
——— haadisc.c

```
970                hagent_addr = (struct in6_addr *)(hap + 1);
971                count = (IPV6_MMTU - sizeof (struct ip6_hdr) -
972                       sizeof (struct mip6_dhaad_rep)) / sizeof (struct in6_addr);
973                /* pick home agent global addresses for this home address */
974                 if ((nhaa = hal_pick(reqaddr, hagent_addr, &src, haif, count)) < 0) {
  ....
976                   goto err;
977                }
978                if (IN6_IS_ADDR_UNSPECIFIED(&src))
979                   goto err;
980                len += nhaa * sizeof (struct in6_addr);
```
——— haadisc.c

970–977 The `hal_pick()` function constructs the payload part of a Dynamic Home Agent Address Discovery reply message.

978–979 The `src` variable is set in the `hal_pick()` function to one of the global home agent addresses that received the request message. If no source address is selected, the home agent aborts replying to the message.

Listing 14-139

haadisc.c

```
982        sndmhdr.msg_name = (caddr_t)coaddr;
983        sndmhdr.msg_namelen = coaddr->sin6_len;
984        sndmhdr.msg_iov[0].iov_base = (caddr_t)buf;
985        sndmhdr.msg_iov[0].iov_len = len;
986
987        cm = CMSG_FIRSTHDR(&sndmhdr);
988        /* specify source address */
989        cm->cmsg_level = IPPROTO_IPV6;
990        cm->cmsg_type = IPV6_PKTINFO;
991        cm->cmsg_len = CMSG_LEN(sizeof(struct in6_pktinfo));
992        pi = (struct in6_pktinfo *)CMSG_DATA(cm);
993        pi->ipi6_addr = src;
994        pi->ipi6_ifindex = 0; /* determined with routing table */
995
996        if ((len = sendmsg(sock, &sndmhdr, 0)) < 0) {
....
998            goto err;
999        }
1000   err:
1001   }
```

haadisc.c

982–985 A destination address of the reply message must be the source address of the request message. The `coaddr` variable is set as the destination address.

993–999 The created Dynamic Home Agent Address Discovery reply message is sent by the `sendmsg()` system call from the source address selected by the `hal_pick()` function.

14.19.5   Constructing a Payload

The `hal_pick()` function constructs the payload of a Dynamic Home Agent Address Discovery reply message.

Listing 14-140

halist.c

```
803    int
804    hal_pick(req_addr, hagent_addrs, src_addr, haif, count)
805        struct in6_addr *req_addr;
806        struct in6_addr *hagent_addrs;
807        struct in6_addr *src_addr;
808        struct hagent_ifinfo *haif;
809        int count;
810    {
811        int naddr;
812        struct hagent_entry *hap, *selfhalp = NULL;
813        struct hagent_gaddr *ha_gaddr;
814        int found_src = 0;
```

halist.c

803–809 The `hal_pick()` function has five parameters: the `req_addr` parameter is the destination address of the request message; the `hagent_addrs` parameter is a pointer to the memory space where the constructed payload is stored; the `src_addr` parameter is the source address of the request message; the `haif` parameter is a pointer to the `hagent_ifinfo{}` instance to which the request message is delivered; and the `count` parameter is the maximum number of addresses to be listed in the payload.

Listing 14-141

_____halist.c

```
816          /* shuffle home agent entries with same preference */
817          hal_shuffle(haif);
818
819          /* lookup self entry from home agent list */
820          if (haif->linklocal)
821              selfhalp = hal_find(haif, &((struct sockaddr_in6 *)(haif->linklocal->
    ifa_addr))->sin6_addr);
822
823          /* list all home agents in the home agent list of this interface */
824          for (naddr = 0, hap = haif->halist_pref.hagent_next_pref;
825              hap && naddr < count; hap = hap->hagent_next_pref) {
826            for (ha_gaddr = hap->hagent_galist.hagent_next_gaddr;
827                (ha_gaddr != NULL) && (naddr < count);
828                ha_gaddr = ha_gaddr->hagent_next_gaddr) {
829              *hagent_addrs = ha_gaddr->hagent_gaddr;
830              if (hap == selfhalp && found_src == 0) {
831                  *src_addr = *hagent_addrs;
832                  found_src++;
833              }
834              hagent_addrs ++;
835              naddr ++;
836            }
837          }
838
839          return naddr;
840      }
```

_____halist.c

Note: Line 821 is broken here for layout reasons. However, it is a single line of code.

817 The `hal_shuffle()` function randomly reorders the `hagent_entries` instances that have the same preference value. A mobile node will use the first address in the home agent address list. If there are multiple home agents in the home network, shuffling the addresses randomly will balance the load between home agents when there are many mobile nodes in the home network.

820–837 The global addresses of home agents, listed in the `halist_pref` variable of the `hagent_ifinfo{}` structure, are listed in the order of preference value. During the process, the `src_addr` variable is set to the global address of the home agent that received the request message.

14.20 Prefix List Management

When a home agent receives a Mobile Prefix Solicitation message, the node needs to reply with a Mobile Prefix Advertisement message as described in Chapter 7. The prefix information of a home network can be taken from the home agent list information that is maintained as a part of Dynamic Home Agent Address Discovery mechanism.

The current implementation only supports solicited request messages. The **had** program does not send an unsolicited Mobile Prefix Advertisement message.

14.21 Sending a Mobile Prefix Advertisement Message

The `mpi_solicit_input()` function is called when the **had** program receives a Mobile Prefix Solicitation message.

Listing 14-142

_____mpa.c
```
122     void
123     mpi_solicit_input(pi, sin6_hoa, mps)
124         struct in6_pktinfo *pi;
125         struct sockaddr_in6 *sin6_hoa;
126         struct mip6_prefix_solicit *mps;
127     {
128         int ifga_index = -1;
129         struct in6_addr ha_addr;
130         struct hagent_ifinfo *haif;
131         struct in6_addr src;
132         int error;
```
_____mpa.c

122–126 The `mpi_solicit_input()` function has three parameters: the `pi` parameter is a pointer to the `in6_pktinfo{}` instance that contains the destination address of the request message; the `sin6_hoa` parameter is the home address of the mobile node, which sent the request message; and the `mps` parameter is a pointer to the head of the message.

Listing 14-143

_____mpa.c
```
137         ha_addr = pi->ipi6_addr;
138         /* determine a home link by the global address */
139         haif = haif_findwithunicast(&pi->ipi6_addr, &ifga_index);
140
141         if (!haif) {
....
143             goto err;
144         }
....
151         src = ha_addr;
152         mpi_advert_output(sin6_hoa, &src, haif, mps->mip6_ps_id);
153     err:
154     }
```
_____mpa.c

137–144 A Mobile Prefix Solicitation message is sent to the global unicast address of the home agent of a mobile node. The `haif_findwithunicast()` function finds the `hagent_ininfo{}` instance, which has the specified unicast global address.

152 The `mpi_advert_output()` function sends a Mobile Prefix Advertisement message.

Listing 14-144

_____mpa.c
```
159     void
160     mpi_advert_output(dst_sa, src, haif, id)
161         struct sockaddr_in6 *dst_sa;        /* home addr of destination MN */
162         struct in6_addr *src;
163         struct hagent_ifinfo *haif;
164         u_int16_t id;
165     {
```

```
166          struct cmsghdr *cm;
167          struct nd_opt_prefix_info *prefix_info;
168          u_int8_t buf[IPV6_MMTU];
169          struct mip6_prefix_advert *map;
170          int len;
171          int count;
172          int npi;
173          struct in6_pktinfo *pi;
```
_____ mpa.c

159–164 The `mpi_advert_output()` function has four parameters. The `dst_sa` parameter
is the home address of the mobile node, which sent the request message. The `src` parameter is the global unicast address of the home agent. The `haif` parameter is a pointer to
the `hagent_ifinfo{}` instance. This parameter represents the network interface on
which the request message arrived. The `id` parameter is the identifier that is contained in
the request message.

Listing 14-145
_____ mpa.c

```
175          /* create ICMPv6 message */
176          map = (struct mip6_prefix_advert *)buf;
177          bzero(map, sizeof (struct mip6_prefix_advert));
178          map->mip6_pa_type = MIP6_PREFIX_ADVERT;
179          map->mip6_pa_code = 0;
180
181          len = sizeof(struct mip6_prefix_advert);
182          prefix_info = (struct nd_opt_prefix_info *)&map[1];
183          /* count number of prefix informations --
   to make assurance (not in spec.) */
184          count = (IPV6_MMTU - sizeof (struct ip6_hdr) - /* XXX: should include the
   size of routing header*/
185                              sizeof (struct mip6_prefix_advert)) / sizeof
   (struct nd_opt_prefix_info);
186
187          /* Pick home agent prefixes */
188          /* -- search by dest. address instead of Home Address */
189          if ((npi = pi_pick(src, prefix_info, haif, count)) < 0) {
....
191              goto err;
192          }
```
_____ mpa.c

Note: Lines 183, 184, and 185 are broken here for layout reasons. However, they are each a single line of code.

176–179 Memory for the reply message is allocated and ICMPv6 header fields are filled in.

182–192 The `pi_pick()` function constructs the payload of the reply message and returns
the number of prefix information entries that have been put into the payload. On lines
184–185, the maximum number of prefix information entries to be included is calculated.
As the comment says, the stack should take into account the size of the Type 2 Routing
Header since a Mobile Prefix Advertisement message always has a Type 2 Routing header;
however, the current code does not count the size at this moment. This is may be a bug
in the code.

Listing 14-146
_____ mpa.c

```
194          if(!npi)
195              return;
196
```

```
197            len += npi * sizeof (struct nd_opt_prefix_info);
198
199            map->mip6_pa_cksum = 0;
200            map->mip6_pa_id = id;
201            sndmhdr.msg_name = (caddr_t)dst_sa;
202            sndmhdr.msg_namelen = dst_sa->sin6_len;
203            sndmhdr.msg_iov[0].iov_base = (caddr_t)buf;
204            sndmhdr.msg_iov[0].iov_len = len;
```
_____ mpa.c

200–201 The identifier field, `mip6_pa_id`, must be copied from the identifier field of the corresponding solicitation message. The destination address of the advertisement message must be the home address of the mobile node, `dst_sa`, which sent the solicitation message.

Listing 14-147

_____ mpa.c
```
206            cm = CMSG_FIRSTHDR(&sndmhdr);
207            /* specify source address */
208            cm->cmsg_level = IPPROTO_IPV6;
209            cm->cmsg_type = IPV6_PKTINFO;
210            cm->cmsg_len = CMSG_LEN(sizeof(struct in6_pktinfo));
211            pi = (struct in6_pktinfo *)CMSG_DATA(cm);
212            pi->ipi6_addr = *src;
213            pi->ipi6_ifindex = 0; /* determined with a routing table */
214
215            if ((len = sendmsg(sock, &sndmhdr, 0)) < 0) {
....
217                goto err;
218            }
219      err:
220      }
```
_____ mpa.c

212–215 The source address must be the destination address of the incoming solicitation message. The created advertisement message is sent by the `sendmsg()` system call.

14.22   Constructing the Payload

The `pi_pick()` function constructs the payload of a Mobile Prefix Advertisement message.

Listing 14-148

_____ mpa.c
```
225      int
226      pi_pick(home_addr, prefix_info, haif, count)
227            struct in6_addr *home_addr;
228            struct nd_opt_prefix_info *prefix_info;
229            struct hagent_ifinfo *haif;
230            int count;
231      {
232            int naddr;
233            struct hagent_entry *hap;
234            struct hagent_gaddr *ha_gaddr;
235            struct nd_opt_prefix_info *h_prefix_info;
236            struct timeval now;
237            u_int32_t vltime, pltime;
```
_____ mpa.c

225–230 The `pi_pick()` function has four parameters: the `home_addr` parameter is the home address of a mobile node; the `prefix_info` parameter is a pointer to the memory space where the payload is stored; the `haif` parameter is a pointer to the `hagent_ifinfo{}` instance that received the solicitation message; and the `count` parameter is the maximum amount of prefix information to be contained.

Listing 14-149

_____ mpa.c

```
239          h_prefix_info = prefix_info;
240          /* search home agent list and pick all prefixes */
241          for (naddr = 0, hap = haif->halist_pref.hagent_next_pref;
242              hap && naddr < count; hap = hap->hagent_next_pref) {
243            for (ha_gaddr = hap->hagent_galist.hagent_next_gaddr;
244              (ha_gaddr != NULL) && (naddr < count);
245              ha_gaddr = ha_gaddr->hagent_next_gaddr) {
246              /* duplication check whether MPA includes duplicated prefixes */
247              if (prefix_dup_check(h_prefix_info, ha_gaddr, naddr))
248                /* duplicated prefix is included */
249                continue;
```
_____ mpa.c

241–245 The prefix information is created from the global addresses of the home agent of the home network. The outer `for` loop checks all `hagent_entry{}` instances related to the home network and the inner `for` loop checks all global addresses of each `hagent_entry{}` instance.

247–249 The `prefix_dup_check()` function checks whether the global address specified by the `ha_gaddr` variable has already been included in the payload that is under construction. If the payload has the global address already, a duplicate prefix information entry is not included.

Listing 14-150

_____ mpa.c

```
251              /* make prefix information */
252              prefix_info->nd_opt_pi_type = ND_OPT_PREFIX_INFORMATION;
253              prefix_info->nd_opt_pi_len = 4;
254              prefix_info->nd_opt_pi_prefix_len = ha_gaddr->hagent_prefixlen;
255              prefix_info->nd_opt_pi_flags_reserved = 0;
256
257              if (ha_gaddr->hagent_flags.onlink)
258                prefix_info->nd_opt_pi_flags_reserved |= ND_OPT_PI_FLAG_ONLINK;
259              if (ha_gaddr->hagent_flags.autonomous)
260                prefix_info->nd_opt_pi_flags_reserved |= ND_OPT_PI_FLAG_AUTO;
261              if (ha_gaddr->hagent_flags.router)
262                prefix_info->nd_opt_pi_flags_reserved |= ND_OPT_PI_FLAG_ROUTER;
```
_____ mpa.c

252–262 A prefix information entry is constructed as a form of the modified prefix information structure described in Section 3.6 of Chapter 3.

Listing 14-151

_____ mpa.c

```
264              if (ha_gaddr->hagent_vltime || ha_gaddr->hagent_pltime)
265                gettimeofday(&now, NULL);
266              if (ha_gaddr->hagent_vltime == 0)
```

```
267                          vltime = ha_gaddr->hagent_expire;
268                 else
269                          vltime = (ha_gaddr->hagent_expire > now.tv_sec) ?
270                                  ha_gaddr->hagent_expire - now.tv_sec : 0;
271                 if (ha_gaddr->hagent_pltime == 0)
272                          pltime = ha_gaddr->hagent_preferred;
273                 else
274                          pltime = (ha_gaddr->hagent_preferred > now.tv_sec) ?
275                                  ha_gaddr->hagent_preferred - now.tv_sec : 0;
276                 if (vltime < pltime) {
277                          /*
278                           * this can happen if vltime is decrement but pltime
279                           * is not.
280                           */
281                          pltime = vltime;
282                 }
283                 prefix_info->nd_opt_pi_valid_time = htonl(vltime);
284                 prefix_info->nd_opt_pi_preferred_time = htonl(pltime);
285                 prefix_info->nd_opt_pi_reserved2 = 0;
286                 prefix_info->nd_opt_pi_prefix = ha_gaddr->hagent_gaddr;
287
288                 prefix_info ++;
289                 naddr ++;
290          }
291      }
292      return naddr;
293  }
```

$\underline{\qquad\qquad\qquad\qquad\qquad\qquad\qquad\qquad\qquad}$mpa.c

264–282 The valid lifetime and the preferred lifetime are calculated based on the lifetime values of the global address from which the prefix information is derived. The lifetime values which are included in the prefix information will be the actual lifetimes. In other words, if the advertised lifetime was 1000 s, but the last Router Advertisement was 400 s ago, the lifetime of the Mobile Prefix Advertisement will be 600 s.

Note that the code when the lifetime values are 0 (that means infinite lifetime in the **had** program) is wrong. Lines 267 and 272 should be as follows:

```
267                          vltime = ND6_INFINITE_LIFETIME;
```

```
272                          pltime = ND6_INFINITE_LIFETIME;
```

15

Mobile Node

A mobile node is an Internet Protocol version 6 (IPv6) node that can change its point of attachment to the Internet without disconnecting existing connections with other nodes. A mobile node has the following capabilities.

- Maintaining binding information to inform its home agent of its current location
- Receiving and sending all the packets using the tunnel connection established between it and its home agent to hide its current location and provide transparent access to its communicating peer nodes
- Performing the return routability procedure to communicate with nodes that support the route optimization function to avoid inefficient tunnel communication

15.1 Files

Table 15-1 shows the files used by a mobile node.

15.2 Binding Update List Entry Management

A mobile node keeps a small amount of information called a binding update list entry, which contains a mobile node's address, communicating node's address, and other information representing the communication status between the two nodes. The information is used to keep the status of home registration or the status of route optimization between a mobile and a correspondent node.

TABLE 15-1

File	Description
`${KAME}/kame/sys/net/if_hif.h`	Home virtual interface structures
`${KAME}/kame/sys/net/if_hif.c`	Implementation of the home virtual interface
`${KAME}/kame/sys/netinet/icmp6.h;`	Dynamic Home Agent Address Discovery and Mobile Prefix Solicitation/Advertisement structures
`${KAME}/kame/sys/netinet/ip6.h`	Home Address option structure
`${KAME}/kame/sys/netinet/ip6mh.h`	Mobility Header structures
`${KAME}/kame/sys/netinet6/mip6_var.h`	All structures that are used in the Mobile IPv6 stack
`${KAME}/kame/sys/netinet6/mip6_mncore.c`	Implementation of mobile node functions
`${KAME}/kame/sys/netinet6/mip6_fsm.c`	The finite state machine of a binding update list entry
`${KAME}/kame/sys/netinet6/mip6_halist.c`	Home agent list management for a mobile node
`${KAME}/kame/sys/netinet6/mip6_prefix.c`	Prefix information management for a mobile node
`${KAME}/kame/sys/netinet6/mip6_icmp6.c`	Implementation of ICMPv6 message related processing
`${KAME}/kame/sys/netinet6/in6_src.c`	Implementation of the default address selection mechanism
`${KAME}/kame/sys/netinet6/ip6_output.c`	Insertion of extension headers for Mobile IPv6 signaling
`${KAME}/kame/sys/netinet6/nd6_rtr.c`	Gathering router information and prefix information
`${KAME}/kame/sys/netinet6/route6.c`	Implementation of the Type 2 Routing Header

Files used by a mobile node.

There are six functions to manage binding update list entries:

- `mip6_bu_create()` — creates a binding update list entry
- `mip6_bu_list_insert()` — inserts a binding update list entry into a binding update list
- `mip6_bu_list_remove()` — removes the specified entry from the list
- `mip6_bu_list_remove_all()` — removes all binding update list entries from the list
- `mip6_bu_list_find_home_registration()` — returns a pointer to the binding update list entry for home registration
- `mip6_bu_list_find_withpaddr()` — returns a pointer to the binding update list entry which has the specified correspondent address

15.2.1 Creating a Binding Update List Entry

A binding update list entry is created by the `mip6_bu_create()` function.

Listing 15-1

```
                                                                    mip6_mncore.c
1482    static struct mip6_bu *
1483    mip6_bu_create(paddr, mpfx, coa, flags, sc)
1484           const struct in6_addr *paddr;
1485           struct mip6_prefix *mpfx;
1486           struct in6_addr *coa;
1487           u_int16_t flags;
1488           struct hif_softc *sc;
1489    {
1490           struct mip6_bu *mbu;
1491           u_int32_t coa_lifetime, cookie;
....
1496           MALLOC(mbu, struct mip6_bu *, sizeof(struct mip6_bu),
1497                   M_TEMP, M_NOWAIT);
1498           if (mbu == NULL) {
....
1502                   return (NULL);
1503           }
                                                                    mip6_mncore.c
```

1482–1488 The `mip6_bu_create()` function has five parameters: the `paddr` parameter is a pointer to the address of a communicating node; the `mpfx` parameter is a pointer to the prefix information of the home address of a mobile node; the `coa` parameter is a pointer to the care-of address of the mobile node; the `flags` parameter is a combination of flags defined in Table 10-13 of Chapter 10; and the `sc` parameter is a pointer to the `hif_softc{}` instance, which indicates the home network of the mobile node.

1496–1503 Memory is allocated for the binding update list entry.

Listing 15-2

```
                                                                    mip6_mncore.c
1505           coa_lifetime = mip6_coa_get_lifetime(coa);
1506
1507           bzero(mbu, sizeof(*mbu));
1508           mbu->mbu_flags = flags;
1509           mbu->mbu_paddr = *paddr;
1510           mbu->mbu_haddr = mpfx->mpfx_haddr;
1511           if (sc->hif_location == HIF_LOCATION_HOME) {
1512                   /* un-registration. */
1513                   mbu->mbu_coa = mpfx->mpfx_haddr;
1514                   mbu->mbu_pri_fsm_state =
1515                       (mbu->mbu_flags & IP6MU_HOME)
1516                       ? MIP6_BU_PRI_FSM_STATE_WAITD
1517                       : MIP6_BU_PRI_FSM_STATE_IDLE;
1518           } else {
1519                   /* registration. */
1520                   mbu->mbu_coa = *coa;
1521                   mbu->mbu_pri_fsm_state =
1522                       (mbu->mbu_flags & IP6MU_HOME)
1523                       ? MIP6_BU_PRI_FSM_STATE_WAITA
1524                       : MIP6_BU_PRI_FSM_STATE_IDLE;
1525           }
1526           if (coa_lifetime < mpfx->mpfx_vltime) {
1527                   mbu->mbu_lifetime = coa_lifetime;
1528           } else {
1529                   mbu->mbu_lifetime = mpfx->mpfx_vltime;
```

```
1530                }
1531                if (mip6ctl_bu_maxlifetime > 0 &&
1532                    mbu->mbu_lifetime > mip6ctl_bu_maxlifetime)
1533                        mbu->mbu_lifetime = mip6ctl_bu_maxlifetime;
```
_____ mip6_mncore.c

1505 The lifetime of a binding update list entry is determined from the lifetime of a care-of and a home address. The `mip6_coa_get_lifetime()` function returns the lifetime of the specified address.

1507–1525 The member variables of the `mip6_bu{}` structure are set. The home address is copied from the prefix information provided by the `mpfx` parameter. The care-of address and registration status are set based on the current location of the mobile node. If the mobile node is at home (`sc->hif_location` is `HIF_LOCATION_HOME`), the care-of address is set to the home address of the mobile node. The registration status is set to the `WAITD` state if the binding update list entry is for home registration (that is, the `IP6MU_HOME` flag is set). This is when the mobile node returns home without a corresponding home registration entry. If the entry is not for home registration, the state is set to `IDLE`. If the mobile node is in a foreign network, the care-of address is set to the address specified as the `coa` parameter. The state is set to the `WAITA` status if the entry is for home registration; otherwise, it is set to `IDLE`.

1526–1533 The lifetime is set to the lifetime of the care-of address if the lifetime of the care-of address is greater than the lifetime of the home address (more precisely, the lifetime of the home prefix); otherwise, it is set to the lifetime of the home address.

1531–1533 `mip6ctl_bu_maxlifetime` is a global variable which is used to limit the lifetime of binding update list entries.

Listing 15-3
_____ mip6_mncore.c

```
1534            mbu->mbu_expire = time_second + mbu->mbu_lifetime;
1535            /* sanity check for overflow */
1536            if (mbu->mbu_expire < time_second)
1537                    mbu->mbu_expire = 0x7fffffff;
1538            mbu->mbu_refresh = mbu->mbu_lifetime;
1539            /* Sequence Number SHOULD start at a random value */
1540            mbu->mbu_seqno = (u_int16_t)arc4random();
1541            cookie = arc4random();
1542            bcopy(&cookie, &mbu->mbu_mobile_cookie[0], 4);
1543            cookie = arc4random();
1544            bcopy(&cookie, &mbu->mbu_mobile_cookie[4], 4);
1545            mbu->mbu_hif = sc;
1546            /* *mbu->mbu_encap = NULL; */
1547            mip6_bu_update_firewallstate(mbu);
1548
1549            return (mbu);
1550    }
```
_____ mip6_mncore.c

1534–1549 The expiration time of the binding update list entry is set. The code checks the overflow of the expiration time, since the time is represented as a 32-bit signed integer. The refresh time is set to the same value as the expiration time. The value may be overwritten by a Binding Acknowledgment message which will arrive later. The sequence number is initialized to a random number and the mobile cookie used for the return routability procedure is initialized to random numbers.

The `mip6_bu_update_firewallstate()` function is not discussed in this book, since the function is used for the KAME-specific experimental code for the firewall traversal mechanism.

15.2.2 Inserting a Binding Update List Entry to List

Binding update list entries are maintained as a list. Each `hif_softc{}` structure, which represents a home network, has a list constructed of binding update list entries that belong to the home network. Figure 10-2 in Chapter 10 shows the relationship between the `hif_softc{}` and `mip6_bu{}` structures.

Listing 15-4

_____ mip6_mncore.c

```
1552    static int
1553    mip6_bu_list_insert(bu_list, mbu)
1554            struct mip6_bu_list *bu_list;
1555            struct mip6_bu *mbu;
1556    {
1557            LIST_INSERT_HEAD(bu_list, mbu, mbu_entry);
1558
1559            if (mip6_bu_count == 0) {
....
1562                    mip6_bu_starttimer();
1563            }
1564            mip6_bu_count++;
1565
1566            return (0);
1567    }
```
_____ mip6_mncore.c

1552–1555 The `mip6_bu_list_insert()` function has two parameters: The `bu_list` parameter is a pointer to the list of binding update list entries kept in a `hif_softc{}` instance and the `mbu` parameter is a pointer to the binding update list entry to insert.

1557–1566 The `mip6_bu_starttimer()` function starts the timer function for binding update list entries when the first binding update list entry is inserted by the `LIST_INSERT_HEAD()` macro. The `mip6_bu_count` variable is the total number of binding update entries currently held by the mobile node. The variable is used to determine when to start and stop the timer function.

15.2.3 Removing a Binding Update List Entry from List

A binding update list entry is removed when its lifetime is expired. The `mip6_bu_list_remove()` function removes a binding update list entry from a list.

Listing 15-5

_____ mip6_mncore.c

```
1569    int
1570    mip6_bu_list_remove(mbu_list, mbu)
1571            struct mip6_bu_list *mbu_list;
1572            struct mip6_bu *mbu;
1573    {
1574            if ((mbu_list == NULL) || (mbu == NULL)) {
1575                    return (EINVAL);
1576            }
```

```
1577
1578                LIST_REMOVE(mbu, mbu_entry);
1579                FREE(mbu, M_TEMP);
1580
1581                mip6_bu_count--;
1582                if (mip6_bu_count == 0) {
1583                        mip6_bu_stoptimer();
....
1587                }
1588
1589                return (0);
1590        }
```

1569–1572 The mip6_bu_list_remove() has two parameters: the mbu_list parameter is
a pointer to the list that includes the entry to remove, and the mbu parameter is a pointer
to the binding update list entry.

1578–1589 The specified entry is removed by the LIST_REMOVE() macro and the memory
used by the entry is released. The mip6_bu_count variable is decremented by 1 and
the mip6_bu_stop_timer() function is called if the variable reaches 0, which means
there is no binding update list entry.

Listing 15-6

```
1592    int
1593    mip6_bu_list_remove_all(mbu_list, all)
1594            struct mip6_bu_list *mbu_list;
1595            int all;
1596    {
1597            struct mip6_bu *mbu, *mbu_next;
1598            int error = 0;
1599
1600            if (mbu_list == NULL) {
1601                    return (EINVAL);
1602            }
1603
1604            for (mbu = LIST_FIRST(mbu_list);
1605                 mbu;
1606                 mbu = mbu_next) {
1607                    mbu_next = LIST_NEXT(mbu, mbu_entry);
1608
1609                    if (!all &&
1610                        (mbu->mbu_flags & IP6MU_HOME) == 0 &&
1611                        (mbu->mbu_state & MIP6_BU_STATE_DISABLE) == 0)
1612                            continue;
1613
1614                    error = mip6_bu_list_remove(mbu_list, mbu);
1615                    if (error) {
....
1619                            continue;
1620                    }
1621            }
1622
1623            return (0);
1624    }
```

1592–1595 The mip6_bu_list_remove_all() function is used to remove all the binding
update list entries. The mbu_list parameter is a pointer to the list that contains the
binding update list entries to remove. All the binding update list entries are removed if the

all parameter is set to true; otherwise, all entries except for correspondent nodes that do not support the Mobile IPv6 function are removed.

1604–1623 All binding update list entries held in the mbu_list are released by the function mip6_bu_list_remove(). If the all variable is set to false, binding update list entries for correspondent nodes that have the MIP6_BU_STATE_DISABLE flag set (the flag means the node does not support Mobile IPv6) are not removed. Keeping the entries of non-Mobile IPv6 aware nodes will avoid unnecessary signaling to those nodes in later communications.

15.2.4 Looking Up a Binding Update List Entry

To retrieve a binding update list entry, the KAME Mobile IPv6 code provides two functions: the mip6_bu_list_find_home_registration() function will find the home registration entry and the mip6_bu_list_find_withpaddr() function will find the entry that matches the specified destination address.

Listing 15-7
_____ mip6_mncore.c

```
1626    struct mip6_bu *
1627    mip6_bu_list_find_home_registration(bu_list, haddr)
1628         struct mip6_bu_list *bu_list;
1629         struct in6_addr *haddr;
1630    {
1631         struct mip6_bu *mbu;
1632
1633         for (mbu = LIST_FIRST(bu_list); mbu;
1634             mbu = LIST_NEXT(mbu, mbu_entry)) {
1635             if (IN6_ARE_ADDR_EQUAL(&mbu->mbu_haddr, haddr) &&
1636                 (mbu->mbu_flags & IP6MU_HOME) != 0)
1637                     break;
1638         }
1639         return (mbu);
1640    }
```
_____ mip6_mncore.c

1626–1629 The mip6_bu_list_find_home_registration() function has two parameters: The bu_list parameter is a pointer to the list of binding update list entries and the haddr parameter is the home address of the target mobile node.

1633–1639 All binding update list entries contained in the bu_list are checked, and the entry of which the home address is the same as the haddr parameter and has the IP6MU_HOME flag set is returned.

Listing 15-8
_____ mip6_mncore.c

```
1648    struct mip6_bu *
1649    mip6_bu_list_find_withpaddr(bu_list, paddr, haddr)
1650         struct mip6_bu_list *bu_list;
1651         struct in6_addr *paddr;
1652         struct in6_addr *haddr;
1653    {
1654         struct mip6_bu *mbu;
1655
1656         /* sanity check. */
1657         if (paddr == NULL)
1658             return (NULL);
```

```
1659
1660            for (mbu = LIST_FIRST(bu_list); mbu;
1661                 mbu = LIST_NEXT(mbu, mbu_entry)) {
1662                if (IN6_ARE_ADDR_EQUAL(&mbu->mbu_paddr, paddr)
1663                    && ((haddr != NULL)
1664                        ? IN6_ARE_ADDR_EQUAL(&mbu->mbu_haddr, haddr)
1665                        : 1))
1666                        break;
1667            }
1668            return (mbu);
1669    }
```
_____ mip6_mncore.c

1648–1652 The mip6_bu_list_find_withpaddr() function has three parameters: the bu_list parameter is a pointer to the list of binding update list entries and the paddr and haddr parameters are the addresses of the correspondent node and the home address of the mobile node being searched. If the mobile node has multiple home addresses, there may be multiple entries for the same node that have different home addresses. To verify that there is at least one entry of the specified peer node, the haddr parameter can be a NULL pointer. In that case, the function returns the first entry whose peer address is the same as the paddr parameter.

1660–1668 All binding update list entries included in the bu_list are checked. If the haddr parameter is NULL, the first entry of which the peer address, mbu_peer, is the same as the paddr parameter is returned. If the haddr parameter is not NULL, the entry of which the peer address, mbu_peer, is the same as paddr and of which the home address, mbu_haddr, is the same as the haddr parameter is returned.

15.2.5   Timer Processing of the mip6_bu{} Structure

Similar to the mip6_bc{} structure, the mip6_bu{} structure has its timer function. The difference between the timer functions of the mip6_bc{} and mip6_bu{} structures is that the mip6_bc{} structure has a timer entry in each instance, while the mip6_bu{} structure shares one timer function among all the binding update list entries. This implementation design is due to historical reasons.

There are three functions related to the timer processing of the mip6_bu{} structure:

(1) mip6_bu_starttimer() — start the timer function

(2) mip6_bu_stoptimer() — stop the timer function

(3) mip6_bu_timeout() — process periodical jobs needed to manage binding update list entries

Listing 15-9

_____ mip6_mncore.c
```
2294    static void
2295    mip6_bu_starttimer()
2296    {
 ....
2298            callout_reset(&mip6_bu_ch,
2299                    MIP6_BU_TIMEOUT_INTERVAL * hz,
2300                    mip6_bu_timeout, NULL);
 ....
2309    }
```
_____ mip6_mncore.c

2294–2309 The `mip6_bu_starttimer()` function is called when the first binding update list entry is created on a mobile node. The function is also called from the `mip6_bu_timeout()` function to reset the timer. The interval to call the timer function is set to 1 s.

Listing 15-10

```
                                                                    mip6_mncore.c
2311    static void
2312    mip6_bu_stoptimer()
2313    {
....
2315            callout_stop(&mip6_bu_ch);
....
2321    }
                                                                    mip6_mncore.c
```

2311–2321 The `mip6_bu_stoptimer()` function stops the timer function. This function is called when all the binding update list entries have been removed.

Listing 15-11

```
                                                                    mip6_mncore.c
2324    mip6_bu_timeout(arg)
2325            void *arg;
2326    {
2327            int s;
2328            struct hif_softc *sc;
2329            int error = 0;
....
2331            struct timeval mono_time;
....
2335            mono_time.tv_sec = time_second;
....
2341            s = splnet();
....
2343            mip6_bu_starttimer();
                                                                    mip6_mncore.c
```

2324–2325 The `mip6_bu_timeout()` function is called periodically to process each binding update list entry kept by a mobile node.

2343 The `mip6_bu_starttimer()` function is called to schedule the next timeout.

Listing 15-12

```
                                                                    mip6_mncore.c
2345            for (sc = LIST_FIRST(&hif_softc_list); sc;
2346                sc = LIST_NEXT(sc, hif_entry)) {
2347                struct mip6_bu *mbu, *mbu_entry;
2348
2349                    for (mbu = LIST_FIRST(&sc->hif_bu_list);
2350                        mbu != NULL;
2351                        mbu = mbu_entry) {
2352                        mbu_entry = LIST_NEXT(mbu, mbu_entry);
2353
2354                            /* check expiration. */
2355                            if (mbu->mbu_expire < mono_time.tv_sec) {
2356                                if ((mbu->mbu_flags & IP6MU_HOME) != 0) {
2357                                    /*
2358                                     * the binding update entry for
2359                                     * the home registration
```

```
2360                                          * should not be removed.
2361                                          */
2362                                         mip6_bu_fsm(mbu,
2363                                             MIP6_BU_PRI_FSM_EVENT_RETRANS_TIMER,
2364                                             NULL);
2365                                 } else {
2366                                     error = mip6_bu_list_remove(
2367                                             &sc->hif_bu_list, mbu);
2368                                     if (error) {
....
2373                                             /* continue anyway... */
2374                                     }
2375                                     continue;
2376                                 }
2377                             }
2378
2379                             /* check if we need retransmit something. */
2380                             if ((mbu->mbu_state & MIP6_BU_STATE_NEEDTUNNEL) != 0)
2381                                 continue;
2382
2383                             /* check timeout. */
2384                             if ((mbu->mbu_retrans != 0)
2385                                 && (mbu->mbu_retrans < mono_time.tv_sec)) {
2386                                 /* order is important. */
2387                                 if(MIP6_IS_BU_RR_STATE(mbu)) {
2388                                         /* retransmit RR signals. */
2389                                         error = mip6_bu_fsm(mbu,
2390                                             MIP6_BU_SEC_FSM_EVENT_RETRANS_TIMER,
2391                                             NULL);
2392                                 } else if (((mbu->mbu_flags & IP6MU_ACK) != 0)
2393                                     && MIP6_IS_BU_WAITA_STATE(mbu)) {
2394                                         /* retransmit a binding update
2395                                          * to register. */
2396                                         error = mip6_bu_fsm(mbu,
2397                                             MIP6_BU_PRI_FSM_EVENT_RETRANS_TIMER,
2398                                             NULL);
2399                                 } else if (MIP6_IS_BU_BOUND_STATE(mbu)) {
2400                                         /* retransmit a binding update
2401                                          * for to refresh binding. */
2402                                         error = mip6_bu_fsm(mbu,
2403                                             MIP6_BU_PRI_FSM_EVENT_REFRESH_TIMER,
2404                                             NULL);
2405                                 }
2406                                 if (error) {
....
2411                                         /* continue, anyway... */
2412                                 }
2413                             }
2414                         }
2415                     }
2416
2417             splx(s);
2418     }
```
—— mip6_mncore.c

2345–2352 Each home network information entry (the instance of the hif_softc{} structure) has a list of binding update list entries. All of the hif_softc{} instances are visited to process all binding update list entries included in each hif_softc{} instance.

2355–2377 The expiration time of the binding update list entry is checked. The timer function will remove the expired entry by calling the mip6_bu_list_remove() function if the entry is not for home registration. In the KAME implementation, home registration entries never expire. A mobile node will try to reregister its information to the home agent forever.

2380–2381 The retransmission procedure is skipped if the entry has been marked with the `MIP6_BU_STATE_NEEDTUNNEL` flag, which means the destination node does not support Mobile IPv6 and requires tunnel communication.

2384–2413 The retransmission timeout of each binding update list entry is checked. Based on the current state of the binding update list entry, a proper event is sent to the state machine operated in each binding update list entry. If the state is `MIP6_IS_BU_RR_STATE`, which means the entry is performing the return routability procedure, the `MIP6_BU_SEC_FSM_EVENT_RETRANS_TIMER` event is sent to retransmit required messages for the return routability procedure. If the state is `MIP6_IS_WAITA_STATE`, which means the entry is waiting for a Binding Acknowledgment message, and the entry has the `IP6MU_ACK` flag set, the `MIP6_BU_PRI_FSM_EVENT_RETRANS_TIMER` event is sent to retransmit a Binding Update message. If the state is `MIP6_IS_BU_BOUND_STATE`, which means the mobile node has successfully registered with the node being described, the `MIP6_BU_PRI_FSM_EVENT_REFRESH_TIMER` event is triggered to send a Binding Update message to extend the lifetime of the registered entry. The state machine will be discussed in Section 15.19.

15.3 Movement Detection

The most interesting part of a mobile node is the mechanism to detect its movement. The movement detection mechanism implemented by the KAME Mobile IPv6 is based on Neighbor Unreachability Detection (NUD) and the status change of care-of addresses. The basic idea is as follows:

(1) Receive a Router Advertisement message

(2) Configure IPv6 addresses based on the received Router Advertisement

(3) Probe all routers by sending Neighbor Solicitation messages

(4) Wait for any status change of assigned IPv6 addresses

(5) Check the available care-of addresses

(6) Recognize movement if the old care-of address is not usable any more

The overview is shown in Figure 15-1. In the KAME IPv6 implementation, the prefix and router information have a close relationship. When a router becomes unreachable, the related prefix is also marked as unusable. The care-of address constructed from the invalid prefix can no longer be used for communication anymore. When a mobile node detects that the current care-of address has become unusable, the node thinks it has moved to another network.

15.3.1 Probing Routers

Router Advertisement messages are processed by the `nd6_ra_input()` function.

FIGURE 15-1

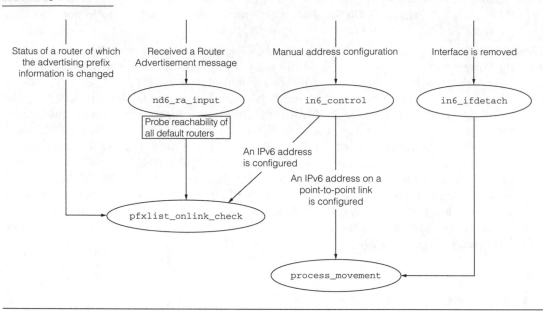

Movement detection overview.

Listing 15-13

_____ nd6_rtr.c

```
229     void
230     nd6_ra_input(m, off, icmp6len)
231             struct   mbuf *m;
232             int off, icmp6len;
233     {
....
436     #if defined(MIP6) && defined(MIP6_MOBILE_NODE)
437             if (MIP6_IS_MN) {
438                     /* check reachability of all routers. */
439                     mip6_probe_routers();
440             }
....
```
_____ nd6_rtr.c

437–440 If a node is acting as a mobile node and receives a Router Advertisement message, the `mip6_probe_routers()` function is called to initiate NUD for all default routers currently kept in the default router list of the mobile node.

Listing 15-14

_____ mip6_mncore.c

```
617     void
618     mip6_probe_routers(void)
619     {
620             struct llinfo_nd6 *ln;
621
622             ln = llinfo_nd6.ln_next;
623             while (ln && ln != &llinfo_nd6) {
624                     if ((ln->ln_router) &&
625                             ((ln->ln_state == ND6_LLINFO_REACHABLE) ||
```

```
626                          (ln->ln_state == ND6_LLINFO_STALE))) {
627                              ln->ln_asked = 0;
628                              ln->ln_state = ND6_LLINFO_DELAY;
629                              nd6_llinfo_settimer(ln, 0);
630                          }
631                      ln = ln->ln_next;
632              }
633      }
```
―――――――――――――――――――――――――――――――――――――― mip6_mncore.c

617–633 The mip6_probe_routers() function changes the neighbor discovery state of routers. The ln_router field of the llinfo_nd6{} structure indicates whether the entry is for a router. If the entry is for a router and the state is either ND6_LLINFO_REACHABLE or ND6_LLINFO_STALE, the state is changed to ND6_LLINFO_DELAY and the timer function for the entry is called by setting the time to the next timeout to 0 s. The timer function will be called soon and a Neighbor Solicitation message will be sent from the timer function because the state is ND6_LLINFO_DELAY. After the Neighbor Solicitation message is sent, the state is changed to ND6_LLINFO_PROBE in the timer function. The detailed state transition of a neighbor cache entry is discussed in Chapter 5 of *IPv6 Core Protocols Implementation*.

If routers are still reachable, the mobile node will receive Neighbor Advertisement messages from the routers. The state of each entry is updated to ND6_LLINFO_REACHABLE when the mobile node receives the messages; otherwise, the entry is removed.

This mechanism works only when the routers of the network to which the mobile node attaches use the Router Advertisement message. If the care-of address of the mobile node is assigned in other ways, such as a PPP link on a point-to-point network, address availability is checked by the in6_control() function described in Listing 15-16.

15.3.2 Updating the Status of Addresses

The status of IPv6 addresses are changed in the following situations:

- A new prefix is advertised and a new address is configured
- The lifetime of an address is expired
- An address is removed manually
- Routers become unreachable and addresses that are generated from the prefixes sent from the routers become detached

When one of these events occurs, the KAME implementation checks the availability of care-of addresses and reregisters a new care-of address if necessary.

Listing 15-15
―――――――――――――――――――――――――――――――――――――― nd6_rtr.c

```
1654    void
1655    pfxlist_onlink_check()
1656    {
....
1827    #if defined(MIP6) && defined(MIP6_MOBILE_NODE)
1828            if (MIP6_IS_MN)
1829                    mip6_process_movement();
1830    #endif /* MIP6 && MIP6_MOBILE_NODE */
1831    }
```
―――――――――――――――――――――――――――――――――――――― nd6_rtr.c

1654–1831 The `pfxlist_onlink_check()` function is called whenever a node needs to check the latest status of the prefix information stored in the node. Based on the status of each prefix, the statuses of addresses assigned to the node are also updated. At the end of this function, the `mip6_process_movement()` function is called to verify whether movement has occurred.

Listing 15-16

```
                                                                                    in6.c
464     int
....
473     in6_control(so, cmd, data, ifp, p)
474             struct  socket *so;
475             u_long cmd;
476             caddr_t data;
477             struct ifnet *ifp;
478             struct proc *p;
....
480     {
....
823             case SIOCAIFADDR_IN6:
824             {
....
870                     if (pr0.ndpr_plen == 128) {
871     #if defined(MIP6) && defined(MIP6_MOBILE_NODE)
872                             if (MIP6_IS_MN)
873                                     mip6_process_movement();
874     #endif /* MIP6 && MIP6_MOBILE_NODE */
875                             break;  /* we don't need to install a host route. */
876                     }
....
                                                                                    in6.c
```

823–876 When an address is manually configured, the `in6_control()` function is called as a part of the address assignment procedure. In that case, the `SIOCAIFADDR_IN6` command and required address information are passed to the function. In the corresponding code, the `mip6_process_movement()` function is called only when the prefix length of the assigned address is 128. Typically, this condition occurs when assigning an IPv6 address on a Point-to-Point interface, such as a gif interface. Addresses that have a prefix length less than 128 are handled by the `pfxlist_onlink_ckech()` function.

Listing 15-17

```
                                                                            in6_ifattach.c
957     void
958     in6_ifdetach(ifp)
959             struct ifnet *ifp;
960     {
....
1095    #if defined(MIP6) && defined(MIP6_MOBILE_NODE)
1096            if (MIP6_IS_MN)
1097                    mip6_process_movement();
1098    #endif /* MIP6 && MIP6_MOBILE_NODE */
1099    }
                                                                            in6_ifattach.c
```

957–1099 The `in6_ifdetach()` function is called when a network interface is deleted, for example, when removing a PCMCIA network card. In this case, all the addresses assigned to the network interface are removed and the mobile node needs to check the latest status of available care-of addresses.

Listing 15-18

```
641     void
642     mip6_process_movement(void)
643     {
644             struct hif_softc *sc;
645             int coa_changed = 0;
```

641–645 The mip6_process_movement() is called whenever a mobile node needs to check the current location. A Binding Update message will be sent if the node detects movement, otherwise, the function has no effect.

Listing 15-19

```
647             for (sc = LIST_FIRST(&hif_softc_list); sc;
648                 sc = LIST_NEXT(sc, hif_entry)) {
649                     hif_save_location(sc);
650                     coa_changed = mip6_select_coa(sc);
651                     if (coa_changed == 1) {
652                             if (mip6_process_pfxlist_status_change(sc)) {
653                                     hif_restore_location(sc);
654                                     continue;
655                             }
656                             if (mip6_register_current_location(sc)) {
657                                     hif_restore_location(sc);
658                                     continue;
659                             }
660                             mip6_bu_list_update_firewallstate(sc);
661                     } else
662                             hif_restore_location(sc);
663             }
664     }
```

647–663 The KAME implementation assigns one care-of address to each hif_softc{} instance (meaning the home network). The loop defined on line 647 checks all the hif_softc{} instances in a mobile node and calls the mip6_select_coa() function to choose the best care-of address. The mip6_select_coa() function returns to 1 if a new care-of address is chosen. In this case, the mip6_process_pfxlist_status_change() function is called to determine the current location and the mip6_register_current_location() function is called subsequently to register the current location by sending a Binding Update message.

Listing 15-20

```
767     int
768     mip6_select_coa(sc)
769             struct hif_softc *sc;
770     {
771             int hoa_scope, ia_best_scope, ia_scope;
772             int ia_best_matchlen, ia_matchlen;
773             struct in6_ifaddr *ia, *ia_best;
```

```
774              struct in6_addr *hoa;
775              struct mip6_prefix *mpfx;
776              int i;
777
778              hoa = NULL;
779              hoa_scope = ia_best_scope = -1;
780              ia_best_matchlen = -1;
```
_____ mip6_mncore.c

767–769 The `mip6_select_coa()` function has one parameter. The `sc` parameter is a pointer
to the `hif_softc{}` instance that indicates one of the home networks of a mobile node.

Listing 15-21
_____ mip6_mncore.c
```
782              /* get the first HoA registered to a certain home network. */
783              for (mpfx = LIST_FIRST(&mip6_prefix_list); mpfx;
784                  mpfx = LIST_NEXT(mpfx, mpfx_entry)) {
785                      if (hif_prefix_list_find_withmpfx(&sc->hif_prefix_list_home,
786                          mpfx) == NULL)
787                          continue;
788                      if (IN6_IS_ADDR_UNSPECIFIED(&mpfx->mpfx_haddr))
789                          continue;
790                      hoa = &mpfx->mpfx_haddr;
791                      hoa_scope = in6_addrscope(hoa);
792              }
```
_____ mip6_mncore.c

782–784 One of the home addresses assigned to the home network, specified by the parameter
of the `mip6_select_coa()` function, is located. The `mip6_prefix_list` variable is
a list that keeps all prefix information that the mobile node currently has. Each prefix
information entry has an IPv6 address generated from the prefix. If the prefix is a home
prefix, then the address kept in the information is a home address.

785–791 If the prefix information currently being checked is not one of the home prefixes of
the `hif_softc{}` instance specified by the parameter, the prefix information is ignored.
The `hif_prefix_list_find_withmpfx()` function searches for a specified prefix
information entry in the prefix list, which is kept in the `hif_softc{}` instance. The
`hif_prefix_list_home` member variable holds all home prefixes of a particular
`hif_softc{}` instance. If a home prefix is found, the home address of the prefix and
its scope identifier are set based on the information stored in the prefix information.

Listing 15-22
_____ mip6_mncore.c
```
794              ia_best = NULL;
795              for (ia = in6_ifaddr; ia; ia = ia->ia_next) {
796                      ia_scope = -1;
797                      ia_matchlen = -1;
```
_____ mip6_mncore.c

794–797 The following code selects a new care-of address for a particular home network.
The algorithm is similar to the default source address selection algorithm implemented
as the `in6_selectsrc()` function discussed in Section 3.13.1 of *IPv6 Core Protocols
Implementation*. In the loop defined on line 795, all IPv6 addresses assigned to the mobile
node are checked to determine if there is a more appropriate address for use as a care-of
address than the current candidate.

Listing 15-23

```
                                                         mip6_mncore.c
799                    /* IFT_HIF has only home addresses. */
800                    if (ia->ia_ifp->if_type == IFT_HIF)
801                         goto next;
802
803                    if (ia->ia6_flags &
804                        (IN6_IFF_ANYCAST
....
808                        /* | IN6_IFF_TENTATIVE */
809                        | IN6_IFF_DETACHED
810                        | IN6_IFF_DUPLICATED))
811                         goto next;
812
813                    /* loopback address cannot be used as a CoA. */
814                    if (IN6_IS_ADDR_LOOPBACK(&ia->ia_addr.sin6_addr))
815                         goto next;
816
817                    /* link-local addr as a CoA is impossible? */
818                    if (IN6_IS_ADDR_LINKLOCAL(&ia->ia_addr.sin6_addr))
819                         goto next;
820
821                    /* tempaddr as a CoA is not supported. */
822                    if (ia->ia6_flags & IN6_IFF_TEMPORARY)
823                         goto next;
                                                         mip6_mncore.c
```

800–801 Addresses that cannot be used as a care-of address are excluded. The interface type IFT_HIF means a virtual interface that represents a home network. The addresses assigned to the virtual interface are home addresses and are never used as care-of addresses.

803–811 An anycast address, detached address, or duplicated address cannot be used as a care-of address.

814–815 A loopback address cannot be used as a care-of address.

818–819 Using a link-local address as a care-of address is possible in theory; however, it is almost meaningless since a packet of which the source address is a link-local address cannot be forwarded by routers.

821–823 An address that is generated based on the privacy extension specification [RFC3041] can be used as a care-of address; however, the KAME implementation does not use such addresses so that we can reduce the frequency of movement. The privacy extension mechanism invalidates the autoconfigured address in a short time and generates a new autoconfigured address to make it difficult to bind the address and its user. If it is used as a care-of address, the stack has to send a new Binding Update message when a new privacy enhanced address is generated.

Listing 15-24

```
                                                         mip6_mncore.c
825                    /* prefer a home address. */
826                    for (mpfx = LIST_FIRST(&mip6_prefix_list); mpfx;
827                        mpfx = LIST_NEXT(mpfx, mpfx_entry)) {
828                         if (hif_prefix_list_find_withmpfx(
829                             &sc->hif_prefix_list_home, mpfx) == NULL)
830                              continue;
831                         if (IN6_ARE_ADDR_EQUAL(&mpfx->mpfx_haddr,
832                             &ia->ia_addr.sin6_addr)) {
```

```
833                                           ia_best = ia;
834                                           goto out;
835                         }
836                 }
```

826–836 If an IPv6 address that is the same as one of the home addresses is assigned to a mobile node, the address is chosen as a care-of address. This means the mobile node has returned home.

Listing 15-25

```
838                 if (ia_best == NULL)
839                         goto replace;
840
841                 /* prefer appropriate scope. */
842                 ia_scope = in6_addrscope(&ia->ia_addr.sin6_addr);
843                 if (IN6_ARE_SCOPE_CMP(ia_best_scope, ia_scope) < 0) {
844                         if (IN6_ARE_SCOPE_CMP(ia_best_scope, hoa_scope) < 0)
845                                 goto replace;
846                         goto next;
847                 } else if (IN6_ARE_SCOPE_CMP(ia_scope, ia_best_scope) < 0) {
848                         if (IN6_ARE_SCOPE_CMP(ia_scope, hoa_scope) < 0)
849                                 goto next;
850                         goto replace;
851                 }
```

842–851 An address that has the same scope identifier as the home address is preferred. If the address being checked has the same scope identifier as the home address, the address is chosen as a new candidate care-of address. If the current candidate has the same scope identifier, the current candidate is kept. If both addresses have (or do not have) the same scope identifier, then other conditions are considered.

Listing 15-26

```
853                 /* avoid a deprecated address. */
854                 if (!IFA6_IS_DEPRECATED(ia_best) && IFA6_IS_DEPRECATED(ia))
855                         goto next;
856                 if (IFA6_IS_DEPRECATED(ia_best) && !IFA6_IS_DEPRECATED(ia))
857                         goto replace;
```

853–857 A deprecated address is not used as a care-of address. If the current candidate is deprecated, then the address currently being examined is chosen as a new candidate. If the address currently being examined is deprecated, then the current candidate is kept. If both address are (or are not) deprecated, then other conditions are checked.

Listing 15-27

```
859                 /* prefer an address on an alive interface. */
860                 if ((ia_best->ia_ifp->if_flags & IFF_UP) &&
861                     !(ia->ia_ifp->if_flags & IFF_UP))
862                         goto next;
863                 if (!(ia_best->ia_ifp->if_flags & IFF_UP) &&
```

```
864                     (ia->ia_ifp->if_flags & IFF_UP))
865                         goto replace;
```
——————————————————————————————————————— mip6_mncore.c

860–865 An address that is assigned to the active interface is preferred.

Listing 15-28
——————————————————————————————————————— mip6_mncore.c
```
867                 /* prefer an address on a preferred interface. */
868                 for (i = 0; i < sizeof(mip6_preferred_ifnames.mip6pi_ifname);
869                     i++) {
870                         if ((strncmp(if_name(ia_best->ia_ifp),
871                             mip6_preferred_ifnames.mip6pi_ifname[i],
872                             IFNAMSIZ) == 0)
873                             && (strncmp(if_name(ia->ia_ifp),
874                             mip6_preferred_ifnames.mip6pi_ifname[i],
875                             IFNAMSIZ) != 0))
876                                 goto next;
877                         if ((strncmp(if_name(ia_best->ia_ifp),
878                             mip6_preferred_ifnames.mip6pi_ifname[i],
879                             IFNAMSIZ) != 0)
880                             && (strncmp(if_name(ia->ia_ifp),
881                             mip6_preferred_ifnames.mip6pi_ifname[i],
882                             IFNAMSIZ) == 0))
883                                 goto replace;
884                 }
```
——————————————————————————————————————— mip6_mncore.c

868–884 The KAME implementation provides a feature to set priorities among interfaces when selecting a care-of address. For example, a user can specify that the ne0 Ethernet interface is more preferable to the wi0 wireless network interface. The mip6_preferred_ifnames variable is a configurable variable for specifying the preference. The value can be set by using the I/O control mechanism (I/O control is discussed in Section 15.27). The variable keeps the interface names ordered by preference.

Listing 15-29
——————————————————————————————————————— mip6_mncore.c
```
886                 /* prefer a longest match address. */
887                 if (hoa != NULL) {
888                         ia_matchlen = in6_matchlen(&ia->ia_addr.sin6_addr,
889                             hoa);
890                         if (ia_best_matchlen < ia_matchlen)
891                                 goto replace;
892                         if (ia_matchlen < ia_best_matchlen)
893                                 goto next;
894                 }
```
——————————————————————————————————————— mip6_mncore.c

887–894 An address that is similar to the home address is preferred. The in6_matchlen() function returns the number of bits that are the same from the MSB side between two addresses. The address that has a longer matching part is selected as a new candidate care-of address.

Listing 15-30
——————————————————————————————————————— mip6_mncore.c
```
896                 /* prefer same CoA. */
897                 if ((ia_best == sc->hif_coa_ifa)
```

```
898                           && (ia != sc->hif_coa_ifa))
899                                   goto next;
900                   if ((ia_best != sc->hif_coa_ifa)
901                           && (ia == sc->hif_coa_ifa))
902                                   goto replace;
```
_____ mip6_mncore.c

897–902 If the care-of address is changed, then the mobile node needs to reregister its care-of address to its home agent. Since a care-of address indicates the current topological location of the mobile node, frequent changes of care-of addresses may cause frequent path changes of traffic flow from the mobile node, which sometimes causes bad performance. To avoid the situation as much as possible, the KAME implementation tries to keep the current care-of address as long as the care-of address is available.

Listing 15-31
_____ mip6_mncore.c

```
904           replace:
905                   ia_best = ia;
906                   ia_best_scope = (ia_scope >= 0 ? ia_scope :
907                           in6_addrscope(&ia_best->ia_addr.sin6_addr));
908                   if (hoa != NULL)
909                           ia_best_matchlen = (ia_matchlen >= 0 ? ia_matchlen :
910                                   in6_matchlen(&ia_best->ia_addr.sin6_addr, hoa));
911           next:
912                   continue;
913           out:
914                   break;
915           }
916
917           if (ia_best == NULL) {
....
921                   return (0);
922           }
```
_____ mip6_mncore.c

904–910 The candidate care-of address is replaced with the address currently being examined.

917–922 If there is no proper care-of address, then the current status is kept until a usable care-of address becomes available.

Listing 15-32
_____ mip6_mncore.c

```
924           /* check if the CoA has been changed. */
925           if (sc->hif_coa_ifa == ia_best) {
926                   /* CoA has not been changed. */
927                   return (0);
928           }
929
930           if (sc->hif_coa_ifa != NULL)
931                   IFAFREE(&sc->hif_coa_ifa->ia_ifa);
932           sc->hif_coa_ifa = ia_best;
933           IFAREF(&sc->hif_coa_ifa->ia_ifa);
....
938           return (1);
939   }
```
_____ mip6_mncore.c

925–928 If the care-of address has not been changed, the current status is kept.

930–938 If the care-of address has been changed, the pointer to the care-of address stored in the `hif_softc{}` instance is updated to the new care-of address. The function returns to 1 to notify the caller that the care-of address has been changed.

Listing 15-33

mip6_mncore.c

```
666     int
667     mip6_process_pfxlist_status_change(sc)
668             struct hif_softc *sc;
669     {
670             struct hif_prefix *hpfx;
671             struct sockaddr_in6 hif_coa;
672             int error = 0;
673
674             if (sc->hif_coa_ifa == NULL) {
....
678                     sc->hif_location = HIF_LOCATION_UNKNOWN;
679                     return (0);
680             }
681             hif_coa = sc->hif_coa_ifa->ia_addr;
682             if (in6_addr2zoneid(sc->hif_coa_ifa->ia_ifp,
683                     &hif_coa.sin6_addr, &hif_coa.sin6_scope_id)) {
684                     /* must not happen. */
685             }
686             if (in6_embedscope(&hif_coa.sin6_addr, &hif_coa)) {
687                     /* must not happen. */
688             }
```

mip6_mncore.c

666–669 The `mip6_process_pfxlist_status_change()` function has one parameter. The `sc` parameter is a pointer to the `hif_softc{}` instance, which indicates the home network of a mobile node.

674–679 If the care-of address, `hif_coa_ifa`, is unknown, or has not been set yet, the location is considered as unknown (`HIF_LOCATION_UNKNOWN`).

682–688 The `hif_coa` variable, which is an instance of the `sockaddr_in6{}` structure, is created from the `hif_coa_ifa` variable, which is an instance of the `in6_ifaddr{}` structure.

Listing 15-34

mip6_mncore.c

```
690             sc->hif_location = HIF_LOCATION_UNKNOWN;
691             for (hpfx = LIST_FIRST(&sc->hif_prefix_list_home); hpfx;
692                 hpfx = LIST_NEXT(hpfx, hpfx_entry)) {
693                     if (in6_are_prefix_equal(&hif_coa.sin6_addr,
694                         &hpfx->hpfx_mpfx->mpfx_prefix,
695                         hpfx->hpfx_mpfx->mpfx_prefixlen)) {
696                             sc->hif_location = HIF_LOCATION_HOME;
697                             goto i_know_where_i_am;
698                     }
699             }
700             sc->hif_location = HIF_LOCATION_FOREIGN;
```

mip6_mncore.c

690–700 The code verifies whether the prefix of the current care-of address is the same as one of the home prefixes of the home network. If the prefix is one of the home prefixes, the mobile node is at home (`HIF_LOCATION_HOME`), otherwise, the mobile node is in a foreign network (`HIF_LOCATION_FOREIGN`).

Listing 15-35

```
                                                      mip6_mncore.c
701      i_know_where_i_am:
....
706              /*
707               * configure home addresses according to the home
708               * prefixes and the current location determined above.
709               */
710              error = mip6_haddr_config(sc);
711              if (error) {
....
715                      return (error);
716              }
717
718              return (0);
719      }
                                                      mip6_mncore.c
```

710–716 The `mip6_haddr_config()` function is called to configure home addresses based on the current location information stored in the `hif_softc{}` instance.

Listing 15-36

```
                                                      mip6_mncore.c
726      static int
727      mip6_register_current_location(sc)
728              struct hif_softc *sc;
729      {
730              int error = 0;
731
732              switch (sc->hif_location) {
733              case HIF_LOCATION_HOME:
734                      /*
735                       * we moved to home.  unregister our home address.
736                       */
737                      error = mip6_home_registration(sc);
738                      break;
739
740              case HIF_LOCATION_FOREIGN:
741                      /*
742                       * we moved to foreign.  register the current CoA to
743                       * our home agent.
744                       */
745                      /* XXX: TODO register to the old subnet's AR. */
746                      error = mip6_home_registration(sc);
747                      break;
748
749              case HIF_LOCATION_UNKNOWN:
750                      break;
751              }
752
753              return (error);
754      }
                                                      mip6_mncore.c
```

726–753 The `mip6_register_current_location()` function has one parameter. The `sc` parameter is a pointer to the `hif_softc{}` instance, which represents the home network of a mobile node. The `mip6_register_current_location()` function performs any necessary tasks to register the current care-of address based on the current location stored in the `sc` parameter. At this moment, the function calls the `mip6_home_registration()` function when the location is either `HIF_LOCATION_HOME` or `HIF_LOCATION_FOREIGN`.

The `mip6_home_registration()` function can handle both registration and deregistration requests. If the location is unknown, the function does nothing.

15.4 Configuring Home Addresses

When a mobile node detects movement, it configures home addresses based on the current location. The address configuration is implemented in the `mip6_haddr_config()` function and subsequent functions listed here.

- `mip6_haddr_config()` — manages all home address configuration
- `mip6_attach_haddrs()` — removes all home addresses assigned to physical interfaces and configures all home addresses on the specified home virtual interface
- `mip6_detach_haddrs()` — removes all home addresses assigned to the specified virtual interface
- `mip6_add_haddrs()` — does the actual job of assigning home addresses
- `mip6_remove_haddrs()` — does the actual job of removing home addresses
- `mip6_remove_addr()` — this generic function removes an IPv6 address from the specified network interface

The call flow of these functions is illustrated in Figure 15-2.

FIGURE 15-2

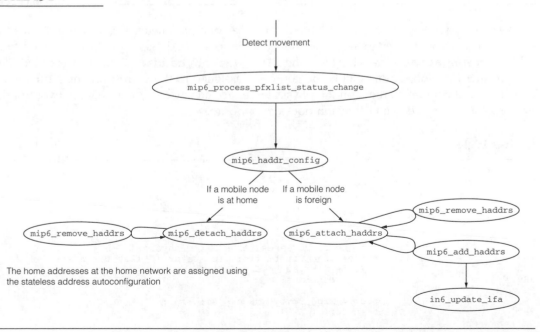

Call flow of home address configuration.

Listing 15-37

_____ mip6_mncore.c

```
941     static int
942     mip6_haddr_config(sc)
943             struct hif_softc *sc;
944     {
945             int error = 0;
946
947             switch (sc->hif_location) {
948             case HIF_LOCATION_HOME:
949                     /*
950                      * remove all home addresses attached to hif.
951                      * all physical addresses are assigned in a
952                      * address autoconfiguration manner.
953                      */
954                     error = mip6_detach_haddrs(sc);
955
956                     break;
957
958             case HIF_LOCATION_FOREIGN:
959                     /*
960                      * attach all home addresses to the hif interface.
961                      * before attach home addresses, remove home addresses
962                      * from physical i/f to avoid the duplication of
963                      * address.
964                      */
965                     error = mip6_attach_haddrs(sc);
966                     break;
967
968             case HIF_LOCATION_UNKNOWN:
969                     break;
970             }
971
972             return (error);
973     }
```

_____ mip6_mncore.c

941–943 The `mip6_haddr_config()` function has one parameter, `sc`, which indicates the home network related to the home address to be configured. Based on the current location, the `mip6_attach_haddrs()` or the `mip6_detach_haddrs()` function is called. The former function removes all home addresses assigned to the corresponding virtual interface and configures them on the physical interface attached to the foreign network of the mobile node. The latter function does the opposite.

Listing 15-38

_____ mip6_mncore.c

```
979     static int
980     mip6_attach_haddrs(sc)
981             struct hif_softc *sc;
982     {
983             struct ifnet *ifp;
984             int error = 0;
985
986             /* remove all home addresses for sc from physical I/F. */
987             for (ifp = ifnet.tqh_first; ifp; ifp = ifp->if_list.tqe_next) {
988                     if (ifp->if_type == IFT_HIF)
989                             continue;
990
991                     error = mip6_remove_haddrs(sc, ifp);
992                     if (error) {
....
997                             return (error);
998                     }
999             }
```

_____ mip6_mncore.c

979–981 The `mip6_attach_haddrs()` has one parameter, `sc`, which represents the home network of which the home addresses will be configured.

987–999 All home addresses related to the home network specified by the `sc` parameter are removed from all the physical interfaces. The `mip6_remove_haddrs()` function removes all home addresses related to the home virtual interface specified as its first parameter from the physical interface specified as the second parameter.

Listing 15-39

```
                                                         _____mip6_mncore.c
1001            /* add home addresses for sc to hif(itself) */
1002            error = mip6_add_haddrs(sc, (struct ifnet *)sc);
1003            if (error) {
....
1008                    return (error);
1009            }
1010
1011            return (0);
1012    }
                                                         _____mip6_mncore.c
```

1002–1009 The `mip6_add_haddrs()` function is called to assign home addresses. The `mip6_add_haddrs()` function configures all home addresses related to the home virtual interface specified as the first parameter at the physical interface specified as the second parameter.

Listing 15-40

```
                                                         _____mip6_mncore.c
1017    int
1018    mip6_detach_haddrs(sc)
1019            struct hif_softc *sc;
1020    {
1021            struct ifnet *hif_ifp = (struct ifnet *)sc;
1022            struct ifaddr *ia, *ia_next;
1023            struct in6_ifaddr *ia6;
1024            int error = 0;
                                                         _____mip6_mncore.c
```

1017–1019 The `mip6_detach_haddrs()` function removes all home addresses assigned to the specified virtual home interface as its parameter. Unlike the `mip6_attach_haddrs()` function, `mip6_detach_haddrs()` only removes the home addresses and does not assign home addresses to physical interfaces. Assigning addresses to physical interfaces is done by the Stateless Address Autoconfiguration mechanism (see Chapter 5 of *IPv6 Core Protocols Implementation*).

Listing 15-41

```
                                                         _____mip6_mncore.c
1027            for (ia = TAILQ_FIRST(&hif_ifp->if_addrhead);
1028                    ia;
1029                    ia = ia_next)
....
1035            {
....
1037                    ia_next = TAILQ_NEXT(ia, ifa_link);
....
1041
```

```
1042                    if (ia->ifa_addr->sa_family != AF_INET6)
1043                            continue;
1044                    ia6 = (struct in6_ifaddr *)ia;
1045                    if (IN6_IS_ADDR_LINKLOCAL(&ia6->ia_addr.sin6_addr))
1046                            continue;
1047
1048                    error = mip6_remove_addr(hif_ifp, ia6);
1049                    if (error) {
....
1054                            return (error);
1055                    }
1056            }
....
1073            return (error);
1074    }
```
—— mip6_mncore.c

1027–1056 The addresses assigned to a particular interface are held in the `if_addrhead` member variable of the `ifnet{}` structure. In the loop defined on line 1027, all addresses of which the address family is IPv6 and of which the type is not link-local are removed by calling the `mip6_remove_addr()` function.

Listing 15-42
—— mip6_mncore.c
```
1079    static int
1080    mip6_add_haddrs(sc, ifp)
1081            struct hif_softc *sc;
1082            struct ifnet *ifp;
1083    {
1084            struct mip6_prefix *mpfx;
1085            struct in6_aliasreq ifra;
1086            struct in6_ifaddr *ia6;
1087            int error = 0;
....
1089            struct timeval mono_time;
....
1093            microtime(&mono_time);
```
—— mip6_mncore.c

1079–1082 The `mip6_add_haddrs()` function has two parameters: The `sc` parameter is a pointer to the virtual home interface that has information of home addresses to be assigned and the `ifp` parameter is a pointer to the network interface to which the home addresses will be assigned.

Listing 15-43
—— mip6_mncore.c
```
1096            if ((sc == NULL) || (ifp == NULL)) {
1097                    return (EINVAL);
1098            }
1099
1100            for (mpfx = LIST_FIRST(&mip6_prefix_list); mpfx;
1101                mpfx = LIST_NEXT(mpfx, mpfx_entry)) {
1102                    if (hif_prefix_list_find_withmpfx(&sc->hif_prefix_list_home,
1103                        mpfx) == NULL)
1104                            continue;
```
—— mip6_mncore.c

1099–1104 The home address configuration will be done on every home prefix information entry kept in the virtual home network interface specified as the `sc` variable.

The `hif_prefix_list_find_withmpfx()` function checks to see if the prefix information specified by the second parameter is included in the list specified by the first parameter. Since the `sc->hif_prefix_list_home` member variable keeps home prefixes of the virtual home network specified as `sc`, we can use the `hif_prefix_list_find_withmpfx()` function to verify whether the prefix information is a home prefix.

Listing 15-44

————————————————————————————————— mip6_mncore.c

```
1106                     /*
1107                      * assign home address to mip6_prefix if not
1108                      * assigned yet.
1109                      */
1110                     if (IN6_IS_ADDR_UNSPECIFIED(&mpfx->mpfx_haddr)) {
1111                             error = mip6_prefix_haddr_assign(mpfx, sc);
1112                             if (error) {
....
1117                                     return (error);
1118                             }
1119                     }
1120
1121                     /* skip a prefix that has 0 lifetime. */
1122                     if (mpfx->mpfx_vltime == 0)
1123                             continue;
```

————————————————————————————————— mip6_mncore.c

1110–1119 The prefix structure (the `mip6_prefix{}` structure) keeps not only prefix information but also a home address that is generated from the prefix information. The home address is kept in the `mpfx_haddr` member variable. If the address has not been generated, the `mip6_prefix_haddr_assign()` function is called to generate a home address.

1122–1123 If the valid lifetime of the prefix is 0, the home address is not assigned because it has expired.

Listing 15-45

————————————————————————————————— mip6_mncore.c

```
1125                     /* construct in6_aliasreq. */
1126                     bzero(&ifra, sizeof(ifra));
1127                     bcopy(if_name(ifp), ifra.ifra_name, sizeof(ifra.ifra_name));
1128                     ifra.ifra_addr.sin6_len = sizeof(struct sockaddr_in6);
1129                     ifra.ifra_addr.sin6_family = AF_INET6;
1130                     ifra.ifra_addr.sin6_addr = mpfx->mpfx_haddr;
1131                     ifra.ifra_prefixmask.sin6_len = sizeof(struct sockaddr_in6);
1132                     ifra.ifra_prefixmask.sin6_family = AF_INET6;
1133                     ifra.ifra_flags = IN6_IFF_HOME | IN6_IFF_AUTOCONF;
1134                     if (ifp->if_type == IFT_HIF) {
1135                             in6_prefixlen2mask(&ifra.ifra_prefixmask.sin6_addr,
1136                                 128);
1137                     } else {
1138                             in6_prefixlen2mask(&ifra.ifra_prefixmask.sin6_addr,
1139                                 mpfx->mpfx_prefixlen);
1140                     }
1141                     ifra.ifra_lifetime.ia6t_vltime = mpfx->mpfx_vltime;
1142                     ifra.ifra_lifetime.ia6t_pltime = mpfx->mpfx_pltime;
1143                     if (ifra.ifra_lifetime.ia6t_vltime == ND6_INFINITE_LIFETIME)
1144                             ifra.ifra_lifetime.ia6t_expire = 0;
1145                     else
```

```
1146                          ifra.ifra_lifetime.ia6t_expire = mono_time.tv_sec
1147                              + ifra.ifra_lifetime.ia6t_vltime;
1148                      if (ifra.ifra_lifetime.ia6t_pltime == ND6_INFINITE_LIFETIME)
1149                          ifra.ifra_lifetime.ia6t_preferred = 0;
1150              else
1151                          ifra.ifra_lifetime.ia6t_preferred = mono_time.tv_sec
1152                              + ifra.ifra_lifetime.ia6t_pltime;
```
─── mip6_mncore.c

1126–1152 The `in6_aliasreq{}` structure is a common structure used to manipulate address assignment. When configuring a home address, the `IN6_IFF_HOME` flag, which indicates that the address is a home address, must be set. The `IN6_IFF_AUTOCONF` flag (which means the address is configured automatically) must be set, since the KAME Mobile IPv6 implementation generates home addresses based on the stateless address autoconfiguration algorithm. The prefix length is set to 128 when assigning a home address to home virtual interfaces; otherwise, it is set to the length specified by the prefix information. The lifetimes (the preferred lifetime and the valid lifetime) of the home address are copied from the lifetime information of the prefix information entry.

Listing 15-46

─── mip6_mncore.c
```
1153                      ia6 = in6ifa_ifpwithaddr(ifp, &ifra.ifra_addr.sin6_addr);
1154                      error = in6_update_ifa(ifp, &ifra, ia6, 0);
1155                      if (error) {
....
1161                          return (error);
1162                      }
1163              }
1164
1165          return (0);
1166      }
```
─── mip6_mncore.c

1153–1163 The `ia6` variable will point to the home address information that matches the information of the home address we are going to assign. If there is no such address, `ia6` will be a NULL pointer. If such a home address has already been assigned, then `ia6` is passed to the `in6_update_ifa()` function so that the function can update the existing entry; otherwise, a new address will be assigned to the specified network interface as a new home address.

Listing 15-47

─── mip6_prefix.c
```
183      int
184      mip6_prefix_haddr_assign(mpfx, sc)
185              struct mip6_prefix *mpfx;
186              struct hif_softc *sc;
187      {
188              struct in6_addr ifid;
189              int error = 0;
190
191              if ((mpfx == NULL) || (sc == NULL)) {
192                      return (EINVAL);
193              }
....
200              {
201                      error = get_ifid((struct ifnet *)sc, NULL, &ifid);
```

```
202                      if (error)
203                              return (error);
204               }
205
206               /* XXX */
207               mpfx->mpfx_haddr = mpfx->mpfx_prefix;
208               mpfx->mpfx_haddr.s6_addr32[2] = ifid.s6_addr32[2];
209               mpfx->mpfx_haddr.s6_addr32[3] = ifid.s6_addr32[3];
210
211               return (0);
212       }
```
——— mip6_prefix.c

183–212 The `mip6_prefix_haddr_assign()` function assigns a home address to the
`mip6_prefix{}` instance based on the prefix information stored in the prefix structure. The home address is copied into the `mpfx_haddr` field of the `mip6_prefix{}`
structure. The prefix part is copied from the `mpfx_prefix` variable that stores prefix
information. The interface identifier part is copied from the result of the `get_ifid()`
function, which returns the interface identifier of a specified network interface.

Listing 15-48
——— mip6_mncore.c
```
1171      static int
1172      mip6_remove_haddrs(sc, ifp)
1173              struct hif_softc *sc;
1174              struct ifnet *ifp;
1175      {
1176              struct ifaddr *ia, *ia_next;
1177              struct in6_ifaddr *ia6;
1178              struct mip6_prefix *mpfx;
1179              int error = 0;
```
——— mip6_mncore.c

1171–1174 The `mip6_remove_haddrs()` function has two parameters. The `sc` parameter is a pointer to the `hif_softc{}` instance, which holds home prefix information,
and the `ifp` parameter is a pointer to the network interface from which the home
addresses related to the `hif_softc{}` instance specified as the first parameter are
removed.

Listing 15-49
——— mip6_mncore.c
```
1182              for (ia = TAILQ_FIRST(&ifp->if_addrhead);
1183                   ia;
1184                   ia = ia_next)
....
1190              {
....
1192                      ia_next = TAILQ_NEXT(ia, ifa_link);
....
1196
1197                      if (ia->ifa_addr->sa_family != AF_INET6)
1198                              continue;
1199                      ia6 = (struct in6_ifaddr *)ia;
```
——— mip6_mncore.c

1182–1199 All IPv6 addresses assigned to the network interface specified by `ifp` will be
checked.

Listing 15-50

```
                                                                  mip6_mncore.c
1201                     for (mpfx = LIST_FIRST(&mip6_prefix_list); mpfx;
1202                         mpfx = LIST_NEXT(mpfx, mpfx_entry)) {
1203                         if (hif_prefix_list_find_withmpfx(
1204                             &sc->hif_prefix_list_home, mpfx) == NULL)
1205                             continue;
1206
1207                         if (!in6_are_prefix_equal(&ia6->ia_addr.sin6_addr,
1208                             &mpfx->mpfx_prefix, mpfx->mpfx_prefixlen)) {
1209                             continue;
1210                         }
1211                         error = mip6_remove_addr(ifp, ia6);
1212                         if (error) {
  ....
1216                             continue;
1217                         }
1218                     }
1219                 }
1220
1221             return (error);
1222     }
                                                                  mip6_mncore.c
```

1201–1205 All home prefixes that belong to the home network specified as sc will be checked.

1207–1217 The home prefix and the prefix part of each address assigned to the network interface specified by ifp are compared. If these prefixes are equal, the mip6_remove_addr() function is called to remove the address that is assigned on the home network.

Listing 15-51

```
                                                                  mip6_mncore.c
1227    static int
1228    mip6_remove_addr(ifp, ia6)
1229            struct ifnet *ifp;
1230            struct in6_ifaddr *ia6;
1231    {
1232            struct in6_aliasreq ifra;
1233            int i = 0, purgeprefix = 0;
1234            struct nd_prefixctl pr0;
1235            struct nd_prefix *pr = NULL;
1236
1237            bcopy(if_name(ifp), ifra.ifra_name, sizeof(ifra.ifra_name));
1238            bcopy(&ia6->ia_addr, &ifra.ifra_addr, sizeof(struct sockaddr_in6));
1239            bcopy(&ia6->ia_prefixmask, &ifra.ifra_prefixmask,
1240                sizeof(struct sockaddr_in6));
                                                                  mip6_mncore.c
```

1227–1230 The mip6_remove_addr() function has two parameters. The ifp parameter is a pointer to the network interface and the ia6 parameter is a pointer to the address information, which is to be removed.

1237–1240 An in6_aliasreq{} instance, which keeps the information of the address to be removed, is created.

Listing 15-52

```
                                                                  mip6_mncore.c
1242            /* address purging code is copied from in6_control(). */
1243
```

```
1244                /*
1245                 * If the address being deleted is the only one that owns
1246                 * the corresponding prefix, expire the prefix as well.
1247                 * XXX: theoretically, we don't have to worry about such
1248                 * relationship, since we separate the address management
1249                 * and the prefix management.  We do this, however, to provide
1250                 * as much backward compatibility as possible in terms of
1251                 * the ioctl operation.
1252                 */
1253                bzero(&pr0, sizeof(pr0));
1254                pr0.ndpr_ifp = ifp;
1255                pr0.ndpr_plen = in6_mask2len(&ia6->ia_prefixmask.sin6_addr, NULL);
1256                if (pr0.ndpr_plen == 128)
1257                        goto purgeaddr;
1258                pr0.ndpr_prefix = ia6->ia_addr;
1259                for (i = 0; i < 4; i++) {
1260                        pr0.ndpr_prefix.sin6_addr.s6_addr32[i] &=
1261                            ia6->ia_prefixmask.sin6_addr.s6_addr32[i];
1262                }
1263                /*
1264                 * The logic of the following condition is a bit complicated.
1265                 * We expire the prefix when
1266                 * 1. the address obeys autoconfiguration and it is the
1267                 *    only owner of the associated prefix, or
1268                 * 2. the address does not obey autoconf and there is no
1269                 *    other owner of the prefix.
1270                 */
1271                if ((pr = nd6_prefix_lookup(&pr0)) != NULL &&
1272                    (((ia6->ia6_flags & IN6_IFF_AUTOCONF) != 0 &&
1273                    pr->ndpr_refcnt == 1) ||
1274                    ((ia6->ia6_flags & IN6_IFF_AUTOCONF) == 0 &&
1275                    pr->ndpr_refcnt == 0)))
1276                        purgeprefix = 1;
```
——mip6_mncore.c

1242 As the comment says, the code is based on the address removal code in the
in6_control() function.

1253–1262 An instance of the nd_prefixctl{} structure, which includes the prefix information of the address to be removed, is prepared.

1271–1276 In the KAME IPv6 implementation, an address information structure and prefix
information structure are linked together. Every IPv6 address of which the prefix length is
less than 128 has a corresponding prefix information entry. When an address is removed,
the corresponding prefix information has to be removed properly. The removal algorithm
differs depending on whether the prefix has an autoconfigured address. If an autoconfigured address is removed and the reference count of related prefix information is 1, then
the prefix information related to the address is removed as well. If the address is not an
autoconfigured address, then only the related prefix is removed if the reference count of
the prefix is 0.

Listing 15-53

——mip6_mncore.c
```
1278    purgeaddr:
1279            in6_purgeaddr(&ia6->ia_ifa);
1280            if (pr && purgeprefix)
1281                    prelist_remove(pr);
1282
1283            return (0);
1284    }
```
——mip6_mncore.c

1278–1281 The `in6_purgeaddr()` function is called to remove the specified address. The `prelist_remove()` function is called if the address has corresponding prefix information and it needs to be removed.

15.5 Sending a Binding Update Message

When a mobile node detects its movement, it needs to send a Binding Update message to update the current care-of address, which is registered to its home agent. There are two kinds of Binding Update messages. One type is used for home registration and the other is sent to correspondent nodes. The `mip6_home_registration()`, the `mip6_home_registration2()`, and the `mip6_bu_send_bu()` functions are used for the former purpose, whereas the `mip6_bu_send_cbu()` function is used for the latter. The overview of the registration procedure is described in Section 4.2 of Chapter 4 and the return routability procedure is described in Section 5.1 of Chapter 5.

15.5.1 Sending a Home Registration Message

Figure 15-3 shows the function call flow when a home registration message is sent.

Listing 15-54

————————————————————————————————————mip6_mncore.c
```
1671    int
1672    mip6_home_registration(sc)
1673            struct hif_softc *sc;
1674    {
1675            struct in6_addr hif_coa;
1676            struct mip6_prefix *mpfx;
1677            struct mip6_bu *mbu;
1678            const struct in6_addr *haaddr;
1679            struct mip6_ha *mha;
```
————————————————————————————————————mip6_mncore.c

1671–1673 The `mip6_home_registration()` function is called from the `mip6_register_current_location()` function to initiate the home registration or home deregistration procedure. The parameter `sc` is a pointer to the `hif_softc{}` instance, which indicates the home network of a mobile node–performing registration.

Listing 15-55

————————————————————————————————————mip6_mncore.c
```
1681            /* get current CoA and recover its scope information. */
1682            if (sc->hif_coa_ifa == NULL) {
....
1686                    return (0);
1687            }
1688            hif_coa = sc->hif_coa_ifa->ia_addr.sin6_addr;
```
————————————————————————————————————mip6_mncore.c

1682–1688 If the current care-of address of the specified home network has not been set yet, there is nothing to do. If there is a valid care-of address, the care-of address stored in the `hif_coa_ifa` variable is copied to the `hif_coa` variable.

FIGURE 15-3

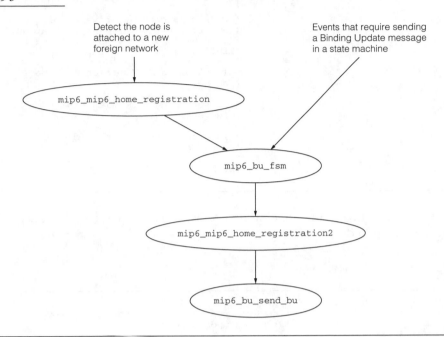

Call flow when sending a home registration message.

Listing 15-56

```
                                                              mip6_mncore.c
1690            for (mpfx = LIST_FIRST(&mip6_prefix_list); mpfx;
1691                mpfx = LIST_NEXT(mpfx, mpfx_entry)) {
1692                if (hif_prefix_list_find_withmpfx(&sc->hif_prefix_list_home,
1693                    mpfx) == NULL)
1694                    continue;
1695
1696                for (mbu = LIST_FIRST(&sc->hif_bu_list); mbu;
1697                    mbu = LIST_NEXT(mbu, mbu_entry)) {
1698                    if ((mbu->mbu_flags & IP6MU_HOME) == 0)
1699                        continue;
1700                    if (IN6_ARE_ADDR_EQUAL(&mbu->mbu_haddr,
1701                        &mpfx->mpfx_haddr))
1702                        break;
1703                }
                                                              mip6_mncore.c
```

1690–1703 The loop searches for home registration entries of home addresses of the mobile node. The home addresses are stored in `mip6_prefix{}` structures. The outer loop finds the home address by checking the `hif_prefix_list_home` field of the `sc` parameter that contains all home prefixes for the virtual home interface. The inner loop looks for home registration entries of the home addresses.

Listing 15-57

```
                                                              mip6_mncore.c
1704            if (mbu == NULL) {
1705                /* not exist */
```

```
1706                              if (sc->hif_location == HIF_LOCATION_HOME) {
1707                                      /*
1708                                       * we are home and we have no binding
1709                                       * update entry for home registration.
1710                                       * this will happen when either of the
1711                                       * following two cases happens.
1712                                       *
1713                                       * 1. enabling MN function at home
1714                                       * subnet.
1715                                       *
1716                                       * 2. returning home with expired home
1717                                       * registration.
1718                                       *
1719                                       * in either case, we should do
1720                                       * nothing.
1721                                       */
1722                                      continue;
1723                              }
```
—— mip6_mncore.c

1704–1723 If there is no existing binding update list entry and the mobile node is at home, there is nothing to do.

Listing 15-58

—— mip6_mncore.c

```
1725                              /*
1726                               * no home registration found.  create a new
1727                               * binding update entry.
1728                               */

1730                              /* pick the preferable HA from the list. */
1731                              mha = hif_find_preferable_ha(sc);

1733                              if (mha == NULL) {
1734                                      /*
1735                                       * if no home agent is found, set an
1736                                       * unspecified address for now.  DHAAD
1737                                       * is triggered when sending a binging
1738                                       * update message.
1739                                       */
1740                                      haaddr = &in6addr_any;
1741                              } else {
1742                                      haaddr = &mha->mha_addr;
1743                              }

1745                              mbu = mip6_bu_create(haaddr, mpfx, &hif_coa,
1746                                      IP6MU_ACK|IP6MU_HOME
....
1748                                      |IP6MU_LINK
....
1750                                      , sc);
1751                              if (mbu == NULL)
1752                                      return (ENOMEM);
1753                              /*
1754                               * for the first registration to the home
1755                               * agent, the ack timeout value should be
1756                               * (retrans * dadtransmits) * 1.5.
1757                               */
1758                              /*
1759                               * XXX: TODO: KAME has different dad retrans
1760                               * values for each interfaces.  which retrans
1761                               * value should be selected ?
1762                               */

1764                              mip6_bu_list_insert(&sc->hif_bu_list, mbu);
```

```
1765
1766                        /* XXX */
1767                        if (sc->hif_location != HIF_LOCATION_HOME)
1768                                mip6_bu_fsm(mbu,
1769                                        MIP6_BU_PRI_FSM_EVENT_MOVEMENT, NULL);
1770                        else
1771                                mip6_bu_fsm(mbu,
1772                                        MIP6_BU_PRI_FSM_EVENT_RETURNING_HOME,
1773                                        NULL);
```
——— mip6_mncore.c

1731–1752 If there is no existing binding update list entry and the mobile node is in a foreign
network, a new binding update list entry is created to register the current location of
the mobile node. The `hif_find_preferable_ha()` function returns the home agent
that has the highest priority from the home agent list maintained in the mobile node. If
the mobile node has not learned any information about its home agents, the unspecified
address is temporarily set as a home agent address. The address will be determined by
the Dynamic Home Agent Address Discovery procedure launched later. When creating
a binding update list entry for home registration, the `IP6MU_ACK`, `IP6MU_HOME` and
`IP6MU_LINK` flags are specified. The `IP6MU_ACK` and `IP6MU_HOME` flags are necessary
for home registration. The `IP6MU_LINK` flag must be set when the interface identifier
part of the home address is the same as the interface identifier of the link-local address of
the mobile node. The flag is always set in the KAME implementation, since it always uses
the interface identifier of the link-local address of the virtual interface when generating a
home address.

1764–1773 The newly created binding update list entry is inserted to the binding update list
and the state machine of the entry is triggered based on the current location. If the mobile
node is at home, the RETURNING_HOME event is sent; otherwise, the MOVEMENT event is
sent. State machines will be discussed in Section 15.19.

Listing 15-59
——— mip6_mncore.c
```
1774                        } else {
1775                                if (sc->hif_location != HIF_LOCATION_HOME)
1776                                        mip6_bu_fsm(mbu,
1777                                                MIP6_BU_PRI_FSM_EVENT_MOVEMENT, NULL);
1778                                else
1779                                        mip6_bu_fsm(mbu,
1780                                                MIP6_BU_PRI_FSM_EVENT_RETURNING_HOME,
1781                                                NULL);
1782                        }
1783                }
1784
1785                return (0);
1786        }
```
——— mip6_mncore.c

1774–1782 If a binding update list entry already exists, the `mip6_bu_fsm()` function is called
to trigger an event of the state machine based on the current location.

Listing 15-60
——— mip6_mncore.c
```
1788    int
1789    mip6_home_registration2(mbu)
```

```
1790              struct mip6_bu *mbu;
1791      {
1792              struct in6_addr hif_coa;
1793              struct mip6_prefix *mpfx;
1794              int32_t coa_lifetime, prefix_lifetime;
1795              int error;
....
1797              struct timeval mono_time;
```
――― mip6_mncore.c

1788–1790 The `mip6_home_registration2()` function is called from the state machine of a binding update list entry as a result of state transition. The `mbu` parameter is a pointer to the binding update list entry that needs to be registered.

Listing 15-61
――― mip6_mncore.c
```
1800              /* sanity check. */
1801              if (mbu == NULL)
1802                      return (EINVAL);
....
1808              /* get current CoA and recover its scope information. */
1809              if (mbu->mbu_hif->hif_coa_ifa == NULL) {
....
1813                      return (0);
1814              }
1815              hif_coa = mbu->mbu_hif->hif_coa_ifa->ia_addr.sin6_addr;
```
――― mip6_mncore.c

1809–1815 The `hif_coa_ifa` variable points to the current care-of address. If there is no available care-of address, the home registration procedure is aborted.

Listing 15-62
――― mip6_mncore.c
```
1817              /*
1818               * a binding update entry exists. update information.
1819               */
1820
1821              /* update CoA. */
1822              if (mbu->mbu_hif->hif_location == HIF_LOCATION_HOME) {
1823                      /* home de-registration. */
1824                      mbu->mbu_coa = mbu->mbu_haddr;
1825              } else {
1826                      /* home registration. */
1827                      mbu->mbu_coa = hif_coa;
1828              }
1829
1830              /* update lifetime. */
1831              coa_lifetime = mip6_coa_get_lifetime(&mbu->mbu_coa);
1832              prefix_lifetime = 0x7fffffff;
1833              for (mpfx = LIST_FIRST(&mip6_prefix_list); mpfx;
1834                  mpfx = LIST_NEXT(mpfx, mpfx_entry)) {
1835                      if (hif_prefix_list_find_withmpfx(
1836                          &mbu->mbu_hif->hif_prefix_list_home, mpfx) == NULL)
1837                              continue;
1838                      if (mpfx->mpfx_vltime < prefix_lifetime)
1839                              prefix_lifetime = mpfx->mpfx_vltime;
1840              }
1841              if (coa_lifetime < prefix_lifetime) {
1842                      mbu->mbu_lifetime = coa_lifetime;
1843              } else {
1844                      mbu->mbu_lifetime = prefix_lifetime;
1845              }
```
――― mip6_mncore.c

1822–1845 The member variables of the existing binding update list entry are updated. The care-of address is set to the home address of the mobile node if the current location is home; otherwise, the care-of address is copied from the care-of address stored in the `hif_softc{}` instance. The lifetime of the entry is calculated from the lifetimes of the care-of and home addresses. The shorter lifetime of the two addresses will be chosen as the lifetime of the binding update list entry.

Listing 15-63

_____ mip6_mncore.c

```
1846            mbu->mbu_expire = mono_time.tv_sec + mbu->mbu_lifetime;
1847            /* sanity check for overflow */
1848            if (mbu->mbu_expire < mono_time.tv_sec)
1849                    mbu->mbu_expire = 0x7fffffff;
1850            mbu->mbu_refresh = mbu->mbu_lifetime;
1851            /* mbu->mbu_flags |= IP6MU_DAD ;*/
1852
1853            /* send a binding update. */
1854            error = mip6_bu_send_bu(mbu);
1855
1856            return (error);
1857    }
```
_____ mip6_mncore.c

1846–1854 The expiration time of the entry is set based on the current time and the lifetime calculated above. The refresh interval is set to the same value with the lifetime for now. The value will be updated when a Binding Acknowledgment is received and processed (see Section 15.6).

15.5.2 Sending a Binding Update Message

The `mip6_bu_send_bu()` creates a Binding Update message based on the information passed as its parameter and sends the message to the home agent of a mobile node.

Listing 15-64

_____ mip6_mncore.c

```
2163    int
2164    mip6_bu_send_bu(mbu)
2165            struct mip6_bu *mbu;
2166    {
2167            struct mbuf *m;
2168            struct ip6_pktopts opt;
2169            int error = 0;
....
2171            /* sanity check. */
2172            if (mbu == NULL)
2173                    return (EINVAL);
2174
2175            if (IN6_IS_ADDR_UNSPECIFIED(&mbu->mbu_paddr)) {
2176                    /* we do not know where to send a binding update. */
2177                    if ((mbu->mbu_flags & IP6MU_HOME) != 0) {
2178                            error = mip6_icmp6_dhaad_req_output(mbu->mbu_hif);
2179                            if (error) {
....
2184                                    /* continue, anyway. */
2185                            }
2186                            /*
2187                             * a binding update will be sent
2188                             * immediately after receiving DHAAD
```

```
2189                         * reply.
2190                         */
2191                        goto bu_send_bu_end;
2192                 }
2193                 panic("a peer address must be known when sending a binding
      update.");
2194         }
```

Note: Line 2193 is broken here for layout reasons. However, it is a single line of code.

2175–2192 If the destination address, `mbu_paddr`, of the Binding Update message to be sent is unspecified and the binding update list entry is for home registration, a Dynamic Home Agent Address Discovery (DHAAD) request message is sent by the `mip6_icmp6_dhaad_req_output()` function. This occurs when a mobile node is turned on at a foreign network and does not know anything about its home agents.

Listing 15-65

——mip6_mncore.c
```
2196         /* create an ipv6 header to send a binding update. */
2197         m = mip6_create_ip6hdr(&mbu->mbu_haddr, &mbu->mbu_paddr,
2198             IPPROTO_NONE, 0);
2199         if (m == NULL) {
....
2202                 error = ENOBUFS;
2203                 goto bu_send_bu_end;
2204         }
```
——mip6_mncore.c

2197–2204 An IPv6 header is prepared based on the source and destination address information stored in the binding update list entry. If the preparation fails, the Binding Update message will not be sent. The code will try to send the message at the next timeout of the retransmission timer of the entry.

Listing 15-66

——mip6_mncore.c
```
2206         /* initialize packet options structure. */
2207         ip6_initpktopts(&opt);
2208
2209         /* create a binding update mobility header. */
2210         error = mip6_ip6mu_create(&opt.ip6po_mh, &mbu->mbu_haddr,
2211             &mbu->mbu_paddr, mbu->mbu_hif);
2212         if (error) {
....
2216                 m_freem(m);
2217                 goto free_ip6pktopts;
2218         }
2219
2220         /* send a binding update. */
....
2222         error = ip6_output(m, &opt, NULL, 0, NULL, NULL
....
2224             , NULL
....
2226                                 );
2227         if (error) {
....
2231                 goto free_ip6pktopts;
2232         }
```

```
2233
2234     free_ip6pktopts:
2235             if (opt.ip6po_mh)
2236                     FREE(opt.ip6po_mh, M_IP6OPT);
2237
2238     bu_send_bu_end:
2239             return (error);
2240     }
```
                                                            _____ mip6_mncore.c

2207 A Binding Update message is passed to the output routine as a packet option. The `ip6_pktopts{}` structure, which keeps the message, is initialized.

2210–2232 A Binding Update message is created by the `mip6_ip6mu_create()` function. The created message, based on the information stored in the binding update list entry, is sent by the `ip6_output()` function.

2234–2236 Memory that is allocated to keep the packet option is released before the function finishes.

15.5.3 Sending a Binding Update Message to a Correspondent Node

The `mip6_bu_send_cbu()` function is used when a mobile node sends a Binding Update message to a correspondent node. The Binding Authorization Data option must be used when a mobile node sends a Binding Update message to correspondent nodes. The value that the option contains will be computed by the return routability procedure, which is discussed in Section 15.23.

Listing 15-67
                                                            _____ mip6_mncore.c

```
2242     int
2243     mip6_bu_send_cbu(mbu)
2244             struct mip6_bu *mbu;
2245     {
2246             struct mbuf *m;
2247             struct ip6_pktopts opt;
2248             int error = 0;
2249
2250             /* sanity check. */
2251             if (mbu == NULL)
2252                     return (EINVAL);
2253
2254             ip6_initpktopts(&opt);
2255
2256             m = mip6_create_ip6hdr(&mbu->mbu_haddr, &mbu->mbu_paddr,
             IPPROTO_NONE, 0);
2257             if (m == NULL) {
....
2260                     return (ENOMEM);
2261             }
2262
2263             error = mip6_ip6mu_create(&opt.ip6po_mh, &mbu->mbu_haddr,
2264                     &mbu->mbu_paddr, mbu->mbu_hif);
2265             if (error) {
....
2270                     m_freem(m);
2271                     goto free_ip6pktopts;
2272             }
2273
2274             mip6stat.mip6s_obu++;
```

```
2275                  error = ip6_output(m, &opt, NULL, 0, NULL, NULL
....
2277                                        , NULL
....
2279                                        );
2280             if (error) {
2281                  mip6log((LOG_ERR,
2282                      "%s:%d: sending a binding update failed. (%d)\n",
2283                      __FILE__, __LINE__, error));
2284                  goto free_ip6pktopts;
2285             }
2286
2287      free_ip6pktopts:
2288             if (opt.ip6po_mh)
2289                  FREE(opt.ip6po_mh, M_IP6OPT);
2290
2291             return (error);
2292      }
```
———————————————————————————————————————mip6_mncore.c

Note: Line 2256 is broken here for layout reasons. However, it is a single line of code.

2254–2289 The procedure is almost the same as the `mip6_bu_send_bu()` function. The only difference between these two functions is that the `mip6_bu_send_bu()` function may send a DHAAD request message if a mobile node does not know the address of its home agent. This never happens when sending a Binding Update message to correspondent nodes because the address of the peer node must be known before the message is sent.

15.5.4   Create a Binding Update Message

The `mip6_ip6mu_create()` function prepares a Binding Update message.

Listing 15-68

———————————————————————————————————————mip6_mncore.c
```
3584      int
3585      mip6_ip6mu_create(pktopt_mobility, src, dst, sc)
3586             struct ip6_mh **pktopt_mobility;
3587             struct in6_addr *src, *dst;
3588             struct hif_softc *sc;
3589      {
3590             struct ip6_mh_binding_update *ip6mu;
3591             struct ip6_mh_opt_nonce_index *mopt_nonce = NULL;
3592             struct ip6_mh_opt_auth_data *mopt_auth = NULL;
3593             struct ip6_mh_opt_altcoa *mopt_altcoa = NULL;
3594             struct in6_addr altcoa;
3595             int ip6mu_size, pad;
3596             int bu_size = 0, nonce_size = 0, auth_size = 0, altcoa_size = 0;
3597             struct mip6_bu *mbu, *hrmbu;
3598             struct mip6_prefix *mpfx;
3599             int need_rr = 0, ignore_co_nonce = 0;
3600             u_int8_t key_bm[MIP6_KBM_LEN]; /* Stated as 'Kbm' in the spec */
```
———————————————————————————————————————mip6_mncore.c

3584–3588 This function can handle both a home registration message and a correspondent registration message. The `pktopt_mobility` parameter is a pointer to the memory in which the created message is stored. The `src` parameter is the home address of the mobile node, the `dst` parameter is the address of the home agent or the correspondent node, and the `sc` parameter is a pointer to the `hif_softc{}` instance, which indicates the home network of the mobile node.

Listing 15-69

_____ mip6_mncore.c
```
3604                *pktopt_mobility = NULL;
3605
3606                mbu = mip6_bu_list_find_withpaddr(&sc->hif_bu_list, dst, src);
3607                hrmbu = mip6_bu_list_find_home_registration(&sc->hif_bu_list, src);
3608                if ((mbu == NULL) &&
3609                    (hrmbu != NULL) &&
3610                    (MIP6_IS_BU_BOUND_STATE(hrmbu))) {
3611                        /* XXX */
3612                        /* create a binding update entry and send CoTI/HoTI. */
3613                        return (0);
3614                }
3615                if (mbu == NULL) {
3616                        /*
3617                         * this is the case that the home registration is on
3618                         * going.  that is, (mbu == NULL) && (hrmbu != NULL)
3619                         * but hrmbu->mbu_fsm_state != STATE_REG.
3620                         */
3621                        return (0);
3622                }
3623                if ((mbu->mbu_state & MIP6_BU_STATE_NEEDTUNNEL) != 0) {
3624                        return (0);
3625                }
```
_____ mip6_mncore.c

3606–3607 The binding update list entry for the destination address and the entry of the home address of the mobile node are identified. The mbu variable will point to the binding update list entry that contains the information between the mobile node and the node to which the message is sent, if it already exists. The hrmbu variable will point to the binding update list entry for home registration between the home address of the mobile node and its home agent, if it exists.

3608–3614 If the mobile node has already registered to its home agent, but it does not have a binding update list entry for the correspondent node specified by the dst parameter, then nothing will be done. Before sending a Binding Update message to the correspondent node, the return routability procedure has to be performed to secure the message. The return routability code will send the message when it completes the procedure.

3615–3622 If the mobile node has not finished home registration, a Binding Update message cannot be sent to the correspondent node.

3623–3625 If the correspondent node is marked as a node that does not support Mobile IPv6 (that is, the MIP6_BU_STATE_NEEDTUNNEL flag is set), a Binding Update message will not be sent. The node will not understand the message and will just reply with an Internet Control Message Protocol version 6 (ICMPv6) error.

Listing 15-70

_____ mip6_mncore.c
```
3626                if (IN6_IS_ADDR_UNSPECIFIED(&mbu->mbu_paddr)) {
3627                        /*
3628                         * the peer addr is unspecified.  this happens when
3629                         * home registration occurs but no home agent address
3630                         * is known.
3631                         */
....
3635                        mip6_icmp6_dhaad_req_output(sc);
3636                        return (0);
3637                }
```
_____ mip6_mncore.c

3626–3637 If the destination address of the binding update list entry for home registration is an unspecified address, the DHAAD procedure is initiated. This only happens when the entry is for home registration.

Listing 15-71

_____mip6_mncore.c

```
3639                if (!(mbu->mbu_flags & IP6MU_HOME)) {
3640                        need_rr = 1;
3641                }
3642
3643                /* check if we have a valid prefix information. */
3644                mpfx = mip6_prefix_list_find_withhaddr(&mip6_prefix_list, src);
3645                if (mpfx == NULL)
3646                        return(EINVAL);
```
_____mip6_mncore.c

3639–3641 If the binding update list entry is for a correspondent node, a Binding Authorization data option has to be included in the Binding Update message. The `need_rr` variable activates the code of the option processing.

3644–3646 The mobile node must have the prefix information related to the source address. The lifetime information of the prefix will be used to calculate the lifetime of binding information later.

Listing 15-72

_____mip6_mncore.c

```
3648                bu_size = sizeof(struct ip6_mh_binding_update);
3649                if (need_rr) {
3650                        /*
3651                         |<- bu_size -> <- nonce_size -> <- auth_size ->
3652                         +------------+---------------+--------------+
3653                         | bind. up.  |  nonce opt.   |  auth. opt.  |
3654                         +------------+---------------+--------------+
3655                          <------->
3656                          sizeof(struct ip6_mh_binding_update)
3657                                 <-->
3658                                 Padding for nonce opt. alignment
3659                         */
3660                        bu_size += MIP6_PADLEN(bu_size, 2, 0);
3661                        nonce_size = sizeof(struct ip6_mh_opt_nonce_index);
3662                        nonce_size += MIP6_PADLEN(bu_size + nonce_size, 8, 2);
3663                        /* (6.2.7)
3664                           The Binding Authorization Data option does not
3665                           have alignment requirements as such.  However,
3666                           since this option must be the last mobility option,
3667                           an implicit alignment requirement is 8n + 2.
3668                         */
3669                        auth_size = IP6MOPT_AUTHDATA_SIZE;
3670                        auth_size += MIP6_PADLEN(bu_size + nonce_size
    + auth_size, 8, 0);
....
3674                        altcoa_size = 0;
3675                } else {
3676                        bu_size += MIP6_PADLEN(bu_size, 8, 6);
3677                        altcoa_size = sizeof(struct ip6_mh_opt_altcoa);
3678                        nonce_size = auth_size = 0;
3679                }
3680                ip6mu_size = bu_size + nonce_size + auth_size + altcoa_size;
```
_____mip6_mncore.c

Note: Line 3670 is broken here for layout reasons. However, it is a single line of code.

3648–3680 The size of the Binding Update message is calculated. The size of the message varies based on the node type of the destination. If the destination is the home agent and the message is for home registration, the message will include the Alternate Care-of Address option; otherwise, the message will include the Nonce Index option and the Binding Authorization Data option.

The alignment requirement of the Nonce Index option is $2n + 0$ and the size of the option is represented by the size of the `ip6_mh_opt_nonce_index{}` structure. The size of the Binding Authorization Data option is defined as the `IP6MOPT_AUTHDATA_SIZE` macro and its implicit alignment requirement is $8n + 2$. If the Binding Update message is for a correspondent node, then these lengths are added to the size of the Binding Update message.

The alignment requirement of the Alternate Care-of Address option is $8n + 2$. The size of the option is the size of the `ip6_mh_opt_altcoa{}` structure. If the Binding Update message is for home registration, the size is added to the size of the Binding Update message.

Listing 15-73

```
                                                                    mip6_mncore.c
3682            MALLOC(ip6mu, struct ip6_mh_binding_update *,
3683                ip6mu_size, M_IP6OPT, M_NOWAIT);
3684        if (ip6mu == NULL)
3685                return (ENOMEM);
3686
3687        if (need_rr) {
3688                mopt_nonce = (struct ip6_mh_opt_nonce_index *)((u_int8_t *)
    ip6mu + bu_size);
3689                mopt_auth = (struct ip6_mh_opt_auth_data *)((u_int8_t *)
    mopt_nonce + nonce_size);
3690        } else {
3691                mopt_altcoa = (struct ip6_mh_opt_altcoa *)((u_int8_t *)ip6mu
    + bu_size);
3692        }
                                                                    mip6_mncore.c
```

Note: Lines 3688, 3689, and 3691 are broken here for layout reasons. However, they are each a single line of code.

3682–3692 The required memory for the Binding Update message is allocated and all pointers (`mopt_once`, `mopt_auth` and `mopt_altcoa`) point to the proper addresses at which the options are located.

Listing 15-74

```
                                                                    mip6_mncore.c
3694        /* update sequence number of this binding update entry. */
3695        mbu->mbu_seqno++;
3696
3697        bzero(ip6mu, ip6mu_size);
3698
3699        ip6mu->ip6mhbu_proto = IPPROTO_NONE;
3700        ip6mu->ip6mhbu_len = (ip6mu_size >> 3) - 1;
3701        ip6mu->ip6mhbu_type = IP6_MH_TYPE_BU;
3702        ip6mu->ip6mhbu_flags = mbu->mbu_flags;
3703        ip6mu->ip6mhbu_seqno = htons(mbu->mbu_seqno);
3704        if (IN6_ARE_ADDR_EQUAL(&mbu->mbu_haddr, &mbu->mbu_coa)) {
3705                /* this binding update is for home un-registration. */
3706                ip6mu->ip6mhbu_lifetime = 0;
3707                if (need_rr) {
```

```
3708                              ignore_co_nonce = 1;
3709                      }
3710              } else {
3711                      u_int32_t haddr_lifetime, coa_lifetime, lifetime;
3712
3713                      haddr_lifetime = mpfx->mpfx_vltime;
3714                      coa_lifetime = mip6_coa_get_lifetime(&mbu->mbu_coa);
3715                      lifetime = haddr_lifetime < coa_lifetime ?
3716                              haddr_lifetime : coa_lifetime;
3717                      if ((mbu->mbu_flags & IP6MU_HOME) == 0) {
3718                              if (mip6ctl_bu_maxlifetime > 0 &&
3719                                  lifetime > mip6ctl_bu_maxlifetime)
3720                                      lifetime = mip6ctl_bu_maxlifetime;
3721                      } else {
3722                              if (mip6ctl_hrbu_maxlifetime > 0 &&
3723                                  lifetime > mip6ctl_hrbu_maxlifetime)
3724                                      lifetime = mip6ctl_hrbu_maxlifetime;
3725                      }
3726                      mbu->mbu_lifetime = lifetime;
3727                      mbu->mbu_expire = time_second + mbu->mbu_lifetime;
3728                      mbu->mbu_refresh = mbu->mbu_lifetime;
3729                      ip6mu->ip6mhbu_lifetime =
3730                          htons((u_int16_t)(mbu->mbu_lifetime >> 2));
    /* units 4 secs */
3731              }
```
_____ mip6_mncore.c

Note: Line 3730 is broken here for layout reasons. However, it is a single line of code.

3695 The sequence number kept in the binding update list entry is incremented to prevent a replay attack.

3699–3730 The packet information is filled based on the binding update list entry. The lifetime is set to 0 when the mobile node is at home. In addition, the `ignore_co_nonce` variable is set to true, since there is no need to differentiate between the home address and the care-of address of the mobile node when it is at home. When the mobile node is in a foreign link, the lifetime is set to the shorter lifetime of the home address or the care-of address of the mobile node. In the KAME implementation, it is possible to limit the maximum lifetime of binding update list entries to prevent a mobile node from creating an entry that has a very long lifetime. For correspondent nodes, the global variable `mip6ctl_bu_maxlifetime` indicates the limit and for home registration, `mip6ctl_hrbu_maxlifetime` limits the lifetime. The value has to be shifted 2 bits when setting it to the lifetime field in the message since the lifetime field is represented in units of 4 s.

Listing 15-75
_____ mip6_mncore.c
```
3733              if ((pad = bu_size - sizeof(struct ip6_mh_binding_update)) >= 2) {
3734                      u_char *p =
3735                              (u_int8_t *)ip6mu + sizeof
    (struct ip6_mh_binding_update);
3736                      *p = IP6_MHOPT_PADN;
3737                      *(p + 1) = pad - 2;
3738              }
```
_____ mip6_mncore.c

Note: Line 3735 is broken here for layout reasons. However, it is a single line of code.

3733–3738 The PadN option between the Binding Update message and the inserted options is filled, if necessary.

Listing 15-76

—— mip6_mncore.c

```
3740            if (need_rr) {
3741                    /* nonce indices and authdata insertion. */
3742                    if (nonce_size) {
3743                            if ((pad = nonce_size - sizeof
     (struct ip6_mh_opt_nonce_index))
3744                                    >= 2) {
3745                                    u_char *p = (u_int8_t *)ip6mu + bu_size
3746                                            + sizeof(struct ip6_mh_opt_nonce_index);
3747                                    *p = IP6_MHOPT_PADN;
3748                                    *(p + 1) = pad - 2;
3749                            }
3750                    }
3751                    if (auth_size) {
3752                            if ((pad = auth_size - IP6MOPT_AUTHDATA_SIZE) >= 2) {
3753                                    u_char *p = (u_int8_t *)ip6mu
3754                                            + bu_size + nonce_size
     + IP6MOPT_AUTHDATA_SIZE;
3755                                    *p = IP6_MHOPT_PADN;
3756                                    *(p + 1) = pad - 2;
3757                            }
3758                    }
```

—— mip6_mncore.c

Note: Lines 3743 and 3754 are broken here for layout reasons. However, they are each a single line of code.

3740–3758 The code from lines 3740–3805 is for Binding Update messages for correspondent nodes. The PadN option is inserted after the Nonce Index option and the Binding Authorization Data option, if these options exist. In this case the padding option is necessary.

Listing 15-77

—— mip6_mncore.c

```
3760                    /* Nonce Indicies */
3761                    mopt_nonce->ip6moni_type = IP6_MHOPT_NONCEID;
3762                    mopt_nonce->ip6moni_len = sizeof
     (struct ip6_mh_opt_nonce_index) - 2;
3763                    SET_NETVAL_S(&mopt_nonce->ip6moni_home_nonce8,
3764                            mbu->mbu_home_nonce_index);
3765                    if (!ignore_co_nonce) {
3766                            SET_NETVAL_S(&mopt_nonce->ip6moni_coa_nonce8,
3767                                    mbu->mbu_careof_nonce_index);
3768                    }
```

—— mip6_mncore.c

Note: Line 3762 is broken here for layout reasons. However, it is a single line of code.

3760–3768 A Nonce Index option is created. The option type is `IP6_MHOPT_NONCEID` and the size is that of the `ip6_mh_opt_nonce_index{}` structure. The nonce index values are kept in the binding update list entry. These values are copied to the option fields and sent to the correspondent node. The care-of nonce is only used when the `ignore_co_nonce` variable is set to false, which means the mobile node is in a foreign network.

Listing 15-78

—— mip6_mncore.c

```
3770                    /* Auth. data */
3771                    mopt_auth->ip6moad_type = IP6_MHOPT_BAUTH;
3772                    mopt_auth->ip6moad_len = IP6MOPT_AUTHDATA_SIZE - 2;
3773
```

```
3774                     if (auth_size > IP6MOPT_AUTHDATA_SIZE) {
3775                             *((u_int8_t *)ip6mu + bu_size + nonce_size
     + IP6MOPT_AUTHDATA_SIZE)
3776                                     = IP6_MHOPT_PADN;
3777                             *((u_int8_t *)ip6mu + bu_size + nonce_size
     + IP6MOPT_AUTHDATA_SIZE + 1)
3778                                     = auth_size - IP6MOPT_AUTHDATA_SIZE - 2;
3779                     }
3780
....
3785                     /* Calculate Kbm */
3786                     mip6_calculate_kbm(&mbu->mbu_home_token,
3787                                     ignore_co_nonce ? NULL :
     &mbu->mbu_careof_token,
3788                                     key_bm);
....
3792
3793                     /* Calculate authenticator (5.2.6) */
3794                     /* First(96, HMAC_SHA1(Kbm, (coa, | cn | BU))) */
3795                     if (mip6_calculate_authenticator(key_bm,
3796                         (u_int8_t *)(mopt_auth + 1), &mbu->mbu_coa,
3797                         dst, (caddr_t)ip6mu,
3798                         bu_size + nonce_size + auth_size,
3799                         bu_size + nonce_size + sizeof(struct ip6_mh_opt_auth_data),
3800                         MIP6_AUTHENTICATOR_LEN)) {
....
3804                             return (EINVAL);
3805                     }
```
_____ mip6_mncore.c

Note: Lines 3775, 3777, and 3787 are broken here for layout reasons. However, they are each a single line of code.

3771–3780 A Binding Authorization Data option is created. The option type is `IP6_MHOPT_BAUTH` and the size is defined as the `IP6MOPT_AUTHDATA_SIZE` macro. The insertion code of the padding option on lines 3774–3779 is a duplicated code. The same process has already been done on lines 3751–3758.

3786–3805 The key value used to compute the hash value of the Binding Update message is prepared by the `mip6_calculate_kbm()` function. The key is used by the `mip6_calculate_authenticator()` function called on line 3795 to get the hash value of the message. The first 96 bits are used as the authenticator value, which will be included in the Binding Authorization Data option.

Listing 15-79

_____ mip6_mncore.c

```
3809             } else {
3810                     if (altcoa_size) {
3811                             if ((pad = altcoa_size
3812                                 - sizeof(struct ip6_mh_opt_altcoa)) >= 2) {
3813                                     u_char *p = (u_int8_t *)ip6mu + bu_size
3814                                         + sizeof(struct ip6_mh_opt_nonce_index);
3815                                     *p = IP6_MHOPT_PADN;
3816                                     *(p + 1) = pad - 2;
3817                             }
3818                     }
3819                     mopt_altcoa->ip6moa_type = IP6_MHOPT_ALTCOA;
3820                     mopt_altcoa->ip6moa_len = sizeof(struct ip6_mh_opt_altcoa) - 2;
3821                     altcoa = mbu->mbu_coa;
3822                     in6_clearscope(&altcoa);
3823                     bcopy(&altcoa, mopt_altcoa->ip6moa_addr,
3824                         sizeof(struct in6_addr));
3825             }
```
_____ mip6_mncore.c

3810–3825 This part of the code is for home registration. The code from lines 3810–3813 aims to put a PadN option after the Alternate Care-of Address option; however, the code is wrong. Line 3814 should add the size of the `ip6_mh_opt_altcoa{}` structure instead of the `ip6_mh_opt_nonce_index{}` structure. Fortunately, the bug does not cause a problem since the PadN option created here is overwritten by the Alternate Care-of Address option value filled on lines 3823–3824.

The type of the Alternate Care-of Address option is `IP6_MHOPT_ALTCOA` and the size is that of the `ip6_mh_opt_altcoa{}` structure. The care-of address of the mobile node is kept in the binding update list entry. The address is copied to the option data field to complete the option.

Listing 15-80

mip6_mncore.c

```
3827            /* calculate checksum. */
3828            ip6mu->ip6mhbu_cksum = mip6_cksum(&mbu->mbu_haddr, dst, ip6mu_size,
3829                IPPROTO_MH, (char *)ip6mu);
3830
3831            *pktopt_mobility = (struct ip6_mh *)ip6mu;
3832
3833            return (0);
3834    }
```

mip6_mncore.c

3828–3833 Finally, the checksum value of the Binding Update message is computed by the `mip6_cksum()` function and the pointer to the created message is returned.

15.6 Receiving a Binding Acknowledgment Message

A mobile node will receive a Binding Acknowledgment message in response to a Binding Update message. The processing of the received Binding Acknowledgment message is implemented as the `mip6_ip6ma_input()` function.

Listing 15-81

mip6_mncore.c

```
2863    int
2864    mip6_ip6ma_input(m, ip6ma, ip6malen)
2865            struct mbuf *m;
2866            struct ip6_mh_binding_ack *ip6ma;
2867            int ip6malen;
2868    {
2869            struct ip6_hdr *ip6;
2870            struct hif_softc *sc;
2871            struct mip6_bu *mbu;
2872            u_int16_t seqno;
2873            u_int32_t lifetime, refresh;
    ....
2877            int error = 0;
2878            struct mip6_mobility_options mopt;
2879            u_int8_t ba_safe = 0;
```

mip6_mncore.c

2863–2867 The `mip6_ip6ma_input()` function has three parameters: the m parameter is a pointer to the mbuf that contains the received Binding Acknowledgment message; the `ip6ma` parameter is a pointer to the address of the message; and the `ip6malen` parameter is the length of the message.

Listing 15-82
_____ mip6_mncore.c

```
2894            ip6 = mtod(m, struct ip6_hdr *);
2895
2896            /* packet length check. */
2897            if (ip6malen < sizeof(struct ip6_mh_binding_ack)) {
....
2905                    /* send ICMP parameter problem. */
2906                    icmp6_error(m, ICMP6_PARAM_PROB, ICMP6_PARAMPROB_HEADER,
2907                        (caddr_t)&ip6ma->ip6mhba_len - (caddr_t)ip6);
2908                    return (EINVAL);
2909            }
```
_____ mip6_mncore.c

2897–2909 If the received size of the message is smaller than the size of the
ip6_mh_binding_ack{} structure, an ICMPv6 Parameter Problem message is sent
to indicate that the length field of the Binding Acknowledgment message is incorrect.
The code value of the ICMPv6 message is set to ICMP6_PARAMPROB_HEADER, which
means there is an error in the header part, and the problem pointer is set to the length
field of the Binding Acknowledgment message.

Listing 15-83
_____ mip6_mncore.c

```
2911    #ifdef M_DECRYPTED      /* not openbsd */
2912            if (((m->m_flags & M_DECRYPTED) != 0)
2913                || ((m->m_flags & M_AUTHIPHDR) != 0)) {
2914                    ba_safe = 1;
2915            }
2916    #endif
....
2923            if ((error = mip6_get_mobility_options((struct ip6_mh *)ip6ma,
2924                    sizeof(*ip6ma), ip6malen, &mopt))) {
2925                    m_freem(m);
....
2927                    return (error);
2928            }
```
_____ mip6_mncore.c

2912–2915 The M_DECRYPTED flag of an mbuf means that the packet is protected by ESP. The
M_AUTHIPHDR flag is set if the packet is protected by Authentication Header (AH) or
ESP. If the input packet has the M_DECRYPTED or M_AUTHIPHDR flag, then the ba_safe
variable is set to true to indicate that the packet is protected by IPsec.

2923–2928 The mip6_get_mobility_options() function parses the options contained in
the received Mobility Header message and stores the option information in the instance
of the mip6_options{} structure specified as the mopt variable.

Listing 15-84
_____ mip6_mncore.c

```
2949            sc = hif_list_find_withhaddr(&ip6->ip6_dst);
2950            if (sc == NULL) {
2951                    /*
2952                     * if we receive a binding ack before sending binding
2953                     * updates(!), sc will be NULL.
2954                     */
....
2958                    /* silently ignore. */
```

```
2959                        m_freem(m);
2960                        return (EINVAL);
2961              }
```

2949–2961 The `hif_list_find_withhaddr()` function returns a pointer to the virtual home network interface related to the home address specified by the function parameter. If there is no related virtual home network interface that relates to the destination address (that is, the home address of the mobile node) of the Binding Acknowledgment message, the packet is dropped.

Listing 15-85

```
2962              mbu = mip6_bu_list_find_withpaddr(&sc->hif_bu_list, &ip6->ip6_src,
2963                      &ip6->ip6_dst);
2964              if (mbu == NULL) {
....
2968                      /* silently ignore */
2969                      m_freem(m);
....
2971                      return (EINVAL);
2972              }
```

2962–2972 The `mip6_bu_list_find_withpaddr()` function returns a pointer to the binding update list entry identified by its function parameters. If there is no binding update list entry related to the received Binding Acknowledgment message, the packet is dropped.

Listing 15-86

```
2974              if (mopt.valid_options & MOPT_AUTHDATA) {
2975                      /* Check Authenticator */
2976                      u_int8_t key_bm[MIP6_KBM_LEN];
2977                      u_int8_t authdata[MIP6_AUTHENTICATOR_LEN];
2978                      u_int16_t cksum_backup;
2979                      int ignore_co_nonce;
2980                      ignore_co_nonce = IN6_ARE_ADDR_EQUAL(&mbu->mbu_haddr,
2981                          &mbu->mbu_coa);
```

2974–2979 A Binding Acknowledgment message from a correspondent node must have the Binding Authorization Data option. If the received Binding Acknowledgment message has the option, the message must be verified using the option value.

2980–2981 The care-of nonce value is not used if the mobile node is at home. The `ignore_co_nonce` variable is set to true if the home address and the care-of address of the mobile node are the same, which means it is at home.

Listing 15-87

```
2983              cksum_backup = ip6ma->ip6mhba_cksum;
2984              ip6ma->ip6mhba_cksum = 0;
2985              /* Calculate Kbm */
```

```
2986                        mip6_calculate_kbm(&mbu->mbu_home_token,
2987                           ignore_co_nonce ? NULL : &mbu->mbu_careof_token, key_bm);
2988                        /* Calculate Authenticator */
2989                        if (mip6_calculate_authenticator(key_bm, authdata,
2990                            &mbu->mbu_coa, &ip6->ip6_dst,
2991                            (caddr_t)ip6ma, ip6malen,
2992                            (caddr_t)mopt.mopt_auth + 2 - (caddr_t)ip6ma,
2993                            min(MOPT_AUTH_LEN(&mopt) + 2,
      MIP6_AUTHENTICATOR_LEN)) == 0) {
2994                                ip6ma->ip6mhba_cksum = cksum_backup;
2995                                if (bcmp(authdata, mopt.mopt_auth + 2,
2996                                       min(MOPT_AUTH_LEN(&mopt) + 2,
      MIP6_AUTHENTICATOR_LEN))
2997                                          == 0)
2998                                        goto accept_binding_ack;
2999                        }
3000                }
```
——— mip6_mncore.c

Note: Lines 2993 and 2996 are broken here for layout reasons. However, they are each a single line of code.

2983–2984 The checksum value of the Mobility Header must be cleared before verifying the header with the Binding Authorization Data option, since the field is assumed to be 0 when computing the checksum on the sender side.

2986–2987 The `mip6_calculate_kbm()` function computes the shared key between the correspondent node and the mobile node using the token values exchanged by the return routability procedure.

2989–2999 The `mip6_calculate_authenticator()` function computes the value of the Binding Authorization Data option using the key computed by the `mip6_calculate_kbm()` function. The result is stored to the `authdata` variable. If the computed value and the value stored in the received Binding Authorization Data option are the same, the message is accepted as a valid message.

Listing 15-88
——— mip6_mncore.c

```
3002            if (!mip6ctl_use_ipsec && (mbu->mbu_flags & IP6MU_HOME)) {
3003                    ba_safe = 1;
3004                    goto accept_binding_ack;
3005            }
3006
3007            if (mip6ctl_use_ipsec
3008                && (mbu->mbu_flags & IP6MU_HOME) != 0
3009                && ba_safe == 1)
3010                    goto accept_binding_ack;
3011
3012            if ((mbu->mbu_flags & IP6MU_HOME) == 0) {
3013                    goto accept_binding_ack;
3014            }
3015
3016            /* otherwise, discard this packet. */
3017            m_freem(m);
3018            mip6stat.mip6s_haopolicy++; /* XXX */
3019            return (EINVAL);
```
——— mip6_mncore.c

3002–3004 The `mip6ctl_use_ipsec` variable is a tunable variable. If the variable is set to false, Binding Acknowledgment messages for home registration will be accepted even if they are not protected by IPsec.

3007–3010 If the `mip6ctl_use_ipsec` variable is set to true and the packet is protected by IPsec (that is, the `ba_safe` variable is set to true), the Binding Acknowledgment message for home registration is accepted.

3012–3014 There is a bug in this part of the code. If there is no Binding Authorization Data option in the incoming Binding Acknowledgment message from the correspondent node, then the message will be incorrectly accepted. However, as discussed, Binding Acknowledgment messages from correspondent nodes must have the Binding Authorization Data option and must be processed on lines 2974–3000. All messages reached at this point must be dropped.

Fortunately, the bug usually does not cause any serious problems since Binding Acknowledgment messages from correspondent nodes have the option as long as the implementation of the peer node follows the specification.

Listing 15-89
```
                                                              mip6_mncore.c
3021    accept_binding_ack:
3022
3023            seqno = htons(ip6ma->ip6mhba_seqno);
3024            if (ip6ma->ip6mhba_status == IP6_MH_BAS_SEQNO_BAD) {
3025                    /*
3026                     * our home agent has a greater sequence number in its
3027                     * binding cache entry of mine.  we should resent
3028                     * binding update with greater than the sequence
3029                     * number of the binding cache already exists in our
3030                     * home agent.  this binding ack is valid though the
3031                     * sequence number doesn't match.
3032                     */
3033                    goto check_mobility_options;
3034            }
3035
3036            if (seqno != mbu->mbu_seqno) {
....
3044                    /* silently ignore. */
3045                    /* discard */
3046                    m_freem(m);
....
3048                    return (EINVAL);
3049            }
                                                              mip6_mncore.c
```

3023–3034 If the sequence number sent from the mobile node was out of date, the peer node will reply with a Binding Acknowledgment message with the `IP6_MH_BAS_SEQNO_BAD` status code. In this case, the mobile node resends a Binding Update message later.

3036–3049 If the received sequence number does not match the sequence number sent before, the Binding Acknowledgment message is dropped.

Listing 15-90
```
                                                              mip6_mncore.c
3051    check_mobility_options:
....
3060            if (ip6ma->ip6mhba_status >= IP6_MH_BAS_ERRORBASE) {
....
3065                    if (ip6ma->ip6mhba_status == IP6_MH_BAS_NOT_HA &&
3066                        mbu->mbu_flags & IP6MU_HOME &&
3067                        mbu->mbu_pri_fsm_state == MIP6_BU_PRI_FSM_STATE_WAITA) {
```

```
3068                          /* XXX no registration? */
3069                          goto success;
3070                  }
```
—— mip6_mncore.c

3065–3070 When the home agent of the mobile node does not have a binding cache
entry for the mobile node and receives a Binding Update message for deregistration, the home agent will reply with a Binding Acknowledgment message with status
IP6_MH_BAS_NOT_HA. This case looks like an error case, however, it occurs even
in normal cases when a Binding Acknowledgment message from the home agent to
the mobile node is lost. The home agent removes the related binding cache entry of
the mobile node after sending the Binding Acknowledgment message. If the mobile
node does not receive the message because the message is lost, it will resend a Binding Update message. The KAME implementation treats the Binding Acknowledgment
message with the IP6_MH_BAS_NOT_HA status as successful when it is at home.

Listing 15-91
—— mip6_mncore.c
```
3071                  if (ip6ma->ip6mhba_status == IP6_MH_BAS_SEQNO_BAD) {
3072                          /* seqno is too small.  adjust it and resend. */
3073                          mbu->mbu_seqno = ntohs(ip6ma->ip6mhba_seqno) + 1;
3074                          /* XXX */
3075                          mip6_bu_send_bu(mbu);
3076                          return (0);
3077                  }
3078
3079                  /* sending binding update failed. */
3080                  error = mip6_bu_list_remove(&sc->hif_bu_list, mbu);
3081                  if (error) {
....
3085                          m_freem(m);
3086                          return (error);
3087                  }
3088                  /* XXX some error recovery process needed. */
3089                  return (0);
3090          }
```
—— mip6_mncore.c

3071–3077 If the recorded sequence number in the binding cache entry of the peer node
is larger than that of the Binding Update message, the peer node will send a Binding
Acknowledgment message with the IP6_MH_BAS_SEQNO_BAD status code. In this case,
the mobile node resends a Binding Update message with the latest sequence number
returned with the Binding Acknowledgment message.

3080–3089 If resending a Binding Update message fails, then the binding update list entry
related to the message is removed.

Listing 15-92
—— mip6_mncore.c
```
3092   success:
3093          /*
3094           * the binding update has been accepted.
3095           */
3096
```

```
3097                /* update lifetime and refresh time. */
3098                lifetime = htons(ip6ma->ip6mhba_lifetime) << 2; /* units of 4 secs */
3099                if (lifetime < mbu->mbu_lifetime) {
3100                        mbu->mbu_expire -= (mbu->mbu_lifetime - lifetime);
3101                        if (mbu->mbu_expire < time_second)
3102                                mbu->mbu_expire = time_second;
3103                }
```
── mip6_mncore.c

3098–3103 Lifetime information is extracted from the received message. If the received lifetime is smaller than that of the related binding update list entry, the expiration time is reduced based on the new lifetime. If the expiration time is past, the expiration time is set to the current time. The entry will be removed in the timer function at the next timeout.

Listing 15-93
── mip6_mncore.c

```
3104                /* binding refresh advice option */
3105                if (mbu->mbu_flags & IP6MU_HOME) {
3106                        if (mopt.valid_options & MOPT_REFRESH) {
3107                                refresh = mopt.mopt_refresh << 2;
3108                                if (refresh > lifetime || refresh == 0) {
3109                                        /*
3110                                         * use default refresh interval for an
3111                                         * invalid binding refresh interval
3112                                         * option.
3113                                         */
3114                                        refresh =
3115                                            MIP6_BU_DEFAULT_REFRESH_INTERVAL(lifetime);
3116                                }
3117                        } else {
3118                                /*
3119                                 * set refresh interval even when a home agent
3120                                 * doesn't specify refresh interval, so that a
3121                                 * mobile node can re-register its binding
3122                                 * before the binding update entry expires.
3123                                 */
3124                                /*
3125                                 * XXX: the calculation algorithm of a default
3126                                 * value must be discussed.
3127                                 */
3128                                refresh = MIP6_BU_DEFAULT_REFRESH_INTERVAL(lifetime);
3129                        }
3130                } else
3131                        refresh = lifetime;
3132                mbu->mbu_refresh = refresh;
```
── mip6_mncore.c

3105–3116 If the binding update list entry is for home registration and the received Binding Acknowledgment message includes a Binding Refresh Advice option, the suggested refresh time is extracted from the option. If the refresh time is greater than the lifetime of the entry or the refresh time is set to 0, the value is ignored. In this case, the refresh interval is calculated based on the lifetime using the `MIP6_BU_DEFAULT_REFRESH_INTERVAL()` macro.

3131 If the binding update list entry is for a correspondent node, the refresh interval is set to the same value as the lifetime, which means the entry is not refreshed for correspondent nodes. The entry will be removed when the lifetime expires.

Listing 15-94

_____ mip6_mncore.c

```
3134              if (mbu->mbu_refresh > mbu->mbu_expire)
3135                      mbu->mbu_refresh = mbu->mbu_expire;
3136
3137              if (ip6ma->ip6mhba_status == IP6_MH_BAS_PRFX_DISCOV) {
3138                      if (mip6_icmp6_mp_sol_output(&mbu->mbu_haddr,
3139                          &mbu->mbu_paddr)) {
....
3144                              /* proceed anyway... */
3145                      }
3146              }
```

_____ mip6_mncore.c

3134–3135 The code is intended to make sure that the refresh time is set to the time before the entry expires; however, there is a bug in the code. The mbu_refresh variable indicates the number of seconds but it does not indicate the time. The mbu_refresh variable must be mbu_retrans in this case.

3137–3146 If the home prefix information has changed, the home agent replies with a Binding Acknowledgment message with the status code IP6_MH_BAS_PRFX_DISCOV, which indicates that the mobile node needs to perform the prefix discovery procedure. In this case, the mobile node sends a Mobile Prefix Solicitation message by calling the mip6_icmp6_mp_sol_output() function.

Listing 15-95

_____ mip6_mncore.c

```
3148              if (mbu->mbu_flags & IP6MU_HOME) {
3149                      /* this is from our home agent. */
3150                      if (mbu->mbu_pri_fsm_state == MIP6_BU_PRI_FSM_STATE_WAITD) {
3151                              struct sockaddr_in6 coa_sa;
3152                              struct sockaddr_in6 daddr; /* XXX */
3153                              struct sockaddr_in6 lladdr;
3154                              struct ifaddr *ifa;
```

_____ mip6_mncore.c

3148–3150 The mobile node will proceed to the home deregistration procedure if the binding update list entry is in the MIP6_BU_PRI_FSM_STATE_WAITD state. The state indicates that the mobile node is waiting for a Binding Acknowledgment message in response to the home deregistration message it sent.

Listing 15-96

_____ mip6_mncore.c

```
3156                              /*
3157                               * home unregistration has completed.
3158                               * send an unsolicited neighbor advertisement.
3159                               */
3160                              bzero(&coa_sa, sizeof(coa_sa));
3161                              coa_sa.sin6_len = sizeof(coa_sa);
3162                              coa_sa.sin6_family = AF_INET6;
3163                              coa_sa.sin6_addr = mbu->mbu_coa;
3164                              /* XXX scope? how? */
3165                              if ((ifa = ifa_ifwithaddr((struct sockaddr *)&coa_sa))
3166                                  == NULL) {
....
3170                                      m_freem(m);
3171                                      return (EINVAL);          /* XXX */
3172                              }
```

_____ mip6_mncore.c

3160–3172 The mobile node sends an unsolicited Neighbor Advertisement message to update neighbor cache entries of the nodes on the home network. An instance of the `sockaddr_in6{}` structure is created, which contains the care-of address of the mobile node. The `ifa_ifwithaddr()` function will return a pointer to the instance of the `in6_ifaddr{}` structure, which is assigned to the node. If the care-of address is not assigned to the mobile node, the procedure is aborted. The pointer is used to recover the scope identifier of the all nodes link-local multicast address in the following procedure.

Listing 15-97

_____ *mip6_mncore.c*

```
3174                    bzero(&daddr, sizeof(daddr));
3175                    daddr.sin6_family = AF_INET6;
3176                    daddr.sin6_len = sizeof(daddr);
3177                    daddr.sin6_addr = in6addr_linklocal_allnodes;
3178                    if (in6_addr2zoneid(ifa->ifa_ifp, &daddr.sin6_addr,
3179                            &daddr.sin6_scope_id)) {
3180                            /* XXX: should not happen */
....
3184                            m_freem(m);
3185                            return (EIO);
3186                    }
3187                    if ((error = in6_embedscope(&daddr.sin6_addr,
3188                            &daddr))) {
3189                            /* XXX: should not happen */
....
3193                            m_freem(m);
3194                            return (error);
3195                    }
3196
3197                    nd6_na_output(ifa->ifa_ifp, &daddr.sin6_addr,
3198                        &mbu->mbu_haddr, ND_NA_FLAG_OVERRIDE, 1, NULL);
```
_____ *mip6_mncore.c*

3174–3195 The `sockaddr_in6{}` instance `daddr`, which contains all nodes' link-local multicast address, is prepared. The scope identifier is recovered from the interface to which the care-of address of the mobile node is assigned.

3197–3198 A Neighbor Advertisement message is sent to the home network. This message overrides all neighbor cache entries stored on the nodes on the home network. While the mobile node is away from home, the neighbor cache entry points to the home agent of the mobile node, since the home agent needs to receive all packets sent to the mobile node to intercept packets. When the mobile node returns home, it needs to update the cache information on all the nodes of the home network.

Listing 15-98

_____ *mip6_mncore.c*

```
3204            /*
3205             * if the binding update entry has the L flag on,
3206             * send unsolicited neighbor advertisement to my
3207             * link-local address.
3208             */
3209            if (mbu->mbu_flags & IP6MU_LINK) {
3210                    bzero(&lladdr, sizeof(lladdr));
3211                    lladdr.sin6_len = sizeof(lladdr);
3212                    lladdr.sin6_family = AF_INET6;
3213                    lladdr.sin6_addr.s6_addr16[0]
```

```
3214                              = IPV6_ADDR_INT16_ULL;
3215                          lladdr.sin6_addr.s6_addr32[2]
3216                              = mbu->mbu_haddr.s6_addr32[2];
3217                          lladdr.sin6_addr.s6_addr32[3]
3218                              = mbu->mbu_haddr.s6_addr32[3];
3219
3220                          if (in6_addr2zoneid(ifa->ifa_ifp,
3221                                  &lladdr.sin6_addr,
3222                                  &lladdr.sin6_scope_id)) {
3223                              /* XXX: should not happen */
....
3227                              m_freem(m);
3228                              return (EIO);
3229                          }
3230                          if ((error = in6_embedscope(&lladdr.sin6_addr,
3231                                  &lladdr))) {
3232                              /* XXX: should not happen */
....
3236                              m_freem(m);
3237                              return (error);
3238                          }
3239
3240                          nd6_na_output(ifa->ifa_ifp, &daddr.sin6_addr,
3241                              &lladdr.sin6_addr, ND_NA_FLAG_OVERRIDE, 1,
3242                              NULL);
....
3248                      }
```
_____ mip6_mncore.c

3209–3242 If the binding update list entry has the L flag (the IP6MU_LINK flag) set, the home
agent of the mobile node will also receive all packets destined to the link-local address of
the mobile node. Sending the Neighbor Advertisement message for the link-local address
of the mobile node will update all the link-local neighbor cache entries stored in the nodes
on the home network.

Listing 15-99
_____ mip6_mncore.c
```
3250                  /* notify all the CNs that we are home. */
3251                  error = mip6_bu_list_notify_binding_change(sc, 1);
3252                  if (error) {
....
3256                      m_freem(m);
3257                      return (error);
3258                  }
```
_____ mip6_mncore.c

3251–3258 The mip6_bu_list_notify_binding_change() function notifies all the
binding update list entries kept in the virtual home network specified as the first parameter to send Binding Update messages to the peer nodes of each of the entries. The
ip6_bu_list_notify_binding_change() function is detailed in Listing 15-104.

Listing 15-100
_____ mip6_mncore.c
```
3260                  /* remove a tunnel to our homeagent. */
3261                  error = mip6_tunnel_control(MIP6_TUNNEL_DELETE,
3262                                              mbu,
3263                                              mip6_bu_encapcheck,
3264                                              &mbu->mbu_encap);
3265                  if (error) {
```

```
         ....
3269                              m_freem(m);
3270                              return (error);
3271                          }
```
_____mip6_mncore.c

3260–3271 The bidirectional tunnel between the mobile node and the home agent is
shut down by the `mip6_tunnel_control()` function.

Listing 15-101
_____mip6_mncore.c
```
3272                    error = mip6_bu_list_remove_all(&sc->hif_bu_list, 0);
3273                    if (error) {
         ....
3277                              m_freem(m);
3278                              return (error);
3279                    }
3280                    mbu = NULL; /* free in mip6_bu_list_remove_all() */
```
_____mip6_mncore.c

3272–3280 All the binding update list entries managed by the mobile node are removed after
the home deregistration procedure has completed.

Listing 15-102
_____mip6_mncore.c
```
3282                 } else if ((mbu->mbu_pri_fsm_state
3283                        == MIP6_BU_PRI_FSM_STATE_WAITA)
3284                        || (mbu->mbu_pri_fsm_state
3285                        == MIP6_BU_PRI_FSM_STATE_WAITAR)) {
         ....
3292                    /* home registration completed */
3293                    error = mip6_bu_fsm
    (mbu, MIP6_BU_PRI_FSM_EVENT_BA, NULL);
3294                    /* create tunnel to HA */
3295                    error = mip6_tunnel_control(MIP6_TUNNEL_CHANGE,
3296                                              mbu,
3297                                              mip6_bu_encapcheck,
3298                                              &mbu->mbu_encap);
3299                    if (error) {
         ....
3303                              m_freem(m);
3304                              return (error);
3305                    }
3306
3307                    /* notify all the CNs that we have a new coa. */
3308                    error = mip6_bu_list_notify_binding_change(sc, 0);
3309                    if (error) {
         ....
3313                              m_freem(m);
3314                              return (error);
3315                    }
```
_____mip6_mncore.c

Note: Line 3293 is broken here for layout reasons. However, it is a single line of code.

3282–3285 If the state of the binding update list entry is either the
`MIP6_BU_PRI_FSM_STATE_WAITA` or the `MIP6_BU_PRI_FSM_STATE_WAITAR`, the
mobile node is waiting for a Binding Acknowledgment message for home registration.

3293 The rest of the Binding Acknowledgment message processing is implemented in the state
machine. The event `MIP6_BU_PRI_FSM_EVENT_BA`, which indicates that the mobile

node receives a Binding Acknowledgment message, is sent to the state machine. The state machine is discussed in Section 15.19.

3295–3305 The `mip6_tunnel_control()` function is called to create a bidirectional tunnel between the mobile node and its home agent.

3308–3315 After the successful home registration procedure, the mobile node sends Binding Update messages to all correspondent nodes for which it has binding information by calling `mip6_bu_list_notify_binding_change()` function.

Listing 15-103

```
                                                                    mip6_mncore.c
3316                    } else if (MIP6_IS_BU_BOUND_STATE(mbu)) {
3317                            /* nothing to do. */
3318                    } else {
....
3322                    }
3323            }
3324
3325            return (0);
3326    }
                                                                    mip6_mncore.c
```

3316–3322 The mobile node does not need to do anything if the state of the binding update list entry is in other states. There is no processing code for a Binding Acknowledgment message from a correspondent node since the KAME implementation never sends a Binding Update message to correspondent nodes with the A (`IP6MU_ACK`) flag set.

Listing 15-104

```
                                                                    mip6_mncore.c
1918    static int
1919    mip6_bu_list_notify_binding_change(sc, home)
1920            struct hif_softc *sc;
1921            int home;
1922    {
1923            struct in6_addr hif_coa;
1924            struct mip6_prefix *mpfx;
1925            struct mip6_bu *mbu, *mbu_next;
1926            int32_t coa_lifetime;
                                                                    mip6_mncore.c
```

1918–1921 The `mip6_bu_list_notify_binding_change()` function has two parameters. The `sc` parameter is a pointer to the virtual home network interface on which the registration procedure has been done and the home parameter is set to true when a mobile node returns to home, otherwise it is set to false.

Listing 15-105

```
                                                                    mip6_mncore.c
1931            /* get current CoA and recover its scope information. */
1932            if (sc->hif_coa_ifa == NULL) {
....
1936                    return (0);
1937            }
1938            hif_coa = sc->hif_coa_ifa->ia_addr.sin6_addr;
                                                                    mip6_mncore.c
```

1931–1938 If the mobile node does not have a valid care-of address, the function returns immediately. If there is a care-of address, the address is written to the `hif_coa` variable.

Listing 15-106

```
1940                 /* for each BU entry, update COA and make them about to send. */
1941                 for (mbu = LIST_FIRST(&sc->hif_bu_list);
1942                     mbu;
1943                     mbu = mbu_next) {
1944                         mbu_next = LIST_NEXT(mbu, mbu_entry);
1945
1946                         if (mbu->mbu_flags & IP6MU_HOME) {
1947                                 /* this is a BU for our home agent */
....
1952                                 continue;
1953                         }
```

1941–1953 All binding update list entries except home registration entries will be processed in this loop.

Listing 15-107

```
1954                         if (IN6_ARE_ADDR_EQUAL(&mbu->mbu_coa, &hif_coa)) {
1955                                 /* XXX no need */
1956                                 continue;
1957                         }
```

1954–1957 If the care-of address of the binding update list entry for a correspondent node is already updated to the latest care-of address, the entry is skipped.

Listing 15-108

```
1958                         mbu->mbu_coa = hif_coa;
1959                         coa_lifetime = mip6_coa_get_lifetime(&mbu->mbu_coa);
1960                         mpfx = mip6_prefix_list_find_withhaddr(&mip6_prefix_list,
1961                             &mbu->mbu_haddr);
1962                         if (mpfx == NULL) {
....
1967                                 mip6_bu_list_remove(&sc->hif_bu_list, mbu);
1968                                 continue;
1969                         }
1970                         if (coa_lifetime < mpfx->mpfx_vltime) {
1971                                 mbu->mbu_lifetime = coa_lifetime;
1972                         } else {
1973                                 mbu->mbu_lifetime = mpfx->mpfx_vltime;
1974                         }
1975                         if (mip6ctl_bu_maxlifetime > 0 &&
1976                             mbu->mbu_lifetime > mip6ctl_bu_maxlifetime)
1977                                 mbu->mbu_lifetime = mip6ctl_bu_maxlifetime;
1978                         mbu->mbu_expire = time_second + mbu->mbu_lifetime;
1979                         /* sanity check for overflow */
1980                         if (mbu->mbu_expire < time_second)
1981                                 mbu->mbu_expire = 0x7fffffff;
1982                         mbu->mbu_refresh = mbu->mbu_lifetime;
```

1958–1982 The care-of address and the lifetime of the binding update list entry are updated. The lifetime of the entry is determined based on the prefix information. The lifetime of the entry must be set to the smaller of either the care-of address or the home address lifetimes. The lifetime is limited to the upper limit defined as the `mip6ctl_bu_maxlifetime` variable, which indicates the maximum lifetime for entries for correspondent nodes. Note that the limit based on the `mip6ctl_bu_maxlifetime` variable is a local policy of the KAME implementation.

The expiration time and refresh time are also updated based on the lifetime.

If there is no related prefix information, the binding update list entry is removed.

Listing 15-109

```
                                                                   mip6_mncore.c
1983                     if (mip6_bu_fsm(mbu,
1984                          (home ?
1985                            MIP6_BU_PRI_FSM_EVENT_RETURNING_HOME :
1986                            MIP6_BU_PRI_FSM_EVENT_MOVEMENT), NULL) != 0) {
....
1991                     }
1992             }
1993
1994         return (0);
1995     }
                                                                   mip6_mncore.c
```

1983–1995 The rest of the procedure is done in the state machine. If the mobile node is at home, the `MIP6_BU_PRI_EVENT_RETURNING_HOME` event is sent to the state machine. If the mobile node is away from home, the `MIP6_BU_PRI_FSM_EVENT_MOVEMENT` event is sent. A more detailed discussion appears in Section 15.19.

15.7 Receiving a Type 2 Routing Header

A mobile node receives a Type 2 Routing Header when it is communicating using route optimization. The Type 2 Routing Header is processed in the `ip6_rthdr2()` function.

Listing 15-110

```
                                                                        route6.c
309     static int
310     ip6_rthdr2(m, ip6, rh2)
311             struct mbuf *m;
312             struct ip6_hdr *ip6;
313             struct ip6_rthdr2 *rh2;
314     {
315             int rh2_has_hoa;
316             struct sockaddr_in6 next_sa;
317             struct hif_softc *sc;
318             struct mip6_bu *mbu;
319             struct in6_addr *nextaddr, tmpaddr;
320             struct in6_ifaddr *ifa;
                                                                        route6.c
```

309–313 The `ip6_rthdr2()` is called from the `route6_input()` function when the type value is 2. The m parameter is a pointer to the mbuf that contains the incoming packet. The `ip6` and `rh2` parameters are pointers to the IPv6 header and the Type 2 Routing Header of the incoming packet, respectively.

Listing 15-111

```
                                                              route6.c
322              rh2_has_hoa = 0;
323
324              /*
325               * determine the scope zone of the next hop, based on the interface
326               * of the current hop.
327               * [draft-ietf-ipngwg-scoping-arch, Section 9]
328               */
329              if ((ifa = ip6_getdstifaddr(m)) == NULL)
330                      goto bad;
331              bzero(&next_sa, sizeof(next_sa));
332              next_sa.sin6_len = sizeof(next_sa);
333              next_sa.sin6_family = AF_INET6;
334              bcopy((const void *)(rh2 + 1), &next_sa.sin6_addr,
335                      sizeof(struct in6_addr));
336              nextaddr = (struct in6_addr *)(rh2 + 1);
337              if (in6_addr2zoneid(ifa->ia_ifp,
338                                  &next_sa.sin6_addr,
339                                  &next_sa.sin6_scope_id)) {
340                      /* should not happen. */
....
342                      goto bad;
343              }
344              if (in6_embedscope(&next_sa.sin6_addr, &next_sa)) {
345                      /* XXX: should not happen */
....
347                      goto bad;
348              }
                                                              route6.c
```

331–348 The address that is listed in a Routing Header does not have a scope identifier. The mobile node has to recover the information based on the interface on which the packet arrived. The `ip6_getdstifaddr()` function returns a pointer to the `in6_ifaddr{}` instance, which indicates the address of the interface from which the packet specified as the parameter was received.

Listing 15-112

```
                                                              route6.c
350              /* check addresses in ip6_dst and rh2. */
351              for (sc = LIST_FIRST(&hif_softc_list); sc;
352                  sc = LIST_NEXT(sc, hif_entry)) {
353                      for (mbu = LIST_FIRST(&sc->hif_bu_list); mbu;
354                          mbu = LIST_NEXT(mbu, mbu_entry)) {
355                              if ((mbu->mbu_flags & IP6MU_HOME) == 0)
356                                      continue;
                                                              route6.c
```

351–356 The address in the Type 2 Routing Header must be a home address or a care-of address of a mobile node. To validate this condition, all binding update list entries kept as home registration entries will be checked to find all the registered home and care-of addresses.

Listing 15-113

```
                                                              route6.c
360                              if (rh2->ip6r2_segleft == 0) {
361                                      struct m_tag *mtag;
362                                      struct ip6aux *ip6a;
363
364                                      /*
```

```
365                              * if segleft == 0, ip6_dst must be
366                              * one of our home addresses.
367                              */
368                             if (!IN6_ARE_ADDR_EQUAL(&ip6->ip6_dst,
369                                 &mbu->mbu_haddr))
370                                 continue;
....
374                             /*
375                              * if the previous hop is the coa that
376                              * is corresponding to the hoa in
377                              * ip6_dst, the route is optimized
378                              * already.
379                              */
380                             if (!IN6_ARE_ADDR_EQUAL(&next_sa.sin6_addr,
381                                 &mbu->mbu_coa)) {
382                                 /* coa mismatch.  discard this. */
383                                 goto bad;
384                             }
```
——— route6.c

360–384 If the segment left field, `ip6r2_segleft`, is 0, that means the header has already been processed. In this case, the destination address of the IPv6 header must be the home address and the address inside the Routing Header must be the care-of address of the mobile node. If the incoming packet does not satisfy these conditions, the packet is dropped.

Listing 15-114

——— route6.c
```
386                                 /*
387                                  * the route is already optimized.
388                                  * set optimized flag in m_aux.
389                                  */
390                                 mtag = ip6_findaux(m);
391                                 if (mtag) {
392                                         ip6a = (struct ip6aux *)(mtag + 1);
393                                         ip6a->ip6a_flags
394                                             |= IP6A_ROUTEOPTIMIZED;
395                                         return (0);
396                                 }
397                                 /* if n == 0 return error. */
398                                 goto bad;
```
——— route6.c

390–395 At this point, it is confirmed that the incoming packet is a valid route-optimized packet. The `IP6A_ROUTEOPTIMIZED` flag is added to the auxiliary mbuf to indicate the packet is route optimized.

Listing 15-115

——— route6.c
```
399                         } else {
400                                 /*
401                                  * if segleft == 1, the specified
402                                  * intermediate node must be one of
403                                  * our home addresses.
404                                  */
405                                 if (!IN6_ARE_ADDR_EQUAL(&next_sa.sin6_addr,
406                                     &mbu->mbu_haddr))
407                                         continue;
408                                 rh2_has_hoa++;
```

```
409                               }
410                         }
411                   }
412                   if (rh2_has_hoa == 0) {
413                         /*
414                          * this rh2 includes an address that is not one of our
415                          * home addresses.
416                          */
417                         goto bad;
418                   }
```
——— route6.c

399–408 If the segment left field is 1, the header has not been processed yet. In this case, the address in the Routing Header must be the home address of the mobile node.

412–417 If the address in the Routing Header is not the home address, the incoming packet is dropped.

Listing 15-116
——— route6.c

```
420                   rh2->ip6r2_segleft--;
421
422                   /*
423                    * reject invalid addresses.  be proactive about malicious use of
424                    * IPv4 mapped/compat address.
425                    * XXX need more checks?
426                    */
427                   if (IN6_IS_ADDR_MULTICAST(&ip6->ip6_dst) ||
428                       IN6_IS_ADDR_UNSPECIFIED(&ip6->ip6_dst) ||
429                       IN6_IS_ADDR_V4MAPPED(&ip6->ip6_dst) ||
430                       IN6_IS_ADDR_V4COMPAT(&ip6->ip6_dst) ||
431                       IN6_IS_ADDR_LOOPBACK(&ip6->ip6_dst)) {
....
433                         goto bad;
434                   }
435
436                   /*
437                    * Swap the IPv6 destination address and nextaddr. Forward the packet.
438                    */
439                   tmpaddr = *nextaddr;
440                   *nextaddr = ip6->ip6_dst;
441                   in6_clearscope(nextaddr);
442                   ip6->ip6_dst = tmpaddr;
443                   ip6_forward(m, 1);
444
445                   return (-1);                      /* m would be freed in ip6_forward() */
446
447       bad:
448                   m_freem(m);
449                   return (-1);
450       }
```
——— route6.c

427–434 The destination address of the incoming IPv6 packet should be the care-of address of the mobile node. A multicast address, the unspecified address, an IPv4-mapped IPv6 address, an IPv4-compatible IPv6 address, and the loopback address cannot be a care-of address.

439–443 The address in the Type 2 Routing Header and the destination address of the IPv6 header are swapped. The packet will be processed as a forwarded packet and is passed to the `ip6_forward()` function.

15.8   Receiving a Binding Refresh Request Message

A mobile node may receive a Binding Refresh Request message from a correspondent node when the lifetime of the binding between the mobile node and the correspondent node becomes small.

Listing 15-117

_____ mip6_mncore.c

```
3328    int
3329    mip6_ip6mr_input(m, ip6mr, ip6mrlen)
3330            struct mbuf *m;
3331            struct ip6_mh_binding_request *ip6mr;
3332            int ip6mrlen;
3333    {
3334            struct ip6_hdr *ip6;
3335            struct hif_softc *sc;
3336            struct mip6_bu *mbu;
3337            int error;
```

_____ mip6_mncore.c

3328–3332 The `mip6_ip6mr_input()` function is called from the `mobility6_input()` function when a mobile node receives a Binding Refresh Request message. The m parameter is a pointer to the mbuf that contains the incoming packet, and the `ip6mr` and `ip6mrlen` parameters are a pointer to the head of the Binding Refresh Request message and its length.

Listing 15-118

_____ mip6_mncore.c

```
3341            ip6 = mtod(m, struct ip6_hdr *);
3342
3343            /* packet length check. */
3344            if (ip6mrlen < sizeof (struct ip6_mh_binding_request)) {
....
3351                    /* send ICMP parameter problem. */
3352                    icmp6_error(m, ICMP6_PARAM_PROB, ICMP6_PARAMPROB_HEADER,
3353                        (caddr_t)&ip6mr->ip6mhbr_len - (caddr_t)ip6);
3354                    return(EINVAL);
3355            }
```

_____ mip6_mncore.c

3343–3355 If the length of the incoming Binding Refresh Request message is smaller than the size of the `ip6_mh_binding_request{}` structure, the mobile node replies with an ICMPv6 Parameter Problem message. The problem pointer is set to point to the length field of the received Binding Refresh Request message.

Listing 15-119

_____ mip6_mncore.c

```
3357            /* find hif corresponding to the home address. */
3358            sc = hif_list_find_withhaddr(&ip6->ip6_dst);
3359            if (sc == NULL) {
3360                    /* we have no such home address. */
....
3362                    goto bad;
3363            }
3364
3365            /* find a corresponding binding update entry. */
3366            mbu = mip6_bu_list_find_withpaddr(&sc->hif_bu_list, &ip6->ip6_src,
3367                &ip6->ip6_dst);
```

```
3368                    if (mbu == NULL) {
3369                            /* we have no binding update entry for dst_sa. */
3370                            return (0);
3371                    }
```
——— mip6_mncore.c

3358–3371 When a mobile node receives a Binding Refresh Request message, it needs to send a Binding Update message to extend the lifetime of the binding information. If the mobile node does not have a binding update list entry related to the incoming Binding Refresh Request message, no action is required.

Listing 15-120
——— mip6_mncore.c

```
3373                    error = mip6_bu_fsm(mbu, MIP6_BU_PRI_FSM_EVENT_BRR, ip6mr);
3374                    if (error) {
....
3378                            goto bad;
3379                    }
3380
3381            return (0);
3382      bad:
3383            m_freem(m);
3384            return (EINVAL);
3385        }
```
——— mip6_mncore.c

3373–3379 If the mobile node has a binding update list entry related to the incoming Binding Refresh Request message, it sends the `MIP6_BU_PRI_FSM_EVENT_BRR` event to its state machine, which will eventually call the sending function for a Binding Update message.

15.9 Receiving a Binding Error Message

A mobile node receives a Binding Error message in the following cases:

- When a mobile node sends a packet when there is no binding information between the mobile node and its peer node
- When a mobile node sends a new (unknown) Mobility Header type to its peer

The former case occurs when a Binding Update message from the mobile node to a correspondent node is lost, but the mobile node sends route-optimized packets. Since the KAME implementation does not require a Binding Acknowledgment message from a correspondent node, the mobile node may send packets with the Home Address option even if there is no binding cache on the peer node. A packet that contains the Home Address option without existing binding information is treated as an error.

The latter case does not occur in the current basic specification since there are no unknown Mobility Header types defined. This error message will be used to detect the future extension of Mobility Header types.

Listing 15-121
——— mip6_mncore.c

```
3387    int
3388    mip6_ip6me_input(m, ip6me, ip6melen)
```

```
3389              struct mbuf *m;
3390              struct ip6_mh_binding_error *ip6me;
3391              int ip6melen;
3392          {
3393              struct ip6_hdr *ip6;
3394              struct sockaddr_in6 hoa;
3395              struct hif_softc *sc;
3396              struct mip6_bu *mbu;
3397              int error = 0;
```
_____ mip6_mncore.c

3387–3391 The `mip6_ip6me_input()` function is called from the `mobility6_input()` function when a mobile node receives a Binding Error message. The m parameter is a pointer to the mbuf that contains the incoming packet and the `ip6me` and `ip6melen` parameters are a pointer to the head of the Binding Error message and its length.

Listing 15-122
_____ mip6_mncore.c
```
3401              ip6 = mtod(m, struct ip6_hdr *);
3402
3403              /* packet length check. */
3404              if (ip6melen < sizeof (struct ip6_mh_binding_error)) {
....
3411                      /* send ICMP parameter problem. */
3412                      icmp6_error(m, ICMP6_PARAM_PROB, ICMP6_PARAMPROB_HEADER,
3413                          (caddr_t)&ip6me->ip6mhbe_len - (caddr_t)ip6);
3414                      return(EINVAL);
3415              }
```
_____ mip6_mncore.c

3404–3415 If the length of the incoming Binding Error message is smaller than the size of the `ip6_mh_binding_error{}` structure, the mobile node replies with an ICMPv6 Parameter Problem message. The problem pointer is set to point to the length field of the incoming Binding Error message.

Listing 15-123
_____ mip6_mncore.c
```
3417              /* extract the home address of the sending node. */
3418              bzero (&hoa, sizeof (hoa));
3419              hoa.sin6_len = sizeof (hoa);
3420              hoa.sin6_family = AF_INET6;
3421              bcopy(&ip6me->ip6mhbe_homeaddr, &hoa.sin6_addr,
3422                  sizeof(struct in6_addr));
3423              if (in6_addr2zoneid(m->m_pkthdr.rcvif, &hoa.sin6_addr,
3424                      &hoa.sin6_scope_id)) {
....
3426                      goto bad;
3427              }
3428              if (in6_embedscope(&hoa.sin6_addr, &hoa)) {
....
3430                      goto bad;
3431              }
```
_____ mip6_mncore.c

3418–3431 The home address of the mobile node may be stored in the Binding Error message if the original packet that caused the error used the home address. The home address is copied to the hoa variable. The scope identifier of the home address is recovered from the interface on which the Binding Error message arrived.

Listing 15-124

_____ mip6_mncore.c

```
3433              /* find hif corresponding to the home address. */
3434              sc = hif_list_find_withhaddr(&hoa.sin6_addr);
3435              if (sc == NULL) {
3436                      /* we have no such home address. */
....
3438                      goto bad;
3439              }
```
_____ mip6_mncore.c

3434–3439 If the mobile node does not have a virtual home interface related to the home address indicated in the incoming Binding Error message, it ignores the error message.

Listing 15-125

_____ mip6_mncore.c

```
3443              switch (ip6me->ip6mhbe_status) {
3444              case IP6_MH_BES_UNKNOWN_HAO:
3445              case IP6_MH_BES_UNKNOWN_MH:
3446                      mbu = mip6_bu_list_find_withpaddr(&sc->hif_bu_list,
3447                          &ip6->ip6_src, &hoa.sin6_addr);
3448                      if (mbu == NULL) {
3449                              /* we have no binding update entry for the CN. */
3450                              goto bad;
3451                      }
3452                      break;
3453
3454              default:
....
3460                      goto bad;
3461                      break;
3462              }
```
_____ mip6_mncore.c

3443–3462 If the error status is either IP6_MH_BES_UNKNOWN_HAO or IP6_MH_BES_UNKNOWN_MH, the mobile node calls the mip6_bu_list_find_withpaddr() function to find the binding update list entry related to the Binding Error message. If the mobile node does not have a related entry, the error message is ignored. Currently, no other error statuses are defined. A message with an unknown error status code is ignored.

Listing 15-126

_____ mip6_mncore.c

```
3464              switch (ip6me->ip6mhbe_status) {
3465              case IP6_MH_BES_UNKNOWN_HAO:
3466                      /* the CN doesn't have a binding cache entry.  start RR. */
3467                      error = mip6_bu_fsm(mbu,
3468                          MIP6_BU_PRI_FSM_EVENT_UNVERIFIED_HAO, ip6me);
3469                      if (error) {
....
3473                              goto bad;
3474                      }
3475
3476                      break;
3477
3478              case IP6_MH_BES_UNKNOWN_MH:
3479                      /* XXX future extension? */
3480                      error = mip6_bu_fsm(mbu,
3481                          MIP6_BU_PRI_FSM_EVENT_UNKNOWN_MH_TYPE, ip6me);
```

```
3482                    if (error) {
....
3486                            goto bad;
3487                    }
3488
3489                    break;
3490
3491            default:
....
3499            }
3500
3501            return (0);
3502
3503     bad:
3504            m_freem(m);
3505            return (EINVAL);
3506      }
```
── mip6_mncore.c

3464–3489 Based on the error status of the Binding Error message, an event is sent to the state machine of the related binding update list entry. If the error status is the IP6_MH_BES_UNKNOWN_HAO, the MIP6_BU_PRI_FSM_EVENT_UNVERIFIED_HAO is sent. If the error status is the IP6_MH_BES_UNKNOWN_MH, the MIP6_BU_PRI_FSM_EVENT_UNKNOWN_MH_TYPE is sent. The error processing and recovery will be done in the state machine. State machines will be discussed in Section 15.19.

15.10   Source Address Selection

A mobile node should prefer to use its home address when sending packets. The source address selection mechanism is modified to meet this recommendation. Source address selection is done in the in6_selectsrc() function. The following code fragments used in this section are quoted from the in6_selectsrc() function. The function is discussed in Section 3.13.1 of *IPv6 Core Protocols Implementation*.

Listing 15-127
── in6_src.c

```
288             /*
289              * a caller can specify IP6PO_USECOA to not to use a home
290              * address.  for example, the case that the neighbor
291              * unreachability detection to the global address.
292              */
293             if (opts != NULL &&
294                 (opts->ip6po_flags & IP6PO_USECOA) != 0) {
295                     usecoa = 1;
296             }
297             /*
298              * a user can specify destination addresses or destination
299              * ports for which he don't want to use a home address when
300              * sending packets.
301              */
302             for (uh = LIST_FIRST(&mip6_unuse_hoa);
303                     uh;
304                     uh = LIST_NEXT(uh, unuse_entry)) {
305                     if ((IN6_IS_ADDR_UNSPECIFIED(&uh->unuse_addr) ||
306                         IN6_ARE_ADDR_EQUAL(dst, &uh->unuse_addr)) &&
307                         (!uh->unuse_port || dstsock->sin6_port == uh->
    unuse_port)) {
308                             usecoa = 1;
```

```
309                                    break;
310                            }
311                    }
```
── in6_src.c

Note: Line 307 is broken here for layout reasons. However, it is a single line of code.

293–311 In some cases, a mobile node may prefer using the care-of address instead of the home address of the node. For example, there is no need to use a home address for very short-lived communications such as domain name server (DNS) queries. The Care-of Test Init message and the Neighbor Solicitation message for the Neighbor Unreachability Detection on a foreign network are other cases that must not use a home address as a source address.

The KAME implementation provides two ways to select a care-of address in the in6_selectsrc() function. One is the IP6PO_USECOA flag. If the packet being sent has the flag in its auxiliary mbuf, the home address will not be chosen as a source address. The other method is using the address and port filter. The global variable mip6_unuse_hoa contains a list of destination addresses and/or port numbers. Packets sent to these addresses or ports should not use a home address.

If the outgoing packet has the IP6PO_USECOA flag, or the destination address or the port number is listed in the mip6_unuse_hoa list, the usecoa variable is set to true.

[RFC3484] defines rule 4 that specifies how a mobile node chooses the source address of packets it sends. Assuming there are two candidates of a source address, SA and SB, the basic rules are as follows:

- If SA is a home address and a care-of address at the same time (that means, the node is at home), and SB is not, prefer SA

- If SB is a home address and a care-of address at the same time (that means, the node is at home), and SA is not, prefer SB

- If SA is just a home address and SB is just a care-of address, prefer SA

- If SB is just a home address and SA is just a care-of address, prefer SB

Lines 384–525 implement the rules.

Listing 15-128

── in6_src.c

```
     ....
384                    /* Rule 4: Prefer home addresses */
     ....
390            {
391                    struct mip6_bu *mbu_ia_best = NULL, *mbu_ia = NULL;
392                    struct sockaddr_in6 ia_addr;
393
394                    /*
395                     * If SA is simultaneously a home address and
396                     * care-of address and SB is not, then prefer
397                     * SA. Similarly, if SB is simultaneously a
398                     * home address and care-of address and SA is
399                     * not, then prefer SB.
400                     */
401                    if (ia_best->ia6_flags & IN6_IFF_HOME) {
402                            /*
403                             * find a binding update entry for ia_best.
```

```
404                                   */
405                                   ia_addr = ia_best->ia_addr;
406                                   if(in6_addr2zoneid(ia_best->ia_ifp,
407                                                   &ia_addr.sin6_addr,
408                                                   &ia_addr.sin6_scope_id)) {
409                                           *errorp = EINVAL; /* XXX */
410                                           return (NULL);
411                                   }
412                                   for (sc = LIST_FIRST(&hif_softc_list); sc;
413                                       sc = LIST_NEXT(sc, hif_entry)) {
414                                           mbu_ia_best =
         mip6_bu_list_find_home_registration(
415                                                   &sc->hif_bu_list,
416                                                   &ia_addr.sin6_addr);
417                                           if (mbu_ia_best)
418                                                   break;
419                                   }
420                           }
```
_____ in6_src.c

Note: Line 414 is broken here for layout reasons. However, it is a single line of code.

401–420 In the `in6_selectsrc()` function, the `ia_best` and the `ia` variables point to
candidate source addresses. The `ia_best` variable is the first candidate and the `ia` variable is the second candidate address. The address structure pointed to by each of these
variables has the `IN6_IFF_HOME` flag set, if the address is a home address.

   If the first candidate address has the flag set and the mobile node has a corresponding
home registration entry, the `mbu_ia_best` variable is set to point to the binding update
list entry.

Listing 15-129

_____ in6_src.c
```
421                           if (ia->ia6_flags & IN6_IFF_HOME) {
422                                   /*
423                                    * find a binding update entry for ia.
424                                    */
425                                   ia_addr = ia->ia_addr;
426                                   if(in6_addr2zoneid(ia->ia_ifp,
427                                                   &ia_addr.sin6_addr,
428                                                   &ia_addr.sin6_scope_id)) {
429                                           *errorp = EINVAL; /* XXX */
430                                           return (NULL);
431                                   }
432                                   for (sc = LIST_FIRST(&hif_softc_list); sc;
433                                       sc = LIST_NEXT(sc, hif_entry)) {
434                                           mbu_ia =
         mip6_bu_list_find_home_registration(
435                                                   &sc->hif_bu_list,
436                                                   &ia_addr.sin6_addr);
437                                           if (mbu_ia)
438                                                   break;
439                                   }
440                           }
```
_____ in6_src.c

Note: Line 434 is broken here for layout reasons. However, it is a single line of code.

421–440 If another candidate address is a home address and has a corresponding home
registration entry, the `mbu_ia` variable points to the entry.

Listing 15-130

```
442                          /*
443                           * even if the address is a home address, we
444                           * do not use them if they are not registered
445                           * (or re-registered) yet.  this condition is
446                           * not explicitly stated in the address
447                           * selection draft.
448                           */
449                          if ((mbu_ia_best &&
450                              (mbu_ia_best->mbu_pri_fsm_state
451                               != MIP6_BU_PRI_FSM_STATE_BOUND))) {
452                                  /* XXX will break stat! */
453                                  REPLACE(0);
454                          }
455                          if ((mbu_ia &&
456                              (mbu_ia->mbu_pri_fsm_state
457                               != MIP6_BU_PRI_FSM_STATE_BOUND))) {
458                                  /* XXX will break stat! */
459                                  NEXT(0);
460                          }
```

449–460 Even if the candidate address is a home address and has the home registration entry related to the home address, the address is not chosen as a source address if the address has not been registered successfully with its home agent. The home address is valid only when it is successfully registered (that is, the state of the entry must be the MIP6_BU_PRI_FSM_STATE_BOUND state). The REPLACE() macro sets the second candidate as the first candidate. The NEXT() macro discards the second candidate and keeps the first candidate. The comparison will continue until all candidate addresses have been checked.

Listing 15-131

```
462                          /*
463                           * if the binding update entry for a certain
464                           * address exists and its registration status
465                           * is MIP6_BU_FSM_STATE_IDLE, the address is a
466                           * home address and a care of address
467                           * simultaneously.
468                           */
469                          if ((mbu_ia_best &&
470                              (mbu_ia_best->mbu_pri_fsm_state
471                               == MIP6_BU_PRI_FSM_STATE_IDLE))
472                              &&
473                              !(mbu_ia &&
474                              (mbu_ia->mbu_pri_fsm_state
475                               == MIP6_BU_PRI_FSM_STATE_IDLE))) {
476                                  NEXT(4);
477                          }
478                          if (!(mbu_ia_best &&
479                              (mbu_ia_best->mbu_pri_fsm_state
480                               == MIP6_BU_PRI_FSM_STATE_IDLE))
481                              &&
482                              (mbu_ia &&
483                              (mbu_ia->mbu_pri_fsm_state
484                               == MIP6_BU_PRI_FSM_STATE_IDLE))) {
485                                  REPLACE(4);
486                          }
```

469–486 This part is actually never executed. The intention of this code is to prefer the address that is a home address and a care-of address simultaneously. The code assumes that the status of the home registration entry becomes the idle state (`MIP6_BU_PRI_FSM_STATE_IDLE`) when a mobile node is at home; however, since the KAME implementation removes a home registration entry when a mobile node returns to home, such a condition never occurs.

Listing 15-132

_____ in6_src.c
```
487                         if (usecoa != 0) {
488                                 /*
489                                  * a sender don't want to use a home
490                                  * address because:
491                                  *
492                                  * 1) we cannot use.   (ex. NS or NA to
493                                  * global addresses.)
494                                  *
495                                  * 2) a user specified not to use.
496                                  * (ex. mip6control -u)
497                                  */
498                                 if ((ia_best->ia6_flags & IN6_IFF_HOME) == 0 &&
499                                     (ia->ia6_flags & IN6_IFF_HOME) != 0) {
500                                         /* XXX will break stat! */
501                                         NEXT(0);
502                                 }
503                                 if ((ia_best->ia6_flags & IN6_IFF_HOME) != 0 &&
504                                     (ia->ia6_flags & IN6_IFF_HOME) == 0) {
505                                         /* XXX will break stat! */
506                                         REPLACE(0);
507                                 }
```
_____ in6_src.c

498–507 If the `usecoa` variable is set to true, a care-of address is preferred. The address that does not have the `IN6_IFF_HOME` flag is chosen as a first candidate. If both addresses do not have the flag, the function will check other conditions to determine a more preferable address.

Listing 15-133

_____ in6_src.c
```
508                         } else {
509                                 /*
510                                  * If SA is just a home address and SB
511                                  * is just a care-of address, then
512                                  * prefer SA. Similarly, if SB is just
513                                  * a home address and SA is just a
514                                  * care-of address, then prefer SB.
515                                  */
516                                 if ((ia_best->ia6_flags & IN6_IFF_HOME) != 0 &&
517                                     (ia->ia6_flags & IN6_IFF_HOME) == 0) {
518                                         NEXT(4);
519                                 }
520                                 if ((ia_best->ia6_flags & IN6_IFF_HOME) == 0 &&
521                                     (ia->ia6_flags & IN6_IFF_HOME) != 0) {
522                                         REPLACE(4);
523                                 }
524                         }
525                 }
```
_____ in6_src.c

508–524 Otherwise, a home address is preferred as a source address. The address that has the IN6_IFF_HOME flag is chosen as a first priority candidate. If both addresses have the flag, the function will check other conditions to determine a more preferable address.

15.11 Home Agent List Management

A mobile node needs to keep the list of its home agents to determine the address to which to send binding information when the node moves to a foreign network.

In the KAME implementation, a mobile node collects the information by two methods. One is listening to Router Advertisement messages while the mobile node is at home before moving to other networks. Since a home agent is an IPv6 router by definition (it forwards packets from/to mobile nodes), it sends Router Advertisement messages periodically as an IPv6 router. The Router Advertisement messages sent from home agents are slightly modified as described in Section 3.6 of Chapter 3. The mobile node can easily distinguish messages of home agents from messages sent from other normal routers.

The other method is using the Dynamic Home Agent Address Discovery mechanism. The Dynamic Home Agent Address Discovery mechanism is used when a mobile node needs to learn the home agent information while in a foreign network. As discussed in Section 14.19 of Chapter 14, the home agent list management is done by a user space program, **had**, on the home agent side; however, on the mobile node side, the home agent list management is done in the kernel.

Home agent information is represented as the mip6_ha{} structure. To manage the list of home agents, the KAME implementation provides the following support functions:

- mip6_ha_create() — create an instance of the mip6_ha{} structure
- mip6_ha_update_lifetime() — update the lifetime information of the specified mip6_ha{} instance
- mip6_ha_list_insert() — insert the specified mip6_ha{} instance to the list
- mip6_ha_list_reinsert() — relocate the specified mip6_ha{} instance based on the preference in the list
- mip6_ha_list_remove() — remove the specified mip6_ha{} instance from the list
- mip6_ha_list_update_hainfo() — update the preference and lifetime of the specified mip6_ha{} instance in the list
- mip6_ha_list_find_withaddr() — find an instance of the mip6_ha{} structure that has the specified address
- mip6_ha_settimer() — set the next timeout of the mip6_ha{} instance
- mip6_ha_timer() — called periodically for each mip6_ha{} instance

15.11.1 Create a Home Agent Entry

The mip6_ha_create() function creates an instance of the mip6_ha{} structure, which represents a home agent in a mobile node.

Listing 15-134

_____ mip6_halist.c

```
94      struct mip6_ha *
95      mip6_ha_create(addr, flags, pref, lifetime)
96              struct in6_addr *addr;
97              u_int8_t flags;
98              u_int16_t pref;
99              int32_t lifetime;
100     {
101             struct mip6_ha *mha = NULL;
....
103             struct timeval mono_time;
....
107             microtime(&mono_time);
```
_____ mip6_halist.c

94–99 The `addr` parameter is the address of the home agent and the `flag` parameter is a copy of the flag value of the Router Advertisement message received by the mobile node. The `pref` and `lifetime` parameters are a preference value and a lifetime value of the home agent.

Listing 15-135

_____ mip6_halist.c

```
110             if (IN6_IS_ADDR_UNSPECIFIED(addr)
111                 || IN6_IS_ADDR_LOOPBACK(addr)
112                 || IN6_IS_ADDR_MULTICAST(addr)) {
....
116                     return (NULL);
117             }
118
119             if (!IN6_IS_ADDR_LINKLOCAL(addr)
120                 && ((flags & ND_RA_FLAG_HOME_AGENT) == 0)) {
....
125                     return (NULL);
126             }
```
_____ mip6_halist.c

110–117 The address of the home agent must not be an unspecified address, the loopback address, or a multicast address.

119–126 The `mip6_ha{}` structure keeps all router information received by the mobile node. If the router is a home agent, the `flags` variable will include the `ND_RA_FLAG_HOME_AGENT` flag. The address of the home agent must be a global address. If the router is just a router and not a home agent, then the address can be a link-local address. This information is used when the mobile node performs movement detection (see Section 15.3).

Listing 15-136

_____ mip6_halist.c

```
128             MALLOC(mha, struct mip6_ha *, sizeof(struct mip6_ha), M_TEMP,
129                 M_NOWAIT);
130             if (mha == NULL) {
....
134                     return (NULL);
135             }
136             bzero(mha, sizeof(*mha));
137             mha->mha_addr = *addr;
138             mha->mha_flags = flags;
139             mha->mha_pref = pref;
```

```
      ....
143              callout_init(&mha->mha_timer_ch);
      ....
147              if (IN6_IS_ADDR_LINKLOCAL(&mha->mha_addr)) {
148                      mha->mha_lifetime = lifetime;
149              } else {
150                      mha->mha_lifetime = 0; /* infinite. */
151              }
152              mip6_ha_update_lifetime(mha, lifetime);
153
154              return (mha);
155      }
```
_____ mip6_halist.c

128–152 Memory for the `mip6_ha{}` instance is allocated and its member variables are set.
The `mha_timer_ch` field is a handle to the timer function of this instance. The lifetime
initialization code on lines 147–151 is actually meaningless. The lifetime information is
updated by the `mip6_ha_update_lifetime()` function called on line 152.

15.11.2   Update the Home Agent Entry

The `mip6_ha_update_lifetime()` function updates the lifetime information of the
specified home agent entry.

Listing 15-137

_____ mip6_halist.c

```
157      void
158      mip6_ha_update_lifetime(mha, lifetime)
159              struct mip6_ha *mha;
160              u_int16_t lifetime;
161      {
      ....
163              struct timeval mono_time;
      ....
167              microtime(&mono_time);
```
_____ mip6_halist.c

157–160 The `mip6_ha_update_lifetime()` function has two parameters. The mha parameter is a pointer to the `mip6_ha{}` instance of which the lifetime is being updated and
the `lifetime` parameter is the new lifetime value.

Listing 15-138

_____ mip6_halist.c

```
170              mip6_ha_settimer(mha, -1);
171              mha->mha_lifetime = lifetime;
172              if (mha->mha_lifetime != 0) {
173                      mha->mha_expire = mono_time.tv_sec + mha->mha_lifetime;
174                      mip6_ha_settimer(mha, mha->mha_lifetime * hz);
175              } else {
176                      mha->mha_expire = 0;
177              }
178      }
```
_____ mip6_halist.c

170–177 The callback timer is reset on line 170 and the new lifetime of the `mip6_ha{}` instance
is set. If the lifetime is specified as 0, the entry is treated as an infinite entry. Otherwise,
the timer function is set to be called when the lifetime expires.

15.11.3   Insert a Home Agent Entry

The `mip6_ha_list_insert()` function inserts a home agent entry into the proper position of the home agent list based on its preference value.

Listing 15-139

```
                                                                    mip6_halist.c
282     void
283     mip6_ha_list_insert(mha_list, mha)
284             struct mip6_ha_list *mha_list;
285             struct mip6_ha *mha;
286     {
287             struct mip6_ha *tgtmha;
288
289             if ((mha_list == NULL) || (mha == NULL)) {
290                     panic("mip6_ha_list_insert: NULL pointer.");
291             }
292
293             /*
294              * insert a new entry in a proper place ordered by preference
295              * value.  if preference value is same, the new entry is placed
296              * at the end of the group which has a same preference value.
297              */
298             for (tgtmha = TAILQ_FIRST(mha_list); tgtmha;
299                 tgtmha = TAILQ_NEXT(tgtmha, mha_entry)) {
300                     if (tgtmha->mha_pref >= mha->mha_pref)
301                             continue;
302                     TAILQ_INSERT_BEFORE(tgtmha, mha, mha_entry);
303                     return;
304             }
305             TAILQ_INSERT_TAIL(mha_list, mha, mha_entry);
306
307             return;
308     }
                                                                    mip6_halist.c
```

282–285 The `mip6_ha_list_insert()` function has two parameters. The `mha_list` parameter is a pointer to the list of `mip6_ha{}` instances and the `mha` parameter is a pointer to the `mip6_ha{}` instance to be inserted.

298–307 The list of the `mip6_ha{}` instances is ordered by the value of the preference field of each `mip6_ha{}` instance. The preference value of each entry, which is already inserted in the list, is checked and the new entry is inserted at the proper position.

15.11.4   Reinsert the Home Agent Entry

Listing 15-140

```
                                                                    mip6_halist.c
310     void
311     mip6_ha_list_reinsert(mha_list, mha)
312             struct mip6_ha_list *mha_list;
313             struct mip6_ha *mha;
314     {
315             struct mip6_ha *tgtmha;
316
317             if ((mha_list == NULL) || (mha == NULL)) {
318                     panic("mip6_ha_list_insert: NULL pointer.");
319             }
320
321             for (tgtmha = TAILQ_FIRST(mha_list); tgtmha;
```

```
322                         tgtmha = TAILQ_NEXT(tgtmha, mha_entry)) {
323                             if (tgtmha == mha)
324                                     break;
325                         }
326
327                 /* insert or move the entry to the proper place of the queue. */
328                 if (tgtmha != NULL)
329                         TAILQ_REMOVE(mha_list, tgtmha, mha_entry);
330                 mip6_ha_list_insert(mha_list, mha);
331
332                 return;
333         }
```
—— mip6_halist.c

310–333 The `mip6_ha_list_reinsert()` function does almost the same things as the
`mip6_ha_list_insert()` does. The difference between these two functions is
`mip6_ha_list_reinsert()` removes the entry specified by the second parameter,
if the entry already exists in the list, and reinserts it at the proper position. This function
is used to reorder the list of entries when the preference value of the entry is changed.

Listing 15-141
—— mip6_halist.c

```
336     int
337     mip6_ha_list_remove(mha_list, mha)
338             struct mip6_ha_list *mha_list;
339             struct mip6_ha *mha;
340     {
341             struct mip6_prefix *mpfx;
342             struct mip6_prefix_ha *mpfxha, *mpfxha_next;
343
344             if ((mha_list == NULL) || (mha == NULL)) {
345                     return (EINVAL);
346             }
347
348             /* remove all references from mip6_prefix entries. */
349             for (mpfx = LIST_FIRST(&mip6_prefix_list); mpfx;
350                 mpfx = LIST_NEXT(mpfx, mpfx_entry)) {
351                     for (mpfxha = LIST_FIRST(&mpfx->mpfx_ha_list); mpfxha;
352                         mpfxha = mpfxha_next) {
353                             mpfxha_next = LIST_NEXT(mpfxha, mpfxha_entry);
354                             if (mpfxha->mpfxha_mha == mha)
355                                     mip6_prefix_ha_list_remove(&mpfx->mpfx_ha_list,
356                                         mpfxha);
357                     }
358             }
359
360             TAILQ_REMOVE(mha_list, mha, mha_entry);
361             mip6_ha_settimer(mha, -1);
362             FREE(mha, M_TEMP);
363
364             return (0);
365     }
```
—— mip6_halist.c

337–339 The `mip6_ha_list_remove()` function has two parameters. The `mha_list` parameter is a pointer to the list of the `mip6_ha{}` instances and the mha parameter is a pointer
to the `mip6_ha{}` instance to be removed.

349–358 When removing an instance of the `mip6_ha{}` structure, the `mip6_prefix{}`
instance related to the entry has to be removed. The `mip6_prefix{}` structure has
a pointer to the `mip6_ha{}` instance that advertises the prefix information. Before the

entry of the `mip6_ha{}` instance is removed, the structure that keeps the pointer has to be removed. The relationship between these structures is illustrated in Figure 10-1 of Chapter 10.

360–362 The entry is removed from the list and the memory block used for the entry is freed.

15.11.5   Update Home Agent Information

The `mip6_ha_list_update_hainfo()` function updates the information of an instance of the `mip6_ha{}` structure based on the information received with Router Advertisement messages.

Listing 15-142

```
                                                                     mip6_halist.c
367     int
368     mip6_ha_list_update_hainfo(mha_list, dr, hai)
369            struct mip6_ha_list *mha_list;
370            struct nd_defrouter *dr;
371            struct nd_opt_homeagent_info *hai;
372     {
373            int16_t pref = 0;
374            u_int16_t lifetime;
375            struct mip6_ha *mha;
                                                                     mip6_halist.c
```

367–371 The `mip6_ha_list_update_hainfo()` function has three parameters: the `mha_list` parameter is a pointer to the list of `mip6_ha{}` instances; the `dr` parameter is a pointer to information about the router that sent the Router Advertisement message; and the `hai` parameter is a pointer to the Home Agent Information option that is included in Router Advertisement messages sent from home agents.

Listing 15-143

```
                                                                     mip6_halist.c
377            if ((mha_list == NULL) ||
378                (dr == NULL) ||
379                !IN6_IS_ADDR_LINKLOCAL(&dr->rtaddr)) {
380                    return (EINVAL);
381            }
382
383            lifetime = dr->rtlifetime;
384            if (hai) {
385                    pref = ntohs(hai->nd_opt_hai_preference);
386                    lifetime = ntohs(hai->nd_opt_hai_lifetime);
387            }
                                                                     mip6_halist.c
```

379 The source address of the Router Advertisement message must be a link-local address according to the specification of Neighbor Discovery.

383–386 The lifetime of the home agent is specified in the Home Agent Information option. If the option does not exist, then the router lifetime specified in the Router Advertisement message is used as the lifetime of the home agent.

Listing 15-144

_____ mip6_halist.c

```
389              /* find an existing entry. */
390              mha = mip6_ha_list_find_withaddr(mha_list, &dr->rtaddr);
391              if (mha == NULL) {
392                      /* an entry must exist at this point. */
393                      return (EINVAL);
394              }
395
396              /*
397               * if received lifetime is 0, delete the entry.
398               * otherwise, update an entry.
399               */
400              if (lifetime == 0) {
401                      mip6_ha_list_remove(mha_list, mha);
402              } else {
403                      /* reset lifetime */
404                      mip6_ha_update_lifetime(mha, lifetime);
405              }
406
407              return (0);
408      }
```
_____ mip6_halist.c

389–405 The `mip6_ha_list_find_withaddr()` function returns an existing entry of the `mip6_ha{}` structure. If there is an existing entry, its lifetime information is updated. The lifetime value 0 means that the router is going to stop functioning as a home agent. In that case, the entry will be removed. In this function, the preference value is not processed at all. This is a bug and the function should consider updating the preference value and reordering the list.

15.11.6   Find the Home Agent Entry

The `mip6_ha_list_find_withaddr()` function returns a pointer to the home agent entry that has the specified address as its function parameter.

Listing 15-145

_____ mip6_halist.c

```
410      struct mip6_ha *
411      mip6_ha_list_find_withaddr(mha_list, addr)
412              struct mip6_ha_list *mha_list;
413              struct in6_addr *addr;
414      {
415              struct mip6_ha *mha;
416
417              for (mha = TAILQ_FIRST(mha_list); mha;
418                  mha = TAILQ_NEXT(mha, mha_entry)) {
419                      if (IN6_ARE_ADDR_EQUAL(&mha->mha_addr, addr))
420                              return (mha);
421              }
422              /* not found. */
423              return (NULL);
424      }
```
_____ mip6_halist.c

410–413 The `mip6_ha_list_find_withaddr()` function has two parameters. The `mha_list` parameter is a pointer to the list of the `mip6_ha{}` instances to be searched and the `addr` parameter is the address of the `mip6_ha{}` instances being searched.

417–423 All entries stored in the list are checked in the `for` loop and the pointer to the entry of which the address is the same as that specified as the `addr` parameter is returned, if it exists.

15.11.7   Set Next Timeout of the Home Agent Entry

The `mip6_ha_settimer()` function sets the next timeout.

Listing 15-146
```
                                                                       mip6_halist.c
180     static void
181     mip6_ha_settimer(mha, tick)
182             struct mip6_ha *mha;
183             long tick;
184     {
....
186             struct timeval mono_time;
....
188             int s;
                                                                       mip6_halist.c
```

180–183 The `mha` parameter is a pointer to the `mip6_ha{}` instance and the `tick` parameter is the time until the callback function is called next.

Listing 15-147
```
                                                                       mip6_halist.c
191             microtime(&mono_time);
....
197             s = splnet();
....
200             if (tick < 0) {
201                     mha->mha_timeout = 0;
202                     mha->mha_ntick = 0;
....
204                     callout_stop(&mha->mha_timer_ch);
....
210             } else {
211                     mha->mha_timeout = mono_time.tv_sec + tick / hz;
212                     if (tick > INT_MAX) {
213                             mha->mha_ntick = tick - INT_MAX;
....
215                             callout_reset(&mha->mha_timer_ch, INT_MAX,
216                                 mip6_ha_timer, mha);
....
222                     } else {
223                             mha->mha_ntick = 0;
....
225                             callout_reset(&mha->mha_timer_ch, tick,
226                                 mip6_ha_timer, mha);
....
232                     }
233             }
234
235             splx(s);
236     }
                                                                       mip6_halist.c
```

200–204 If the specified time is a minus value, the timer is stopped.

211–232 The next timeout is set. The timer handle `mha_timer_ch` cannot count periods longer than the maximum integer value can represent. If we need to set a longer timeout, the mha_ntick variable is used to divide the timeout period into several pieces. If the specified time (tick) is longer than the limit of the maximum integer value (INT_MAX), the next timeout is set to the maximum time that an integer variable can represent and mha_ntick is set to the rest of the time. Otherwise, set the next timeout to the specified time as tick.

15.11.8 Periodical Tasks of the Home Agent Entry

The mip6_ha_timer() function is called when the lifetime of the entry has expired.

Listing 15-148

_____ mip6_halist.c

```
238      static void
239      mip6_ha_timer(arg)
240              void *arg;
241      {
242              int s;
243              struct mip6_ha *mha;
....
245              struct timeval mono_time;
....
249              microtime(&mono_time);
....
255              s = splnet();
....
258              mha = (struct mip6_ha *)arg;
```
_____ mip6_halist.c

238–240 The arg parameter is a pointer to the mip6_ha{} instance to which the timer function is called.

Listing 15-149

_____ mip6_halist.c

```
260              if (mha->mha_ntick > 0) {
261                      if (mha->mha_ntick > INT_MAX) {
262                              mha->mha_ntick -= INT_MAX;
263                              mip6_ha_settimer(mha, INT_MAX);
264                      } else {
265                              mha->mha_ntick = 0;
266                              mip6_ha_settimer(mha, mha->mha_ntick);
267                      }
268                      splx(s);
269                      return;
270              }
271
272              /*
273               * XXX reset all home agent addresses in the binding update
274               * entries.
275               */
276
277              mip6_ha_list_remove(&mip6_ha_list, mha);
278
279              splx(s);
280      }
```
_____ mip6_halist.c

260–277 If the mha_ntick variable has a positive value and it is greater than the maximum integer value (INT_MAX), then the value is reduced by INT_MAX and the next callback time is set after `INT_MAX` time; otherwise, the next timeout is set after `mha_ntick` time. In these cases, the entry will not be removed, since the expiration time has not come yet.

If the `mha_ntick` variable is set to 0 and the timer function is called, the `mip6_ha{}` instance is removed from the list. In this case, all home agent addresses currently used in the list of binding update list entries of the mobile node have to be updated. However, the function is not implemented.

This will not cause any serious problems because when the mobile node tries to use the binding update list entry for which the home agent address is no longer valid, the mobile node will perform the Dynamic Home Agent Address Discovery mechanism.

15.12   Prefix Information Management

A mobile node maintains a prefix list as a clue to check its location. Each prefix information entry is bound to the router information entry which advertised the prefix. The virtual home network structure (the `hif_softc{}` structure) has two lists of the prefix information entry. One is used to keep all prefix information entries that are announced at home. The other is used to keep all foreign prefix information entries. The lists do not keep the actual values of the prefixes; they keep pointers to the entries in the global prefix list explained below.

The KAME Mobile IPv6 implementation has one global prefix list. Every structure that requires prefix information keeps pointers to the entries in the list. The following functions are provided to manage the prefix information:

- `mip6_prefix_create()` — creates an instance of a prefix information structure (the `mip6_prefix{}` structure)

- `mip6_prefix_update_lifetime()` — updates the lifetime information of the `mip6_prefix{}` instance

- `mip6_prefix_settimer()` — sets the next timeout of the prefix information

- `mip6_prefix_timer()` — the function that is called when the timer set by the `mip6_prefix_settimer()` function expires

- `mip6_prefix_send_mps()` — sends a Mobile Prefix Solicitation message

- `mip6_prefix_list_insert()` — inserts the specified instance of the `mip6_prefix{}` structure into the list

- `mip6_prefix_list_remove()` — removes the specified instance of the `mip6_prefix{}` structure from the list

- `mip6_prefix_list_find_withprefix()` — searches for an `mip6_prefix{}` instance with prefix information

- `mip6_prefix_list_find_withhaddr()` — searches for an `mip6_prefix{}` instance with home address information derived from the prefix

- `mip6_prefix_ha_list_insert()` — inserts the pointer structure that points to the `mip6_ha{}` instance that includes a router or home agent information

- `mip6_prefix_ha_list_remove()` — removes the pointer structure of the `mip6_ha{}` instance

- `mip6_prefix_ha_list_find_withaddr()` — searches for a pointer structure with a router or a home agent address

- `mip6_prefix_ha_list_find_withmha()` — searches for a pointer structure with the value of the pointer to the `mip6_ha{}` instance

15.12.1 Create a Prefix Entry

The `mip6_prefix_create()` function creates a new prefix information entry for the Mobile IPv6 stack.

Listing 15-150

_____ mip6_prefix.c

```
95      struct mip6_prefix *
96      mip6_prefix_create(prefix, prefixlen, vltime, pltime)
97              struct in6_addr *prefix;
98              u_int8_t prefixlen;
99              u_int32_t vltime;
100             u_int32_t pltime;
101     {
102             struct in6_addr mask;
103             struct mip6_prefix *mpfx;
104
105             MALLOC(mpfx, struct mip6_prefix *, sizeof(struct mip6_prefix),
106                     M_TEMP, M_NOWAIT);
107             if (mpfx == NULL) {
....
111                     return (NULL);
112             }
113             bzero(mpfx, sizeof(*mpfx));
114             in6_prefixlen2mask(&mask, prefixlen);
115             mpfx->mpfx_prefix = *prefix;
116             mpfx->mpfx_prefix.s6_addr32[0] &= mask.s6_addr32[0];
117             mpfx->mpfx_prefix.s6_addr32[1] &= mask.s6_addr32[1];
118             mpfx->mpfx_prefix.s6_addr32[2] &= mask.s6_addr32[2];
119             mpfx->mpfx_prefix.s6_addr32[3] &= mask.s6_addr32[3];
120             mpfx->mpfx_prefixlen = prefixlen;
121             /* XXX mpfx->mpfx_haddr; */
122             LIST_INIT(&mpfx->mpfx_ha_list);
....
126             callout_init(&mpfx->mpfx_timer_ch);
....
130
131             /* set initial timeout. */
132             mip6_prefix_update_lifetime(mpfx, vltime, pltime);
133
134             return (mpfx);
135     }
```

_____ mip6_prefix.c

95–100 The `mip6_prefix_create()` function has four parameters: The `prefix` parameter is a pointer to the instance of the `in6_addr{}` structure that holds prefix information; the `prefixlen`, `vltime`, and `pltime` parameters are the prefix length, the valid lifetime, and the preferred lifetime of the new prefix information, respectively.

105–122 Memory space is allocated for the new `mip6_prefix{}` instance and each member variable is filled. The `mpfx_prefix{}` structure keeps only the prefix information. The interface identifier part will be filled with 0 using the mask value created from the prefix length information. The `mpfx_ha_list` field is a list on which pointer structures are kept. It points to the `mip6_ha{}` instances advertising the prefix. If multiple routers are advertising the same prefix, the list will include all of them.

126–132 The timer handle (`mpfx_timer_ch`) is initialized and the lifetime is updated based on the parameters passed to the function.

15.12.2   Update Lifetime of a Prefix Entry

The `mip6_prefix_update_lifetime()` function updates the information of an existing prefix entry.

Listing 15-151

```
                                                            mip6_prefix.c
138      #define MIP6_PREFIX_EXPIRE_TIME(ltime) ((ltime) / 4 * 3) /* XXX */
139      void
140      mip6_prefix_update_lifetime(mpfx, vltime, pltime)
141              struct mip6_prefix *mpfx;
142              u_int32_t vltime;
143              u_int32_t pltime;
144      {
....
146              struct timeval mono_time;
....
150              microtime(&mono_time);
....
153              if (mpfx == NULL)
154                      panic("mip6_prefix_update_lifetime: mpfx == NULL");
155
156              mip6_prefix_settimer(mpfx, -1);
157
158              mpfx->mpfx_vltime = vltime;
159              mpfx->mpfx_pltime = pltime;
160
161              if (mpfx->mpfx_vltime == ND6_INFINITE_LIFETIME) {
162                      mpfx->mpfx_vlexpire = 0;
163              } else {
164                      mpfx->mpfx_vlexpire = mono_time.tv_sec + mpfx->mpfx_vltime;
165              }
166              if (mpfx->mpfx_pltime == ND6_INFINITE_LIFETIME) {
167                      mpfx->mpfx_plexpire = 0;
168              } else {
169                      mpfx->mpfx_plexpire = mono_time.tv_sec + mpfx->mpfx_pltime;
170              }
171
172              if (mpfx->mpfx_pltime != ND6_INFINITE_LIFETIME) {
173                      mip6_prefix_settimer(mpfx,
174                          MIP6_PREFIX_EXPIRE_TIME(mpfx->mpfx_pltime) * hz);
175                      mpfx->mpfx_state = MIP6_PREFIX_STATE_PREFERRED;
176              } else if (mpfx->mpfx_vltime != ND6_INFINITE_LIFETIME) {
177                      mip6_prefix_settimer(mpfx,
178                          MIP6_PREFIX_EXPIRE_TIME(mpfx->mpfx_vltime) * hz);
179                      mpfx->mpfx_state = MIP6_PREFIX_STATE_PREFERRED;
180              }
181      }
                                                            mip6_prefix.c
```

139–143 The `mip6_prefix_update_lifetime()` function has three parameters. The `mpfx` parameter is a pointer to the instance of the `mip6_prefix{}` structure, and the `vltime` and `pltime` parameters are a valid lifetime and a preferred lifetime, respectively.

161–164 If the specified valid lifetime indicates the infinite lifetime, the `mpfx_vlexpire` variable is set to 0, which represents infinity; otherwise, the expiration time is set to `vltime` seconds after the current time.

166–170 The same procedure is performed on the preferred lifetime.

172–179 The next timeout is set to call the timer function. If the preferred lifetime is not set to infinity, the timeout is set based on the preferred lifetime. If the preferred lifetime is set to infinity and the valid lifetime is not infinity, the timeout is set based on the valid lifetime. If both lifetime values are set to infinity, the prefix never expires. The `MIP6_PREFIX_EXPIRE_TIME()` macro returns a value that is three-fourths of the parameter passed to it. The timeout value is set to three-fourths of the lifetime so that the timer function is called before the lifetime of the prefix expires.

15.12.3 Set Next Timeout of the Prefix Entry

The `mip6_prefix_settimer()` function sets the next timeout of the specified prefix entry.

Listing 15-152

_____ mip6_prefix.c
```
237     void
238     mip6_prefix_settimer(mpfx, tick)
239             struct mip6_prefix *mpfx;
240             long tick;
241     {
....
243             struct timeval mono_time;
....
245             int s;
....
248             microtime(&mono_time);
....
254             s = splnet();
```
_____ mip6_prefix.c

237–240 The `mip6_prefix_settimer()` function has two parameters. The `mpfx` parameter is a pointer to the `mip6_prefix{}` instance for which a timer is being set and the `tick` parameter indicates the timeout value.

Listing 15-153

_____ mip6_prefix.c
```
257             if (tick < 0) {
258                     mpfx->mpfx_timeout = 0;
259                     mpfx->mpfx_ntick = 0;
....
261                     callout_stop(&mpfx->mpfx_timer_ch);
....
267             } else {
268                     mpfx->mpfx_timeout = mono_time.tv_sec + tick / hz;
269                     if (tick > INT_MAX) {
270                             mpfx->mpfx_ntick = tick - INT_MAX;
....
272                             callout_reset(&mpfx->mpfx_timer_ch, INT_MAX,
273                                 mip6_prefix_timer, mpfx);
....
279                     } else {
280                             mpfx->mpfx_ntick = 0;
....
282                             callout_reset(&mpfx->mpfx_timer_ch, tick,
283                                 mip6_prefix_timer, mpfx);
....
289                     }
290             }
```

```
291
292            splx(s);
293     }
```
                                                                 ———————————— mip6_prefix.c

257–261 If `tick` is set to a negative value, the timer is stopped.

268–289 If `tick` is set to a positive value and it is greater than the maximum value of an integer
        variable (`INT_MAX`), `INT_MAX` is set as a timeout value and the rest of the time is set in
        the `mpfx_ntick` variable. If `tick` is smaller than `INT_MAX`, the `tick` value is set as a
        timeout value.

15.12.4   Periodic Tasks of a Prefix Entry

The `mip6_prefix_timer()` function is called when the timer set by the
`mip6_prefix_settimer()` function expires.

Listing 15-154
                                                                 ———————————— mip6_prefix.c
```
295     #define MIP6_MOBILE_PREFIX_SOL_INTERVAL 10 /* XXX */
296     static void
297     mip6_prefix_timer(arg)
298            void *arg;
299     {
300            int s;
301            struct mip6_prefix *mpfx;
....
303            struct timeval mono_time;
....
307            microtime(&mono_time);
....
313            s = splnet();
```
                                                                 ———————————— mip6_prefix.c

296–297 The timer is maintained separately by each `mip6_prefix{}` instance. The parameter
        is a pointer to the instance of the `mip6_prefix{}` structure of which the timer has
        expired.

Listing 15-155
                                                                 ———————————— mip6_prefix.c
```
316            mpfx = (struct mip6_prefix *)arg;
317
318            if (mpfx->mpfx_ntick > 0) {
319                   if (mpfx->mpfx_ntick > INT_MAX) {
320                          mpfx->mpfx_ntick -= INT_MAX;
321                          mip6_prefix_settimer(mpfx, INT_MAX);
322                   } else {
323                          mpfx->mpfx_ntick = 0;
324                          mip6_prefix_settimer(mpfx, mpfx->mpfx_ntick);
325                   }
326                   splx(s);
327                   return;
328            }
```
                                                                 ———————————— mip6_prefix.c

318–328 `mpfx_ntick` is set if the period to the next timeout is greater than the maximum size
        of an integer value (`INT_MAX`). In this case, if `mpfx_ntick` is still greater than `INT_MAX`,
        the next timeout is set to `INT_MAX`; otherwise, the next timeout is set to `mpfx_ntick`.

Listing 15-156

mip6_prefix.c

```
330                 switch (mpfx->mpfx_state) {
331                 case MIP6_PREFIX_STATE_PREFERRED:
332                         if (mip6_prefix_send_mps(mpfx)) {
....
337                         }
338
339                         if (mpfx->mpfx_vlexpire >
340                             mono_time.tv_sec + MIP6_MOBILE_PREFIX_SOL_INTERVAL) {
341                                 mip6_prefix_settimer(mpfx,
342                                     MIP6_MOBILE_PREFIX_SOL_INTERVAL * hz);
343                         } else {
344                                 mip6_prefix_settimer(mpfx,
345                                     (mpfx->mpfx_vlexpire - mono_time.tv_sec) * hz);
346                         }
347                         mpfx->mpfx_state = MIP6_PREFIX_STATE_EXPIRING;
348                         break;
```

mip6_prefix.c

330–348 Each prefix information has a state field. The field can have either of two states (as described in Table 10-15 of Chapter 10). If the state is MIP6_PREFIX_STATE_PREFERRED, a Mobile Prefix Solicitation message is sent by calling the mip6_prefix_send_mps() function to extend the lifetime of the prefix. The next timeout is set to the time after MIP6_MOBILE_PREFIX_SOL_INTERVAL seconds for retransmission. The state is changed to the MIP6_PREFIX_STATE_EXPIRING state.

Listing 15-157

mip6_prefix.c

```
350                 case MIP6_PREFIX_STATE_EXPIRING:
351                         if (mpfx->mpfx_vlexpire < mono_time.tv_sec) {
352                                 mip6_prefix_list_remove(&mip6_prefix_list, mpfx);
353                                 break;
354                         }
355
356                         if (mip6_prefix_send_mps(mpfx)) {
....
361                         }
362
363                         if (mpfx->mpfx_vlexpire >
364                             mono_time.tv_sec + MIP6_MOBILE_PREFIX_SOL_INTERVAL) {
365                                 mip6_prefix_settimer(mpfx,
366                                     MIP6_MOBILE_PREFIX_SOL_INTERVAL * hz);
367                         } else {
368                                 mip6_prefix_settimer(mpfx,
369                                     (mpfx->mpfx_vlexpire - mono_time.tv_sec) * hz);
370                         }
371                         mpfx->mpfx_state = MIP6_PREFIX_STATE_EXPIRING;
372                         break;
373                 }
374
375             splx(s);
376     }
```

mip6_prefix.c

350–354 If the state is MIP6_PREFIX_STATE_EXPIRING and the valid lifetime is expired, the prefix information is removed.

356–372 A Mobile Prefix Solicitation message is sent by the mip6_prefix_send_mps() function. The message is resent every MIP6_MOBILE_PREFIX_SOL_INTERVAL seconds (10 s) until the lifetime of the prefix is expired. If a corresponding advertisement message is received, the retransmission is stopped. The state remains in the `MIP6_PREFIX_STATE_EXPIRING` state.

15.12.5 Requesting Updated Prefix Information

When a mobile node needs to retrieve the latest prefix information of its home network, the `mip6_prefix_send_mps()` function is called to send a Mobile Prefix Solicitation message.

Listing 15-158

```
                                                                         mip6_prefix.c
214     static int
215     mip6_prefix_send_mps(mpfx)
216             struct mip6_prefix *mpfx;
217     {
218             struct hif_softc *hif;
219             struct mip6_bu *mbu;
220             int error = 0;
221
222             for (hif = LIST_FIRST(&hif_softc_list); hif;
223                  hif = LIST_NEXT(hif, hif_entry)) {
224                     if (!IN6_IS_ADDR_UNSPECIFIED(&mpfx->mpfx_haddr)) {
225                             mbu = mip6_bu_list_find_home_registration(
226                                     &hif->hif_bu_list, &mpfx->mpfx_haddr);
227                             if (mbu != NULL) {
228                                     error = mip6_icmp6_mp_sol_output(
229                                         &mbu->mbu_haddr, &mbu->mbu_paddr);
230                                     break;
231                             }
232                     }
233             }
234             return (error);
235     }
                                                                         mip6_prefix.c
```

215–216 The `mip6_prefix_send_mps()` function takes a pointer to the instance of the `mip6_prefix{}` structure.

222–233 A mobile node needs to update the information about a prefix when the lifetime of the prefix is about to expire. The information can be retrieved by exchanging the Mobile Prefix Solicitation and Advertisement messages. The solicitation message must be sent from the home address of the mobile node to the address of its home agent. In this loop, we check to see if the specified `mip6_prefix{}` instance has a valid home address and find the home agent address to which the home address is registered. The `mip6_icmp6_mp_sol_output()` function sends a Mobile Prefix Solicitation message.

15.12.6 Insert Prefix Entry

The `mip6_prefix_list_insert()` function inserts the specified prefix entry into the prefix list.

Listing 15-159

```
                                                                         mip6_prefix.c
378     int
379     mip6_prefix_list_insert(mpfx_list, mpfx)
380             struct mip6_prefix_list *mpfx_list;
381             struct mip6_prefix *mpfx;
382     {
```

```
383                if ((mpfx_list == NULL) || (mpfx == NULL)) {
384                        return (EINVAL);
385                }
386
387                LIST_INSERT_HEAD(mpfx_list, mpfx, mpfx_entry);
388
389                return (0);
390        }
```
_____ mip6_prefix.c

378–381 The `mip6_prefix_list_insert()` function has two parameters. The
`mpfx_list` is a pointer to the list of instances of the `mip6_prefix{}` structure
and the `mpfx` parameter is a pointer to the instance of the `mip6_prefix{}` structure
to be inserted.

387 The function just calls a macro function that manipulates a list structure.

15.12.7   Remove Prefix Entry

The `mip6_prefix_list_remove()` function removes the specified prefix entry from the
prefix list.

Listing 15-160
_____ mip6_prefix.c

```
392        int
393        mip6_prefix_list_remove(mpfx_list, mpfx)
394                struct mip6_prefix_list *mpfx_list;
395                struct mip6_prefix *mpfx;
396        {
397                struct hif_softc *hif;
398                struct mip6_prefix_ha *mpfxha;
399
400                if ((mpfx_list == NULL) || (mpfx == NULL)) {
401                        return (EINVAL);
402                }
403
404                /* remove all references from hif interfaces. */
405                for (hif = LIST_FIRST(&hif_softc_list); hif;
406                    hif = LIST_NEXT(hif, hif_entry)) {
407                        hif_prefix_list_remove(&hif->hif_prefix_list_home,
408                            hif_prefix_list_find_withmpfx(&hif->hif_prefix_list_home,
409                                mpfx));
410                        hif_prefix_list_remove(&hif->hif_prefix_list_foreign,
411                                hif_prefix_list_find_withmpfx
  (&hif->hif_prefix_list_foreign,
412                                mpfx));
413                }
414
415                /* remove all references to advertising routers. */
416                while (!LIST_EMPTY(&mpfx->mpfx_ha_list)) {
417                        mpfxha = LIST_FIRST(&mpfx->mpfx_ha_list);
418                        mip6_prefix_ha_list_remove(&mpfx->mpfx_ha_list, mpfxha);
419                }
420
421                LIST_REMOVE(mpfx, mpfx_entry);
422                mip6_prefix_settimer(mpfx, -1);
423                FREE(mpfx, M_TEMP);
424
425                return (0);
426        }
```
_____ mip6_prefix.c

Note: Line 411 is broken here for layout reasons. However, it is a single line of code.

392–395 The `mip6_prefix_list_remove()` function has two parameters. The `mpfx_list` parameter is a pointer to the list of instances of the `mip6_prefix{}` structure and the `mpfx` parameter is a pointer to one of the elements in the list to be removed.

405–413 All references from the virtual home interface (the `hif_softc{}` structure) to the `mip6_prefix{}` instance specified as a parameter are removed. As described in Figure 10-2 of Chapter 10, the `hif_softc{}` structure may have references to the prefix information.

415–419 All references to the router information (the `mip6_ha{}` structure) are removed. Each prefix information entry has at least one reference to the instance of the `mip6_ha{}` structure as described in Figure 10-1 of Chapter 10.

421–423 The specified entry is removed from the prefix list. The timer handler is reset and the memory space used by the `mip6_prefix{}` instance is released.

15.12.8   Find the Prefix Entry with Prefix Information

The `mip6_prefix_list_find_withprefix()` function returns the prefix entry that has the prefix information specified by its function parameters.

Listing 15-161
———————————————————————————————— mip6_prefix.c

```
428      struct mip6_prefix *
429      mip6_prefix_list_find_withprefix(prefix, prefixlen)
430              struct in6_addr *prefix;
431              int prefixlen;
432      {
433              struct mip6_prefix *mpfx;
434
435              for (mpfx = LIST_FIRST(&mip6_prefix_list); mpfx;
436                  mpfx = LIST_NEXT(mpfx, mpfx_entry)) {
437                      if (in6_are_prefix_equal(prefix, &mpfx->mpfx_prefix, prefixlen)
438                          && (prefixlen == mpfx->mpfx_prefixlen)) {
439                              /* found. */
440                              return (mpfx);
441                      }
442              }
443
444              /* not found. */
445              return (NULL);
446      }
```
———————————————————————————————— mip6_prefix.c

428–431 The `mip6_prefix_list_find_withprefix()` function has two parameters. The `prefix` parameter is a pointer to the `in6_addr{}` structure, which holds the key information used as an index, and the `prefixlen` parameter is the length of the prefix specified as the first parameter.

435–442 All prefix information is stored in the list called `mip6_prefix_list`. The prefix information (the `mpfx_prefix` member variable of the `mip6_prefix{}` structure) and the information passed as parameters are compared and the pointer to the entry that has the same information is returned.

15.12.9   Find the Prefix Entry with Home Address

The `mip6_prefix_list_find_withhaddr()` function returns the prefix entry of which the home address is the same address as the second parameter of the function.

Listing 15-162

mip6_prefix.c

```
448     struct mip6_prefix *
449     mip6_prefix_list_find_withhaddr(mpfx_list, haddr)
450          struct mip6_prefix_list *mpfx_list;
451          struct in6_addr *haddr;
452     {
453             struct mip6_prefix *mpfx;
454
455             for (mpfx = LIST_FIRST(mpfx_list); mpfx;
456                 mpfx = LIST_NEXT(mpfx, mpfx_entry)) {
457                     if (IN6_ARE_ADDR_EQUAL(haddr, &mpfx->mpfx_haddr)) {
458                             /* found. */
459                             return (mpfx);
460                     }
461             }
462
463             /* not found. */
464             return (NULL);
465     }
```

mip6_prefix.c

448–451 The `mip6_prefix_list_find_withhaddr()` function has two parameters. The `mpfx_list` parameter is a pointer to the list of `mip6_prefix{}` instances and the `haddr` parameter is a home address that is used as a key when searching for an `mip6_prefix{}` instance.

455–461 Each `mip6_prefix{}` instance has a home address, which is generated from its prefix information. In this loop, we compare the home address of the `mip6_prefix{}` instance (the `mpfx_haddr` member variable) and the specified parameter. If we find an entry that has the same home address as the `haddr` parameter, a pointer to it is returned.

15.12.10   Insert Home Agent Information to the Prefix Entry

The `mip6_prefix_ha_list_insert()` function inserts the pointer structure that points to the `mip6_ha{}` instance that includes router or home agent information.

Listing 15-163

mip6_prefix.c

```
467     struct mip6_prefix_ha *
468     mip6_prefix_ha_list_insert(mpfxha_list, mha)
469             struct mip6_prefix_ha_list *mpfxha_list;
470             struct mip6_ha *mha;
471     {
472             struct mip6_prefix_ha *mpfxha;
473
474             if ((mpfxha_list == NULL) || (mha == NULL))
475                     return (NULL);
476
477             mpfxha = mip6_prefix_ha_list_find_withmha(mpfxha_list, mha);
478             if (mpfxha != NULL)
```

```
479                       return (mpfxha);
480
481            MALLOC(mpfxha, struct mip6_prefix_ha *, sizeof(struct mip6_prefix_ha),
482                M_TEMP, M_NOWAIT);
483            if (mpfxha == NULL) {
....
486                       return (NULL);
487            }
488            mpfxha->mpfxha_mha = mha;
489            LIST_INSERT_HEAD(mpfxha_list, mpfxha, mpfxha_entry);
490            return (mpfxha);
491    }
```
_____ mip6_prefix.c

467–470 The `mip6_prefix_ha_list_insert()` function has two parameters. The `mpfxha_list` parameter is a pointer to the list of instances of the `mip6_prefix_ha{}` structure. The list is one of the member variables of the `mip6_prefix{}` structure. The `mha` parameter is a pointer to an instance of the `mip6_ha{}` structure. The function inserts the structure with a pointer to the `mip6_prefix{}` instance of which the prefix information is advertised by the router (or the home agent) specified by the `mha` parameter.

477–479 The `mip6_prefix_ha_list_find_withmha()` function is called to find an existing entry of the `mip6_prefix_ha{}` instance, which has the specified prefix information. If the entry exists, the pointer is returned to the caller.

481–490 If there is no existing `mip6_prefix_ha{}` instance advertising the prefix information, memory for the new `mip6_prefix_ha{}` structure is allocated and inserted into the list.

Listing 15-164
_____ mip6_prefix.c

```
493    void
494    mip6_prefix_ha_list_remove(mpfxha_list, mpfxha)
495            struct mip6_prefix_ha_list *mpfxha_list;
496            struct mip6_prefix_ha *mpfxha;
497    {
498            LIST_REMOVE(mpfxha, mpfxha_entry);
499            FREE(mpfxha, M_TEMP);
500    }
```
_____ mip6_prefix.c

493–500 The `mip6_prefix_ha_list_remove()` function has two parameters. The `mpfxha_list` parameter is a pointer to the list of instances of the `mip6_prefix_ha{}` structure, which is kept in the `mip6_prefix{}` structure. The `mpfxha` parameter is a pointer to the instance of the `mip6_prefix_ha{}` structure to be removed. The function removes the specified entry from the list and releases the memory space used by the entry.

15.12.11   Find the Prefix Entry with Home Agent Address

The `mip6_prefix_ha_list_find_withaddr()` function returns the pointer to the instance of the `mip6_prefix_ha{}` structure of which the router address is the same as the specified address.

Listing 15-165

── mip6_prefix.c
```
502     struct mip6_prefix_ha *
503     mip6_prefix_ha_list_find_withaddr(mpfxha_list, addr)
504             struct mip6_prefix_ha_list *mpfxha_list;
505             struct in6_addr *addr;
506     {
507             struct mip6_prefix_ha *mpfxha;
508
509             for (mpfxha = LIST_FIRST(mpfxha_list); mpfxha;
510                 mpfxha = LIST_NEXT(mpfxha, mpfxha_entry)) {
511                     if (mpfxha->mpfxha_mha == NULL)
512                             continue;
513
514                     if (IN6_ARE_ADDR_EQUAL(&mpfxha->mpfxha_mha->mha_addr, addr))
515                             return (mpfxha);
516             }
517             return (NULL);
518     }
```
── mip6_prefix.c

502–505 The `mip6_prefix_ha_list_find_withaddr()` function has two parameters. The `mpfxha_list` parameter is a pointer to the list of instances of the `mip6_prefix_ha{}` structure and the `addr` parameter is a pointer to the address information of the router or the home agent being searched for.

509–517 Each `mip6_prefix_ha{}` instance in the list is compared to the address of the `mip6_ha{}` instance to which the `mip6_prefix_ha{}` instance points. If there is an entry that has the same address information as the address specified as the second parameter of this function, the pointer to that `mip6_prefix_ha{}` instance is returned.

15.12.12   Find the Prefix Entry with Home Agent Information

The `mip6_prefix_ha_list_find_withmha()` function returns the pointer to the instance of the `mip6_prefix_ha{}` structure that has the specified home agent entry information.

Listing 15-166

── mip6_prefix.c
```
520     struct mip6_prefix_ha *
521     mip6_prefix_ha_list_find_withmha(mpfxha_list, mha)
522             struct mip6_prefix_ha_list *mpfxha_list;
523             struct mip6_ha *mha;
524     {
525             struct mip6_prefix_ha *mpfxha;
526
527             for (mpfxha = LIST_FIRST(mpfxha_list); mpfxha;
528                 mpfxha = LIST_NEXT(mpfxha, mpfxha_entry)) {
529                     if (mpfxha->mpfxha_mha && (mpfxha->mpfxha_mha == mha))
530                             return (mpfxha);
531             }
532             return (NULL);
533     }
```
── mip6_prefix.c

520–532 The   `mip6_prefix_ha_list_find_withmha()`   function   finds   an `mip6_prefix_ha{}` entry by using the pointer to an `mip6_ha{}` instance as a key.

15.13 Receiving Prefix Information by Router Advertisement Messages

A mobile node will receive prefix information by two methods: listening to Router Advertisement messages at home and using Mobile Prefix Solicitation and Advertisement messages in foreign networks.

Listing 15-167

```
                                                                    mip6_mncore.c
241     int
242     mip6_prelist_update(saddr, ndopts, dr, m)
243             struct in6_addr *saddr; /* the addr that sent this RA. */
244             union nd_opts *ndopts;
245             struct nd_defrouter *dr; /* NULL in case of a router shutdown. */
246             struct mbuf *m; /* the received router adv. packet. */
247     {
248             struct mip6_ha *mha;
249             struct hif_softc *sc;
250             int error = 0;
                                                                    mip6_mncore.c
```

241–246 The `mip6_prelist_update()` function updates the prefix information that is kept in a mobile node based on the Router Advertisement message that the mobile node receives. This function has four parameters: the `saddr` parameter is an address of the node, which sent the Router Advertisement message; the `ndopts` parameter is a pointer to the `nd_opts{}` structure, which keeps option values included in the incoming message; the `dr` parameter is a pointer to the `nd_defrouter{}` structure, which represents an entry of the default router list in the mobile node; and the `m` parameter is a pointer to the mbuf, which contains the incoming packet.

Listing 15-168

```
                                                                    mip6_mncore.c
252             /* sanity check. */
253             if (saddr == NULL)
254                     return (EINVAL);
255
256             /* advertising router is shutting down. */
257             if (dr == NULL) {
258                     mha = mip6_ha_list_find_withaddr(&mip6_ha_list, saddr);
259                     if (mha) {
260                             error = mip6_ha_list_remove(&mip6_ha_list, mha);
261                     }
262                     return (error);
263             }
264
265             /* if no prefix information is included, we have nothing to do. */
266             if ((ndopts == NULL) || (ndopts->nd_opts_pi == NULL)) {
267                     return (0);
268             }
                                                                    mip6_mncore.c
```

253–254 The address of the router that sent the Advertisement message must not be NULL.

257–263 When the router is going to shut down, it sends a Router Advertisement message with its lifetime set to 0. In this case, the pointer to the router information becomes NULL. If the router is in the home agent list, the entry is removed from the home agent list (the `mip6_ha_list` global variable) by calling the `mip6_ha_list_remove()` function.

266–268 If the incoming message does not contain any prefix options, there is nothing to do.

Listing 15-169

_____ mip6_mncore.c
```
270             for (sc = LIST_FIRST(&hif_softc_list); sc;
271                 sc = LIST_NEXT(sc, hif_entry)) {
272                     /* reorganize subnet groups. */
273                     error = mip6_prelist_update_sub(sc, saddr, ndopts, dr, m);
274                     if (error) {
....
278                             return (error);
279                     }
280             }
281
282             return (0);
283     }
```
_____ mip6_mncore.c

270–280 All the prefix information enclosed in the incoming Router Advertisement message is passed to each virtual home network interface to update the prefix information of each home interface.

Listing 15-170

_____ mip6_mncore.c
```
285     static int
286     mip6_prelist_update_sub(sc, rtaddr, ndopts, dr, m)
287             struct hif_softc *sc;
288             struct in6_addr *rtaddr;
289             union nd_opts *ndopts;
290             struct nd_defrouter *dr;
291             struct mbuf *m;
292     {
293             int location;
294             struct nd_opt_hdr *ndopt;
295             struct nd_opt_prefix_info *ndopt_pi;
296             struct sockaddr_in6 prefix_sa;
297             int is_home;
298             struct mip6_ha *mha;
299             struct mip6_prefix *mpfx;
300             struct mip6_prefix *prefix_list[IPV6_MMTU/sizeof
    (struct nd_opt_prefix_info)];
301             int nprefix = 0;
302             struct hif_prefix *hpfx;
303             struct sockaddr_in6 haaddr;
304             int i;
305             int error = 0;
```
_____ mip6_mncore.c

Note: Line 300 is broken here for layout reasons. However, it is a single line of code.

285–291 The `mip6_prelist_update_sub()` function updates the prefix information and the router information of each home virtual interface. The function has five parameters. The `sc` parameter is a pointer to the `hif_softc{}` instance, which indicates one home network. The `rtaddr`, `ndopts`, `dr`, and `m` parameters are the same as the parameters passed to the `mip6_prelist_update()` function described previously.

Listing 15-171

```
——————————————————————————————————————————— mip6_mncore.c
307                /* sanity check. */
308                if ((sc == NULL) || (rtaddr == NULL) || (dr == NULL)
309                    || (ndopts == NULL) || (ndopts->nd_opts_pi == NULL))
310                        return (EINVAL);
311
312                /* a router advertisement must be sent from a link-local address. */
313                if (!IN6_IS_ADDR_LINKLOCAL(rtaddr)) {
....
318                        /* ignore. */
319                        return (0);
320                }
——————————————————————————————————————————— mip6_mncore.c
```

308–310 An error is returned if any of the required parameters is NULL.

313–320 The source address of the Router Advertisement message must be a link-local address as specified in the Neighbor Discovery specification.

Listing 15-172

```
——————————————————————————————————————————— mip6_mncore.c
322                location = HIF_LOCATION_UNKNOWN;
323                is_home = 0;
324
325                for (ndopt = (struct nd_opt_hdr *)ndopts->nd_opts_pi;
326                     ndopt <= (struct nd_opt_hdr *)ndopts->nd_opts_pi_end;
327                     ndopt = (struct nd_opt_hdr *)((caddr_t)ndopt
328                         + (ndopt->nd_opt_len << 3))) {
329                    if (ndopt->nd_opt_type != ND_OPT_PREFIX_INFORMATION)
330                        continue;
331                    ndopt_pi = (struct nd_opt_prefix_info *)ndopt;
332
333                    /* sanity check of prefix information. */
334                    if (ndopt_pi->nd_opt_pi_len != 4) {
```
.... (output warning logs)
```
339                    }
340                    if (128 < ndopt_pi->nd_opt_pi_prefix_len) {
....
345                        continue;
346                    }
347                    if (IN6_IS_ADDR_MULTICAST(&ndopt_pi->nd_opt_pi_prefix)
348                        || IN6_IS_ADDR_LINKLOCAL(&ndopt_pi->nd_opt_pi_prefix)) {
....
353                        continue;
354                    }
355                    /* aggregatable unicast address, rfc2374 */
356                    if ((ndopt_pi->nd_opt_pi_prefix.s6_addr8[0] & 0xe0) == 0x20
357                        && ndopt_pi->nd_opt_pi_prefix_len != 64) {
....
363                        continue;
364                    }
——————————————————————————————————————————— mip6_mncore.c
```

325–328 The prefix information is stored between the address space pointed to by the `nd_opts_pi` and the `nd_opts_pi_end` pointers of the `nd_opts{}` structure as a Neighbor Discovery option. In this loop, the prefix information included in the incoming Advertisement message is checked to determine whether it is valid.

329–330 Options other than the prefix information option may exist in the space. If the option is not a prefix information option (the option type is not `ND_OPT_PREFIX_INFORMATION`), the option is skipped.

334–339 The length of the prefix information option must be 4. If the length is invalid, the error is logged. The procedure continues.

340–346 If the prefix of which the prefix length is greater than 128 cannot be processed, the option is ignored.

347–354 If the prefix is a multicast or a link-local prefix, the option is skipped.

356–364 If the prefix information delivered by the Router Advertisement message is not a prefix of an IPv6 global unicast address, the prefix is not processed. In the KAME Mobile IPv6, only IPv6 global unicast addresses are used.

Listing 15-173
_____ mip6_mncore.c
```
366                       bzero(&prefix_sa, sizeof(prefix_sa));
367                       prefix_sa.sin6_family = AF_INET6;
368                       prefix_sa.sin6_len = sizeof(prefix_sa);
369                       prefix_sa.sin6_addr = ndopt_pi->nd_opt_pi_prefix;
370                       if (in6_addr2zoneid(m->m_pkthdr.rcvif, &prefix_sa.sin6_addr,
371                           &prefix_sa.sin6_scope_id))
372                               continue;
373                       if (in6_embedscope(&prefix_sa.sin6_addr, &prefix_sa))
374                               continue;
375                       hpfx = hif_prefix_list_find_withprefix(
376                           &sc->hif_prefix_list_home, &prefix_sa.sin6_addr,
377                           ndopt_pi->nd_opt_pi_prefix_len);
378                       if (hpfx != NULL)
379                               is_home++;
```
_____ mip6_mncore.c

366–373 The prefix information delivered by the Router Advertisement message is restored as a `sockaddr_in6{}` instance. The scope identifier of the prefix is restored from the interface on which the Advertisement message has arrived.

375–379 The `hif_prefix_list_find_withprefix()` is called to see if the received prefix information is registered as one of the home prefixes of the virtual home interface we are now processing. If the prefix is a home prefix, the `is_home` variable is set to true. The `is_home` variable indicates the current location of the mobile node.

Listing 15-174
_____ mip6_mncore.c
```
381                       /*
382                        * since the global address of a home agent is stored
383                        * in a prefix information option, we can reuse
384                        * prefix_sa as a key to search a mip6_ha entry.
385                        */
386                       if (ndopt_pi->nd_opt_pi_flags_reserved
387                           & ND_OPT_PI_FLAG_ROUTER) {
388                               hpfx = hif_prefix_list_find_withhaaddr(
389                                   &sc->hif_prefix_list_home, &prefix_sa.sin6_addr);
390                               if (hpfx != NULL)
391                                       is_home++;
392                       }
393               }
```
_____ mip6_mncore.c

386–392 A prefix information option may contain an address of a home agent. If the flag field in the prefix information option has the `ND_OPT_PI_FLAG_ROUTER` flag set, the contents of the option includes not only the prefix information but also the address information.

If the mobile node receives address information, it checks to see whether the received address is registered as a home agent of its home network. If the mobile node has a home agent entry of which the address is the same as the address contained in the prefix information, the `is_home` variable is set to true.

Listing 15-175

```
                                                              mip6_mncore.c
395             /* check if the router's lladdr is on our home agent list. */
396             if (hif_prefix_list_find_withhaaddr(&sc->hif_prefix_list_home, rtaddr))
397                     is_home++;
398
399             if (is_home != 0) {
400                     /* we are home. */
401                     location = HIF_LOCATION_HOME;
402             } else {
403                     /* we are foreign. */
404                     location = HIF_LOCATION_FOREIGN;
405             }
                                                              mip6_mncore.c
```

396–397 The `hif_prefix_list_find_withhaaddr()` function is called to check whether the link-local address of the router that sent this Router Advertisement message is registered as a router of the home network of the mobile node. If the router is registered, the `is_home` variable is set to true.

399–405 The `location` variable is set based on the value of the `is_home` variable.

Listing 15-176

```
                                                              mip6_mncore.c
407             for (ndopt = (struct nd_opt_hdr *)ndopts->nd_opts_pi;
408                  ndopt <= (struct nd_opt_hdr *)ndopts->nd_opts_pi_end;
409                  ndopt = (struct nd_opt_hdr *)((caddr_t)ndopt
410                      + (ndopt->nd_opt_len << 3))) {
411                     if (ndopt->nd_opt_type != ND_OPT_PREFIX_INFORMATION)
412                             continue;
413                     ndopt_pi = (struct nd_opt_prefix_info *)ndopt;
....
450                     bzero(&prefix_sa, sizeof(prefix_sa));
451                     prefix_sa.sin6_family = AF_INET6;
452                     prefix_sa.sin6_len = sizeof(prefix_sa);
453                     prefix_sa.sin6_addr = ndopt_pi->nd_opt_pi_prefix;
454                     if (in6_addr2zoneid(m->m_pkthdr.rcvif, &prefix_sa.sin6_addr,
455                         &prefix_sa.sin6_scope_id))
456                             continue;
457                     if (in6_embedscope(&prefix_sa.sin6_addr, &prefix_sa))
458                             continue;
                                                              mip6_mncore.c
```

407–413 All prefix information options are checked again to update the prefix and router (or home agent) information stored in the virtual home network structure.

450–458 An instance of the `sockaddr_in6{}` structure that contains the prefix information is constructed.

Listing 15-177

```
                                                              mip6_mncore.c
460                     /* update mip6_prefix_list. */
461                     mpfx = mip6_prefix_list_find_withprefix(&prefix_sa.sin6_addr,
462                         ndopt_pi->nd_opt_pi_prefix_len);
```

```
463                     if (mpfx) {
464                             /* found an existing entry.  just update it. */
465                             mip6_prefix_update_lifetime(mpfx,
466                                 ntohl(ndopt_pi->nd_opt_pi_valid_time),
467                                 ntohl(ndopt_pi->nd_opt_pi_preferred_time));
468                             /* XXX mpfx->mpfx_haddr; */
469                     } else {
470                             /* this is a new prefix. */
471                             mpfx = mip6_prefix_create(&prefix_sa.sin6_addr,
472                                 ndopt_pi->nd_opt_pi_prefix_len,
473                                 ntohl(ndopt_pi->nd_opt_pi_valid_time),
474                                 ntohl(ndopt_pi->nd_opt_pi_preferred_time));
475                             if (mpfx == NULL) {
....
480                                     goto skip_prefix_update;
481                             }
482                             error = mip6_prefix_list_insert(&mip6_prefix_list,
483                                 mpfx);
484                             if (error) {
....
489                                     goto skip_prefix_update;
490                             }
....
496                     }
```
———————————————————————————————————— mip6_mncore.c

461 The `mip6_prefix_list_find_withprefix()` function is called to look up an existing prefix information entry (the `mip6_prefix{}` structure), which has the same information as the received prefix information.

463–496 If the mobile node has the prefix information already, it does not need to create a new entry. The mobile node just updates the preferred lifetime and the valid lifetime of the prefix information to the latest value. If the mobile node does not have a prefix information entry that matches the received prefix information, a new `mip6_prefix{}` instance is created and inserted into the list by calling the `mip6_prefix_list_insert()` function.

Listing 15-178
———————————————————————————————————— mip6_mncore.c

```
498                     /*
499                      *  insert this prefix information to hif structure
500                      *  based on the current location.
501                      */
502                     if (location == HIF_LOCATION_HOME) {
503                             hpfx = hif_prefix_list_find_withmpfx(
504                                 &sc->hif_prefix_list_foreign, mpfx);
505                             if (hpfx != NULL)
506                                     hif_prefix_list_remove(
507                                         &sc->hif_prefix_list_foreign, hpfx);
508                             if (hif_prefix_list_find_withmpfx(
509                                 &sc->hif_prefix_list_home, mpfx) == NULL)
510                                     hif_prefix_list_insert_withmpfx(
511                                         &sc->hif_prefix_list_home, mpfx);
512                     } else {
513                             hpfx = hif_prefix_list_find_withmpfx(
514                                 &sc->hif_prefix_list_home, mpfx);
515                             if (hpfx != NULL)
516                                     hif_prefix_list_remove(
517                                         &sc->hif_prefix_list_home, hpfx);
518                             if (hif_prefix_list_find_withmpfx(
519                                 &sc->hif_prefix_list_foreign, mpfx) == NULL)
520                                     hif_prefix_list_insert_withmpfx(
521                                         &sc->hif_prefix_list_foreign, mpfx);
```

```
522                     }
523
524                     /* remember prefixes advertised with this ND message. */
525                     prefix_list[nprefix] = mpfx;
526                     nprefix++;
527             skip_prefix_update:
528                     }
```
—— mip6_mncore.c

502–511 Based on the current location of the mobile node, the prefix information stored in the virtual home network structure is updated. If the mobile node is at home, the prefix should be added/updated as a home prefix. The hif_prefix_list_find_withmpfx() function finds a specified prefix from the list kept in the virtual home network. If the prefix is stored as a foreign prefix, it is removed from the list of foreign prefixes and added to the list of home prefixes.

513–522 If the mobile node is in a foreign network, the received prefix will be added/updated as a foreign prefix. If the received prefix is stored in the list of home prefixes, it is removed and added to the list of foreign prefixes.

525–526 Each pointer to the prefix information structure is recorded in the prefix_list[] array. These pointers are needed when we update the advertising router information of each prefix later.

Listing 15-179
—— mip6_mncore.c
```
530             /* update/create mip6_ha entry with an lladdr. */
531             mha = mip6_ha_list_find_withaddr(&mip6_ha_list, rtaddr);
532             if (mha) {
533                     /* the entry for rtaddr exists.  update information. */
534                     if (mha->mha_pref == 0 /* XXX */) {
535                             /* XXX reorder by pref. */
536                     }
537                     mha->mha_flags = dr->flags;
538                     mip6_ha_update_lifetime(mha, dr->rtlifetime);
539             } else {
540                     /* this is a lladdr mip6_ha entry. */
541                     mha = mip6_ha_create(rtaddr, dr->flags, 0, dr->rtlifetime);
542                     if (mha == NULL) {
....
546                             goto haddr_update;
547                     }
548                     mip6_ha_list_insert(&mip6_ha_list, mha);
549             }
550             for (i = 0; i < nprefix; i++) {
551                     mip6_prefix_ha_list_insert(&prefix_list[i]->mpfx_ha_list, mha);
552             }
```
—— mip6_mncore.c

531–549 The mip6_ha_list_find_withaddr() function returns a pointer to the existing entry of the mip6_ha{} instance that has the specified address. If the entry exists, its flag and router lifetime information is updated. As the comment on line 535 says, the mobile node needs to reorder the list when preference information is specified in the Router Advertisement message. However, the current KAME implementation does not implement this feature at this time. If the router that sent the Router Advertisement we are processing is a new router, a new instance of the mip6_ha{} structure is created and inserted into the router list.

550–552 Each prefix information entry has a pointer to the router or the home agent that advertises that prefix information. In this loop, the mobile node updates the pointer information
of each prefix information entry. The pointer information contains the pointer to the entry
of the `mip6_ha{}` instance, which is created or updated in lines 532–549.

Listing 15-180

—— mip6_mncore.c

```
554     haaddr_update:
555             /* update/create mip6_ha entry with a global addr. */
556             for (ndopt = (struct nd_opt_hdr *)ndopts->nd_opts_pi;
557                  ndopt <= (struct nd_opt_hdr *)ndopts->nd_opts_pi_end;
558                  ndopt = (struct nd_opt_hdr *)((caddr_t)ndopt
559                  + (ndopt->nd_opt_len << 3))) {
560                     if (ndopt->nd_opt_type != ND_OPT_PREFIX_INFORMATION)
561                             continue;
562                     ndopt_pi = (struct nd_opt_prefix_info *)ndopt;
563
564                     if ((ndopt_pi->nd_opt_pi_flags_reserved
565                        & ND_OPT_PI_FLAG_ROUTER) == 0)
566                             continue;
```

—— mip6_mncore.c

556–566 In this loop, the mobile node creates or updates the `mip6_ha{}` instance that has a
global address of the home agent of the mobile node. The global address of the home
agent is stored in a prefix information option with the `ND_OPT_PI_FLAG_ROUTER` flag
set. Any prefix information option that does not have the flag set is skipped.

Listing 15-181

—— mip6_mncore.c

```
568             bzero(&haaddr, sizeof(haaddr));
569             haaddr.sin6_len = sizeof(haaddr);
570             haaddr.sin6_family = AF_INET6;
571             haaddr.sin6_addr = ndopt_pi->nd_opt_pi_prefix;
572             if (in6_addr2zoneid(m->m_pkthdr.rcvif, &haaddr.sin6_addr,
573                 &haaddr.sin6_scope_id))
574                     continue;
575             if (in6_embedscope(&haaddr.sin6_addr, &haaddr))
576                     continue;
```

—— mip6_mncore.c

568–576 An instance of the `sockaddr_in6{}` structure that contains the global address of
the home agent is created. The scope identifier is recovered from the interface on which
the incoming Router Advertisement message has been received.

Listing 15-182

—— mip6_mncore.c

```
577             mha = mip6_ha_list_find_withaddr(&mip6_ha_list,
578                 &haaddr.sin6_addr);
579             if (mha) {
580                     if (mha->mha_pref == 0 /* XXX */) {
581                             /* XXX reorder by pref. */
582                     }
583                     mha->mha_flags = dr->flags;
584                     mip6_ha_update_lifetime(mha, 0);
585             } else {
586                     /* this is a new home agent . */
```

```
587                          mha = mip6_ha_create(&haaddr.sin6_addr, dr->flags, 0,
588                              0);
589                          if (mha == NULL) {
....
593                                  goto skip_ha_update;
594                          }
595                          mip6_ha_list_insert(&mip6_ha_list, mha);
....
602                      }
603                      for (i = 0; i < nprefix; i++) {
604                          mip6_prefix_ha_list_insert(
605                              &prefix_list[i]->mpfx_ha_list, mha);
606                      }
607              skip_ha_update:
608              }
609              return (0);
610      }
```
———————————————————————————————————— mip6_mncore.c

577–606 The existing or a newly created `mip6_ha{}` instance is updated in the same manner as done on lines 531–549.

15.14 Sending a Mobile Prefix Solicitation Message

A mobile node manages prefix information of its home network to keep its home address up-to-date. A mobile node can receive prefix information by listening to Router Advertisement messages when it is at home. However, when it is away from home, it needs other mechanisms. Sending a Mobile Prefix Solicitation is implemented as the `mip6_icmp6_mp_sol_output()` function.

Listing 15-183
———————————————————————————————————— mip6_icmp6.c
```
693      int
694      mip6_icmp6_mp_sol_output(haddr, haaddr)
695              struct in6_addr *haddr, *haaddr;
696      {
697              struct hif_softc *sc;
698              struct mbuf *m;
699              struct ip6_hdr *ip6;
700              struct mip6_prefix_solicit *mp_sol;
701              int icmp6len;
702              int maxlen;
703              int error;
....
705              struct timeval mono_time;
....
709              microtime(&mono_time);
```
———————————————————————————————————— mip6_icmp6.c

694–695 The `mip6_icmp6_mp_sol_output()` function has two parameters. The `haddr` parameter is a pointer to the home address of a mobile node and the `haaddr` parameter is a pointer to the address of the home agent of the mobile node.

Listing 15-184
———————————————————————————————————— mip6_icmp6.c
```
712              sc = hif_list_find_withhaddr(haddr);
713              if (sc == NULL) {
....
```

```
719                     return (0);
720                 }
721
722             /* rate limitation. */
723             if (sc->hif_mps_lastsent + 1 > mono_time.tv_sec) {
724                     return (0);
725             }
```
─── mip6_icmp6.c

712–720 If we do not have a virtual home network interface that is related to the home address of the mobile node, there is nothing to do.

723–725 To avoid flooding with solicitation messages, the mobile node must limit the number of messages sent to one per second. The `hif_mps_lastsent` variable holds the time when the mobile node sent the last solicitation message.

Listing 15-185
─── mip6_icmp6.c

```
727             /* estimate the size of message. */
728             maxlen = sizeof(*ip6) + sizeof(*mp_sol);
729             /* XXX we must determine the link type of our home address
730                 instead using hardcoded '6' */
731             maxlen += (sizeof(struct nd_opt_hdr) + 6 + 7) & ~7;
732             if (max_linkhdr + maxlen >= MCLBYTES) {
....
736                     return (EINVAL);
737             }
738
739             /* get packet header. */
740             MGETHDR(m, M_DONTWAIT, MT_HEADER);
741             if (m && max_linkhdr + maxlen >= MHLEN) {
742                 MCLGET(m, M_DONTWAIT);
743                 if ((m->m_flags & M_EXT) == 0) {
744                         m_free(m);
745                         m = NULL;
746                 }
747             }
748             if (m == NULL)
749                     return (ENOBUFS);
750             m->m_pkthdr.rcvif = NULL;
```
─── mip6_icmp6.c

728–737 The packet size that is needed to create a Mobile Prefix Solicitation is calculated. `maxlen` will include the size of an IPv6 header and the Mobile Prefix Solicitation message and the length of a link-layer header. Including the length of a link-layer header will avoid an additional mbuf allocation when the node prepends a link-layer header when sending the packet to a physical link.

740–750 An mbuf to store the solicitation message is allocated. If the requested length cannot be allocated with a single mbuf, the mobile node will try to allocate the same size with a cluster mbuf. If it fails to allocate an mbuf, an error is returned.

Listing 15-186
─── mip6_icmp6.c

```
752             icmp6len = sizeof(*mp_sol);
753             m->m_pkthdr.len = m->m_len = sizeof(*ip6) + icmp6len;
754             m->m_data += max_linkhdr;
755
756             sc->hif_mps_id = mip6_mps_id++;
```
─── mip6_icmp6.c

752–754 The total size of the allocated mbuf is set to the size of the IPv6 header and the ICMPv6 length, which is the length of the Mobile Prefix Solicitation to be sent. The data pointer is set to `max_linkhdr` to reserve a space to prepare a link-layer header.

756 The `hif_mps_id` variable keeps a unique identifier of the solicitation message. When the mobile node receives a Mobile Prefix Advertisement message, it compares the identifier of the received advertisement message to the recorded identifier to check whether the received advertisement message is addressed to the mobile node.

Listing 15-187

```
                                                                    mip6_icmp6.c
759              /* fill the mobile prefix solicitation. */
760              ip6 = mtod(m, struct ip6_hdr *);
761              ip6->ip6_flow = 0;
762              ip6->ip6_vfc &= ~IPV6_VERSION_MASK;
763              ip6->ip6_vfc |= IPV6_VERSION;
764              /* ip6->ip6_plen will be set later */
765              ip6->ip6_nxt = IPPROTO_ICMPV6;
766              ip6->ip6_hlim = ip6_defhlim;
767              ip6->ip6_src = *haddr;
768              ip6->ip6_dst = *haaddr;
769              mp_sol = (struct mip6_prefix_solicit *)(ip6 + 1);
770              mp_sol->mip6_ps_type = MIP6_PREFIX_SOLICIT;
771              mp_sol->mip6_ps_code = 0;
772              mp_sol->mip6_ps_id = htons(sc->hif_mps_id);
773              mp_sol->mip6_ps_reserved = 0;
774
775              /* calculate checksum. */
776              ip6->ip6_plen = htons((u_int16_t)icmp6len);
777              mp_sol->mip6_ps_cksum = 0;
778              mp_sol->mip6_ps_cksum = in6_cksum(m, IPPROTO_ICMPV6, sizeof(*ip6),
779                  icmp6len);
780
781              error = ip6_output(m, 0, 0, 0, 0 ,NULL
....
783                  , NULL
....
785                  );
786              if (error) {
....
790              }
791
792              /* update rate limitation factor. */
793              sc->hif_mps_lastsent = mono_time.tv_sec;
794
795              return (error);
796      }
                                                                    mip6_icmp6.c
```

760–773 All fields of the IPv6 header and the Mobile Prefix Solicitation message are filled. The message type of the ICMPv6 header is set to `MIP6_PREFIX_SOLICIT` and the code field is set to 0.

776–793 After the checksum value for this ICMPv6 message is computed by the `in6_cksum()` function, the packet is sent by the `ip6_output()` function. After sending the packet, the mobile node updates `hif_mps_lastsent`, which indicates the time when the last message was sent.

15.15   Receiving a Mobile Prefix Advertisement Message

A mobile node will receive a Mobile Prefix Advertisement message in response to the Mobile Prefix Solicitation message sent from the mobile node. In addition to the solicited messages, the mobile node may receive an unsolicited Mobile Prefix Advertisement message from its home agent. An unsolicited message is sent when the condition of home prefixes changes. A home agent needs to notify the mobile nodes it is serving of such changes. However, the current KAME implementation does not support unsolicited advertisement messages at this moment; it only processes solicited messages. Receiving a Mobile Prefix Advertisement is implemented as the `mip6_icmp6_mp_adv_input()` function. The function is called from the `icmp6_input()` function.

Listing 15-188

_____mip6_icmp6.c

```
798      static int
799      mip6_icmp6_mp_adv_input(m, off, icmp6len)
800              struct mbuf *m;
801              int off;
802              int icmp6len;
803      {
804              struct ip6_hdr *ip6;
805              struct m_tag *mtag;
806              struct ip6aux *ip6a;
807              struct mip6_prefix_advert *mp_adv;
808              union nd_opts ndopts;
809              struct nd_opt_hdr *ndopt;
810              struct nd_opt_prefix_info *ndopt_pi;
811              struct sockaddr_in6 prefix_sa;
812              struct in6_aliasreq ifra;
813              struct in6_ifaddr *ia6;
814              struct mip6_prefix *mpfx;
815              struct mip6_ha *mha;
816              struct hif_softc *hif, *tmphif;
817              struct mip6_bu *mbu;
818              struct ifaddr *ifa;
819              struct hif_prefix *hpfx;
820              int error = 0;
....
822              struct timeval mono_time;
....
826              microtime(&mono_time);
```

_____mip6_icmp6.c

798–802 The `mip6_icmp6_mp_adv_input()` function has three parameters: the `m` parameter is a pointer to the mbuf that contains the advertisement message; the `off` parameter is an offset from the head of the IPv6 packet to the head of the advertisement message; and the `icmp6len` parameter is the length of the Mobile Prefix Advertisement message.

Listing 15-189

_____mip6_icmp6.c

```
833              ip6 = mtod(m, struct ip6_hdr *);
....
835              IP6_EXTHDR_CHECK(m, off, icmp6len, EINVAL);
836              mp_adv = (struct mip6_prefix_advert *)((caddr_t)ip6 + off);
```

_____mip6_icmp6.c

833–836 The contents of the packet must be located in a contiguous memory space so that we can access each field of the message using offsets from the head of the message.

Listing 15-190

```
                                                            mip6_icmp6.c
848              /* find mip6_ha instance. */
849              mha = mip6_ha_list_find_withaddr(&mip6_ha_list, &ip6->ip6_src);
850              if (mha == NULL) {
851                      error = EINVAL;
852                      goto freeit;
853              }
854
855              /* find relevant hif interface. */
856              hif = hif_list_find_withaddr(&ip6->ip6_dst);
857              if (hif == NULL) {
858                      error = EINVAL;
859                      goto freeit;
860              }
861
862              /* sanity check. */
863              if (hif->hif_location != HIF_LOCATION_FOREIGN) {
864                      /* MPA is processed only we are foreign. */
865                      error = EINVAL;
866                      goto freeit;
867              }
                                                            mip6_icmp6.c
```

849–853 If the mobile node does not have a home agent information entry (the `mip6_ha{}` instance), which is related to the source address of the incoming advertisement message, the mobile node drops the packet. The message must be sent from the home agent of the mobile node.

856–860 If the mobile node does not have a virtual home network interface (the `hif_softc{}` instance), which is related to the destination address of the received message, the node drops the packet. The advertisement message must be sent to the home address of the mobile node.

863–867 If the mobile node is at home, it drops the incoming advertisement message. This behavior is not specified in the Mobile IPv6 specification, and some implementation may accept the message. The KAME implementation uses only the Router Advertisement message to get prefix information when a mobile node is home.

Listing 15-191

```
                                                            mip6_icmp6.c
869              mbu = mip6_bu_list_find_home_registration(&hif->hif_bu_list,
870                  &ip6->ip6_dst);
871              if (mbu == NULL) {
872                      error = EINVAL;
873                      goto freeit;
874              }
875              if (!IN6_ARE_ADDR_EQUAL(&mbu->mbu_paddr, &ip6->ip6_src)) {
....
883                      error = EINVAL;
884                      goto freeit;
885              }
                                                            mip6_icmp6.c
```

869–885 The `mip6_bu_list_find_home_registration()` function will return a home registration entry for the home address specified as its second parameter. The advertisement message must be sent from the address of the home agent that is serving the home address. If the home agent address in the binding update list entry (`mbu_paddr`) is different from the source address of the advertisement message, the mobile node drops the packet.

Listing 15-192

<code>mip6_icmp6.c</code>

```
887                /* check type2 routing header. */
888                mtag = ip6_findaux(m);
889                if (mtag == NULL) {
890                        /* this packet doesn't have a type 2 RTHDR. */
891                        error = EINVAL;
892                        goto freeit;
893                } else {
894                        ip6a = (struct ip6aux *)(mtag + 1);
895                        if ((ip6a->ip6a_flags & IP6A_ROUTEOPTIMIZED) == 0) {
896                                /* this packet doesn't have a type 2 RTHDR. */
897                                error = EINVAL;
898                                goto freeit;
899                        }
900                }
```

<code>mip6_icmp6.c</code>

888–900 The Mobile Prefix Advertisement message must have the Type 2 Routing Header since the Mobile Prefix Advertisement message is sent to the home address of the mobile node from its home agent. A message that does not have the Routing Header will be dropped.

Listing 15-193

<code>mip6_icmp6.c</code>

```
903                /* check id.  if it doesn't match, send mps. */
904                if (hif->hif_mps_id != ntohs(mp_adv->mip6_pa_id)) {
905                        mip6_icmp6_mp_sol_output(&mbu->mbu_haddr, &mbu->mbu_paddr);
906                        error = EINVAL;
907                        goto freeit;
908                }
```

<code>mip6_icmp6.c</code>

904–908 The identifier stored in the advertisement message (`mip6_pa_id`) must be the same as the identifier that the mobile node specified in the solicitation message. If the identifier is different from the identifier stored in a virtual home network interface that was recorded when the solicitation message was sent, the advertisement message is dropped.

Listing 15-194

<code>mip6_icmp6.c</code>

```
910                icmp6len -= sizeof(*mp_adv);
911                nd6_option_init(mp_adv + 1, icmp6len, &ndopts);
912                if (nd6_options(&ndopts) < 0) {
   ....
916                        /* nd6_options have incremented stats */
917                        error = EINVAL;
918                        goto freeit;
919                }
```

<code>mip6_icmp6.c</code>

911–919 The contents of the Mobile Prefix Solicitation is prefix information of which the format is the same as the Router Advertisement message. The `nd6_option_init()` function is called to parse the options included in the incoming advertisement message.

Listing 15-195

```
921                 for (ndopt = (struct nd_opt_hdr *)ndopts.nd_opts_pi;
922                      ndopt <= (struct nd_opt_hdr *)ndopts.nd_opts_pi_end;
923                      ndopt = (struct nd_opt_hdr *)((caddr_t)ndopt
924                           + (ndopt->nd_opt_len << 3))) {
925                     if (ndopt->nd_opt_type != ND_OPT_PREFIX_INFORMATION)
926                             continue;
927                     ndopt_pi = (struct nd_opt_prefix_info *)ndopt;
928
929                     /* sanity check of prefix information. */
930                     if (ndopt_pi->nd_opt_pi_len != 4) {
....
936                     }
937                     if (128 < ndopt_pi->nd_opt_pi_prefix_len) {
....
943                             continue;
944                     }
945                     if (IN6_IS_ADDR_MULTICAST(&ndopt_pi->nd_opt_pi_prefix)
946                         || IN6_IS_ADDR_LINKLOCAL(&ndopt_pi->nd_opt_pi_prefix)) {
....
952                             continue;
953                     }
954                     /* aggregatable unicast address, rfc2374 */
955                     if ((ndopt_pi->nd_opt_pi_prefix.s6_addr8[0] & 0xe0) == 0x20
956                         && ndopt_pi->nd_opt_pi_prefix_len != 64) {
....
963                             continue;
964                     }
```

921–924 All prefix information options included in the advertisement message will be processed.

930–964 The same sanity checks are done as when the mobile node does the sanity checks against an incoming Router Advertisement message in the `mip6_prelist_update_sub()` function as discussed in Section 14.19 of Chapter 14.

Listing 15-196

```
966                     bzero(&prefix_sa, sizeof(prefix_sa));
967                     prefix_sa.sin6_family = AF_INET6;
968                     prefix_sa.sin6_len = sizeof(prefix_sa);
969                     prefix_sa.sin6_addr = ndopt_pi->nd_opt_pi_prefix;
970                     /* XXX scope? */
971                     mpfx = mip6_prefix_list_find_withprefix(&prefix_sa.sin6_addr,
972                         ndopt_pi->nd_opt_pi_prefix_len);
```

966–972 An instance of the `sockaddr_in6{}` structure that contains the received prefix information is constructed. The `mip6_prefix_list_find_withprefix()` function will return the pointer to the instance of the `mip6_prefix{}` structure that contains the received prefix if it already exists.

Listing 15-197

```
973                     if (mpfx == NULL) {
974                         mpfx = mip6_prefix_create(&prefix_sa.sin6_addr,
975                             ndopt_pi->nd_opt_pi_prefix_len,
```

```
976                                    ntohl(ndopt_pi->nd_opt_pi_valid_time),
977                                    ntohl(ndopt_pi->nd_opt_pi_preferred_time));
978                     if (mpfx == NULL) {
979                             error = EINVAL;
980                             goto freeit;
981                     }
982                     mip6_prefix_ha_list_insert(&mpfx->mpfx_ha_list, mha);
983                     mip6_prefix_list_insert(&mip6_prefix_list, mpfx);
984                     for (tmphif = LIST_FIRST(&hif_softc_list); tmphif;
985                         tmphif = LIST_NEXT(tmphif, hif_entry)) {
986                             if (hif == tmphif)
987                                     hif_prefix_list_insert_withmpfx(
988                                         &tmphif->hif_prefix_list_home,
989                                         mpfx);
990                             else
991                                     hif_prefix_list_insert_withmpfx(
992                                         &tmphif->hif_prefix_list_foreign,
993                                         mpfx);
994                     }
                                                                      ─── mip6_icmp6.c
```

973–981 If the mobile node does not have the prefix information received by the advertisement message, a new prefix information entry is created.

982–983 The prefix information is associated with the home agent information, which sent the advertisement message, and is inserted into the list of `mip6_prefix{}` instances.

984–994 The received prefix is a home prefix of the virtual home network, which is represented by the `hif` variable. At the same time, the prefix is foreign prefix information of other virtual home networks other than the `hif_softc{}` instances. The newly created prefix information is inserted into all `hif_softc{}` instances.

Listing 15-198
                                                                      ─── mip6_icmp6.c
```
996                      mip6_prefix_haddr_assign(mpfx, hif); /* XXX */
997
998                      /* construct in6_aliasreq. */
999                      bzero(&ifra, sizeof(ifra));
1000                     bcopy(if_name((struct ifnet *)hif), ifra.ifra_name,
1001                         sizeof(ifra.ifra_name));
1002                     ifra.ifra_addr.sin6_len = sizeof(struct sockaddr_in6);
1003                     ifra.ifra_addr.sin6_family = AF_INET6;
1004                     ifra.ifra_addr.sin6_addr = mpfx->mpfx_haddr;
1005                     ifra.ifra_prefixmask.sin6_len
1006                         = sizeof(struct sockaddr_in6);
1007                     ifra.ifra_prefixmask.sin6_family = AF_INET6;
1008                     ifra.ifra_flags = IN6_IFF_HOME | IN6_IFF_AUTOCONF;
1009                     in6_prefixlen2mask(&ifra.ifra_prefixmask.sin6_addr,
1010                         128);
1011                     ifra.ifra_lifetime.ia6t_vltime = mpfx->mpfx_vltime;
1012                     ifra.ifra_lifetime.ia6t_pltime = mpfx->mpfx_pltime;
1013                     if (ifra.ifra_lifetime.ia6t_vltime
1014                         == ND6_INFINITE_LIFETIME)
1015                             ifra.ifra_lifetime.ia6t_expire = 0;
1016                     else
1017                             ifra.ifra_lifetime.ia6t_expire
1018                                 = mono_time.tv_sec
1019                                 + ifra.ifra_lifetime.ia6t_vltime;
1020                     if (ifra.ifra_lifetime.ia6t_pltime
1021                         == ND6_INFINITE_LIFETIME)
1022                             ifra.ifra_lifetime.ia6t_preferred = 0;
1023                     else
1024                             ifra.ifra_lifetime.ia6t_preferred
1025                                 = mono_time.tv_sec
```

```
1026                                 +  ifra.ifra_lifetime.ia6t_pltime;
1027                     ia6 = in6ifa_ifpwithaddr((struct ifnet *)hif,
1028                             &ifra.ifra_addr.sin6_addr);
```
_____mip6_icmp6.c

996 A new home address is generated from the prefix information.

999–1028 The mobile node needs to configure a new home address since it has received a
new home prefix. An `in6_aliasreq{}` instance that contains information of the new
home address is constructed.

The new home address created while the mobile node is away from home is assigned
to the virtual home network. The `IN6_IFF_HOME` and the `IN6_IFF_AUTOCONF` flags
are set to the new address, which indicates that the address is a home address and it is
configured using the stateless autoconfiguration mechanism. The prefix length of a home
address is always 128 when a mobile node is away from home. The preferred lifetime and
the valid lifetime are set based on the lifetime information of the prefix information. If the
lifetime is infinite, the address will not expire.

Listing 15-199
_____mip6_icmp6.c
```
1030                     /* assign a new home address. */
1031                     error = in6_update_ifa((struct ifnet *)hif, &ifra,
1032                             ia6, 0);
1033                     if (error) {
....
1039                             goto freeit;
1040                     }
1041
1042                     mip6_home_registration(hif); /* XXX */
```
_____mip6_icmp6.c

1031–1042 The new home address is assigned by the `in6_update_ifa()` function and
the home registration procedure is triggered for the newly created home address by the
`mip6_home_registration()` function.

Listing 15-200
_____mip6_icmp6.c
```
1043                 } else {
1044                     mip6_prefix_update_lifetime(mpfx,
1045                         ntohl(ndopt_pi->nd_opt_pi_valid_time),
1046                         ntohl(ndopt_pi->nd_opt_pi_preferred_time));
1047
```
_____mip6_icmp6.c

1043–1047 If the received prefix information already exists, the lifetime is updated based on
the received information.

Listing 15-201
_____mip6_icmp6.c
```
1049                     TAILQ_FOREACH(ifa, &((struct ifnet *)hif)->if_addrlist,
1050                         ifa_list)
....
1055                     {
1056                             struct in6_ifaddr *ifa6;
```

```
1057
1058                                if (ifa->ifa_addr->sa_family != AF_INET6)
1059                                        continue;
1060
1061                                ifa6 = (struct in6_ifaddr *)ifa;
1062
1063                                if ((ifa6->ia6_flags & IN6_IFF_HOME) == 0)
1064                                        continue;
1065
1066                                if ((ifa6->ia6_flags & IN6_IFF_AUTOCONF) == 0)
1067                                        continue;
1068
1069                                if (!IN6_ARE_ADDR_EQUAL(&mpfx->mpfx_haddr,
1070                                    &ifa6->ia_addr.sin6_addr))
1071                                        continue;
1072
1073                                ifa6->ia6_lifetime.ia6t_vltime
1074                                    = mpfx->mpfx_vltime;
1075                                ifa6->ia6_lifetime.ia6t_pltime
1076                                    = mpfx->mpfx_pltime;
1077                                if (ifa6->ia6_lifetime.ia6t_vltime ==
1078                                    ND6_INFINITE_LIFETIME)
1079                                        ifa6->ia6_lifetime.ia6t_expire = 0;
1080                                else
1081                                        ifa6->ia6_lifetime.ia6t_expire =
1082                                            mono_time.tv_sec
1083                                            + mpfx->mpfx_vltime;
1084                                if (ifa6->ia6_lifetime.ia6t_pltime ==
1085                                    ND6_INFINITE_LIFETIME)
1086                                        ifa6->ia6_lifetime.ia6t_preferred = 0;
1087                                else
1088                                        ifa6->ia6_lifetime.ia6t_preferred =
1089                                            mono_time.tv_sec
1090                                            + mpfx->mpfx_pltime;
1091                                ifa6->ia6_updatetime = mono_time.tv_sec;
1092                        }
1093                }
1094        }
```
── mip6_icmp6.c

1049–1050 The lifetime of a prefix affects the lifetime of the home address derived from that prefix. Each home address generated from the prefix is checked and its lifetime is updated.

1058–1071 The mobile node only checks addresses whose address family is IPv6 (AF_INET6) and which have the IN6_IFF_HOME and the IN6_IFF_AUTOCONF flags. The home address generated from the prefix information is stored in the mpfx_haddr member variable of the mip6_prefix{} structure. The mobile node checks to see if it has the home address in its virtual home network interfaces. If it has, the lifetime of the home address is updated.

1073–1091 The lifetime of an address is stored in the ia6_lifetime member variable of the in6_ifaddr{} structure. The preferred and the valid lifetimes of the received prefix information are copied to the ia6_lifetime{} structure. If the preferred or valid lifetimes are infinite, the ia6t_preferred or the ia6t_expire variables of the ia6_lifetime{} structure are set to 0, indicating an infinite lifetime. The ia6t_preferred and the ia6t_valid variables indicate the time that the preferred/ valid lifetime will expire.

Finally, the ia6_updatetime variable, which indicates the time that the address is modified, is updated.

Listing 15-202

_____mip6_icmp6.c

```
1096            for (ndopt = (struct nd_opt_hdr *)ndopts.nd_opts_pi;
1097                 ndopt <= (struct nd_opt_hdr *)ndopts.nd_opts_pi_end;
1098                 ndopt = (struct nd_opt_hdr *)((caddr_t)ndopt
1099                     + (ndopt->nd_opt_len << 3))) {
1100                if (ndopt->nd_opt_type != ND_OPT_PREFIX_INFORMATION)
1101                        continue;
1102                ndopt_pi = (struct nd_opt_prefix_info *)ndopt;
1103
1104                if ((ndopt_pi->nd_opt_pi_flags_reserved
1105                    & ND_OPT_PI_FLAG_ROUTER) == 0)
1106                        continue;
```
_____mip6_icmp6.c

1096–1099 The prefix options are processed again to update home agent addresses, which may be embedded in the prefix information options.

1104–1106 When the ND_OPT_PI_FLAG_ROUTER flag is set in the prefix information option, the home agent address is embedded in the prefix value.

Listing 15-203

_____mip6_icmp6.c

```
1108                bzero(&prefix_sa, sizeof(prefix_sa));
1109                prefix_sa.sin6_family = AF_INET6;
1110                prefix_sa.sin6_len = sizeof(prefix_sa);
1111                prefix_sa.sin6_addr = ndopt_pi->nd_opt_pi_prefix;
1112                /* XXX scope. */
1113                mha = mip6_ha_list_find_withaddr(&mip6_ha_list,
1114                    &prefix_sa.sin6_addr);
1115                if (mha == NULL) {
1116                        mha = mip6_ha_create(&prefix_sa.sin6_addr,
1117                            ND_RA_FLAG_HOME_AGENT, 0, 0);
1118                        mip6_ha_list_insert(&mip6_ha_list, mha);
1119                } else {
1120                        if (mha->mha_pref != 0) {
1121                                /*
1122                                 * we have no method to know the
1123                                 * preference of this home agent.
1124                                 * assume pref = 0.
1125                                 */
1126                                mha->mha_pref = 0;
1127                                mip6_ha_list_reinsert(&mip6_ha_list, mha);
1128                        }
1129                        mip6_ha_update_lifetime(mha, 0);
1130                }
```
_____mip6_icmp6.c

1108–1111 An instance of the sockaddr_in6{} structure, which contains the home agent address delivered with the prefix information option, is constructed.

1113–1118 If the mobile node does not have information about the home agent, then the mip6_ha_create() function is called to create an instance of an mip6_ha{} structure, which keeps the new home agent information.

1120–1129 If the mobile node already has information about the home agent, its lifetime is updated. However, the mobile node cannot know the lifetime of the home agent since the Mobile Prefix Advertisement message does not include any lifetime information of home agents. The mobile node assumes that the home agent has an infinite lifetime. Also, if the existing home agent has a preference value other than 0, the mobile node resets the preference value to 0 because the Mobile Prefix Solicitation message does not include preference information either.

Listing 15-204

```
                                                                        mip6_icmp6.c
1131                    for (hpfx = LIST_FIRST(&hif->hif_prefix_list_home); hpfx;
1132                        hpfx = LIST_NEXT(hpfx, hpfx_entry)) {
1133                            mip6_prefix_ha_list_insert(
1134                                &hpfx->hpfx_mpfx->mpfx_ha_list, mha);
1135                        }
1136                }
1137
1138                return (0);
1139
1140        freeit:
1141                m_freem(m);
1142                return (error);
1143        }
                                                                        mip6_icmp6.c
```

1131–1135 Finally, the list of pointers to the home agent information of each home prefix stored in the virtual home network interface is updated.

15.16   Sending a Dynamic Home Agent Address Discovery Request Message

A mobile node sends a DHAAD message when the node needs to know the address of its home agent. Usually, the mobile node sends the message when it is turned on in a foreign network. Sending a DHAAD request message is implemented as the `mip6_icmp6_dhaad_req_output()` function.

Listing 15-205

```
                                                                        mip6_icmp6.c
563     int
564     mip6_icmp6_dhaad_req_output(sc)
565            struct hif_softc *sc;
566     {
567            struct in6_addr hif_coa;
568            struct in6_addr haanyaddr;
569            struct mip6_prefix *mpfx;
570            struct mbuf *m;
571            struct ip6_hdr *ip6;
572            struct mip6_dhaad_req *hdreq;
573            u_int32_t icmp6len, off;
574            int error;
                                                                        mip6_icmp6.c
```

563–565 The `mip6_icmp6_dhaad_req_output()` function takes one parameter that points to the virtual home network interface to which a mobile node sends a DHAAD request message.

Listing 15-206

```
                                                                        mip6_icmp6.c
583            /* rate limitation. */
584            if (sc->hif_dhaad_count != 0) {
```

```
585                    if (sc->hif_dhaad_lastsent + (1 << sc->hif_dhaad_count)
586                        > time_second)
587                            return (0);
588            }
```
—— mip6_icmp6.c

583–588 The number of DHAAD request messages sent to the home network of a mobile
node is limited to avoid flooding the network. The KAME implementation performs
exponential backoff when resending a message. The initial timeout is set to 1 s. Strictly
speaking, the behavior does not satisfy the specification. The specification says the initial
timeout is 3 s.

Listing 15-207
—— mip6_icmp6.c

```
590            /* get current CoA and recover its scope information. */
591            if (sc->hif_coa_ifa == NULL) {
  ....
595                    return (0);
596            }
597            hif_coa = sc->hif_coa_ifa->ia_addr.sin6_addr;
```
—— mip6_icmp6.c

591–597 If the mobile node does not have a valid care-of address, the node cannot send a
DHAAD request message since the message must be sent from the care-of address of the
mobile node.

Listing 15-208
—— mip6_icmp6.c

```
599            /*
600             * we must determine the home agent subnet anycast address.
601             * to do this, we pick up one home prefix from the prefix
602             * list.
603             */
604            for (mpfx = LIST_FIRST(&mip6_prefix_list); mpfx;
605                mpfx = LIST_NEXT(mpfx, mpfx_entry)) {
606                    if (hif_prefix_list_find_withmpfx(&sc->hif_prefix_list_home,
607                        mpfx))
608                            break;
609            }
610            if (mpfx == NULL) {
611                    /* we must have at least one home subnet. */
612                    return (EINVAL);
613            }
614            if (mip6_icmp6_create_haanyaddr(&haanyaddr, mpfx))
615                    return (EINVAL);
```
—— mip6_icmp6.c

604–614 The destination address of a DHAAD request message is the Home Agent Anycast
Address of its home network. The address can be computed from the home prefix of the
mobile node. The `hif_prefix_list_find_withmpfx()` function will check to see
if the prefix information passed as the second parameter belongs to the list specified as
the first parameter. The mobile node searches for at least one prefix information entry that
belongs to the `hif_prefix_list_home` variable, which contains all home prefix information of a virtual home network. The `mip6_icmp6_create_haanyaddr()` function

creates the Home Agent Anycast Address from the prefix information passed as the second parameter of the function.

Listing 15-209

——— mip6_icmp6.c
```
617                 /* allocate the buffer for the ip packet and DHAAD request. */
618                 icmp6len = sizeof(struct mip6_dhaad_req);
619                 m = mip6_create_ip6hdr(&hif_coa, &haanyaddr,
620                                     IPPROTO_ICMPV6, icmp6len);
621                 if (m == NULL) {
....
627                         return (ENOBUFS);
628                 }
629
630                 sc->hif_dhaad_id = mip6_dhaad_id++;
631
632                 ip6 = mtod(m, struct ip6_hdr *);
633                 hdreq = (struct mip6_dhaad_req *)(ip6 + 1);
634                 bzero((caddr_t)hdreq, sizeof(struct mip6_dhaad_req));
635                 hdreq->mip6_dhreq_type = MIP6_HA_DISCOVERY_REQUEST;
636                 hdreq->mip6_dhreq_code = 0;
637                 hdreq->mip6_dhreq_id = htons(sc->hif_dhaad_id);
638
639                 /* calculate checksum for this DHAAD request packet. */
640                 off = sizeof(struct ip6_hdr);
641                 hdreq->mip6_dhreq_cksum = in6_cksum(m, IPPROTO_ICMPV6, off, icmp6len);
```
——— mip6_icmp6.c

618–641 A DHAAD request message is constructed. An mbuf for the packet is prepared by the `mip6_create_ip6hdr()` function with source and destination addresses. The protocol number of the IPv6 packet is ICMPv6. The `hif_dhaad_id` variable is a unique identifier that distinguishes the reply message to be received. The mobile node drops any DHAAD reply message that does not match the identifier sent with the DHAAD request message. The ICMPv6 type number is set to `MIP6_HA_DISCOVERY_REQUEST` and the code value is set to 0. The checksum is computed by the `in6_cksum()` function in the same manner as other ICMPv6 packets.

Listing 15-210

——— mip6_icmp6.c
```
643                 /* send the DHAAD request packet to the home agent anycast address. */
644                 error = ip6_output(m, NULL, NULL, 0, NULL, NULL
....
646                                         , NULL
....
648                                         );
649                 if (error) {
....
654                         return (error);
655                 }
656
657                 /* update rate limitation factor. */
658                 sc->hif_dhaad_lastsent = time_second;
659                 if (sc->hif_dhaad_count++ > MIP6_DHAAD_RETRIES) {
660                         /*
661                          * XXX the spec says that the number of retires for
662                          * DHAAD request is restricted to DHAAD_RETRIES(=3).
663                          * But, we continue retrying until we receive a reply.
664                          */
```

```
665                                 sc->hif_dhaad_count = MIP6_DHAAD_RETRIES;
666                    }
667
668                    return (0);
669        }
```
——— mip6_icmp6.c

644–655 The created message is sent by the `ip6_output()` function.

658–666 The Mobile IPv6 specification says that a mobile node must not send a DHAAD request message over three times. However, the KAME implementation ignores this rule. When a mobile node is disconnected from the Internet, it soon reaches the maximum transmission limit. Continuing to send DHAAD messages may recover the registration status when the mobile node acquires an Internet connection again.

15.16.1 Create a Home Agent Anycast Address

The `mip6_icmp6_create_haanyaddr()` function creates a home agent anycast address from the specified prefix information.

Listing 15-211
——— mip6_icmp6.c
```
109    static const struct in6_addr haanyaddr_ifid64 = {
110            {{ 0x00, 0x00, 0x00, 0x00, 0x00, 0x00, 0x00, 0x00,
111               0xfd, 0xff, 0xff, 0xff, 0xff, 0xff, 0xff, 0xfe }}
112    };
113    static const struct in6_addr haanyaddr_ifidnn = {
114            {{ 0xff, 0xff, 0xff, 0xff, 0xff, 0xff, 0xff, 0xff,
115               0xff, 0xff, 0xff, 0xff, 0xff, 0xff, 0xff, 0xfe }}
116    };
....
671    static int
672    mip6_icmp6_create_haanyaddr(haanyaddr, mpfx)
673            struct in6_addr *haanyaddr;
674            struct mip6_prefix *mpfx;
675    {
676            struct nd_prefix ndpr;
677
678            if (mpfx == NULL)
679                    return (EINVAL);
680
681            bzero(&ndpr, sizeof(ndpr));
682            ndpr.ndpr_prefix.sin6_addr = mpfx->mpfx_prefix;
683            ndpr.ndpr_plen = mpfx->mpfx_prefixlen;
684
685            if (mpfx->mpfx_prefixlen == 64)
686                    mip6_create_addr(haanyaddr, &haanyaddr_ifid64, &ndpr);
687            else
688                    mip6_create_addr(haanyaddr, &haanyaddr_ifidnn, &ndpr);
689
690            return (0);
691    }
```
——— mip6_icmp6.c

671–674 The `mip6_icmp6_create_haanyaddr()` function has two parameters. The `haanyaddr` parameter is a pointer to store the computed Home Agent Anycast Address and the `mpfx` parameter is a pointer to the home prefix information.

681–688 There are two computation rules for the Home Agent Anycast Address. One is for the address of which the prefix length is 64. The other is for the address of which the prefix length is not 64. Figure 3-24 in Chapter 3 shows the algorithm. The `mip6_create_addr()` function creates an IPv6 address using the second parameter as an interface identifier part and the third parameter as a prefix part.

15.17 Receiving a Dynamic Home Agent Address Discovery Reply Message

A mobile node receives a Dynamic Home Agent Address Discovery reply message from its home agent in response to a Dynamic Home Agent Address Discovery request message. Receiving the message is implemented as the `mip6_icmp6_dhaad_rep_input()` function. The function is called from the `icmp6_input()` function.

Listing 15-212

─── mip6_icmp6.c

```
382     static int
383     mip6_icmp6_dhaad_rep_input(m, off, icmp6len)
384             struct mbuf *m;
385             int off;
386             int icmp6len;
387     {
388             struct ip6_hdr *ip6;
389             struct mip6_dhaad_rep *hdrep;
390             u_int16_t hdrep_id;
391             struct mip6_ha *mha, *mha_preferred = NULL;
392             struct in6_addr *haaddrs, *haaddrptr;
393             struct sockaddr_in6 haaddr_sa;
394             int i, hacount = 0;
395             struct hif_softc *sc;
396             struct mip6_bu *mbu;
....
398             struct timeval mono_time;
....
402             microtime(&mono_time);
```

─── mip6_icmp6.c

382–386 The `mip6_icmp6_dhaad_rep_input()` function is called from the `icmp6_input()` function and it has three parameters. The `m` parameter is a pointer to the mbuf that contains a DHAAD reply message, the `off` parameter is an offset from the head of the packet to the address of the DHAAD message part, and the `icmp6len` parameter is the length of the ICMPv6 part.

Listing 15-213

─── mip6_icmp6.c

```
405             ip6 = mtod(m, struct ip6_hdr *);
....
407             IP6_EXTHDR_CHECK(m, off, icmp6len, EINVAL);
408             hdrep = (struct mip6_dhaad_rep *)((caddr_t)ip6 + off);
....
418             haaddrs = (struct in6_addr *)(hdrep + 1);
```

─── mip6_icmp6.c

405–418 The packet must be located in a contiguous memory space so that we can access the contents by casting the address to each structure. If the packet is not located properly, we drop the packet. Otherwise, `haaddrs` is set to the end of the DHAAD reply message structure to point to the head of the list of home agent addresses. The list of addresses of home agents immediately follows the message part.

Listing 15-214

```
420                 /* sainty check. */
421                 if (hdrep->mip6_dhrep_code != 0) {
422                         m_freem(m);
423                         return (EINVAL);
424                 }
425
426                 /* check the number of home agents listed in the message. */
427                 hacount = (icmp6len - sizeof(struct mip6_dhaad_rep))
428                         / sizeof(struct in6_addr);
429                 if (hacount == 0) {
....
434                         m_freem(m);
435                         return (EINVAL);
436                 }
```

421–424 The ICMPv6 code field of the reply message must be 0. The mobile node drops any packet that does not have the correct code value.

427–436 The number of addresses stored at the end of the reply message can be calculated from the length of the reply message and the size of an IPv6 address, which is 128 bytes. If there is no address information, the packet is dropped.

Listing 15-215

```
438                 /* find hif that matches this receiving hadiscovid of DHAAD reply. */
439                 hdrep_id = hdrep->mip6_dhrep_id;
440                 hdrep_id = ntohs(hdrep_id);
441                 for (sc = LIST_FIRST(&hif_softc_list); sc;
442                     sc = LIST_NEXT(sc, hif_entry)) {
443                         if (sc->hif_dhaad_id == hdrep_id)
444                                 break;
445                 }
446                 if (sc == NULL) {
447                         /*
448                          * no matching hif.  maybe this DHAAD reply is too late.
449                          */
450                         return (0);
451                 }
452
453                 /* reset rate limitation factor. */
454                 sc->hif_dhaad_count = 0;
```

439–451 The identifier value (hdrep_id) of the reply message is copied from the request message. The mobile node searches for the virtual home network interface relevant to the identifier. If there is no matching interface, the mobile node ignores the reply message.

454 The hif_dhaad_count variable is used when a mobile node decides the sending rate of DHAAD messages. After receiving a correct reply message, hif_dhaad_count is reset to 0, which means the rate limitation state is in its initial state.

Listing 15-216

_____mip6_icmp6.c

```
456                /* install addresses of a home agent specified in the message */
457                haaddrptr = haaddrs;
458                for (i = 0; i < hacount; i++) {
459                        bzero(&haaddr_sa, sizeof(haaddr_sa));
460                        haaddr_sa.sin6_len = sizeof(haaddr_sa);
461                        haaddr_sa.sin6_family = AF_INET6;
462                        haaddr_sa.sin6_addr = *haaddrptr++;
463                        /*
464                         * XXX we cannot get a correct zone id by looking only
465                         * in6_addr structure.
466                         */
467                        if (in6_addr2zoneid(m->m_pkthdr.rcvif, &haaddr_sa.sin6_addr,
468                            &haaddr_sa.sin6_scope_id))
469                                continue;
470                        if (in6_embedscope(&haaddr_sa.sin6_addr, &haaddr_sa))
471                                continue;
```

_____mip6_icmp6.c

457–471 All home agent addresses contained in the reply message are installed. The scope identifier can be recovered from the interface on which the reply message is received. However, we need not worry about the scope identifier much since the address of a home agent is usually a global address.

Listing 15-217

_____mip6_icmp6.c

```
472                        mha = mip6_ha_list_find_withaddr(&mip6_ha_list,
473                            &haaddr_sa.sin6_addr);
474                        if (mha) {
475                                /*
476                                 * if this home agent already exists in the list,
477                                 * update its lifetime.
478                                 */
479                                if (mha->mha_pref == 0) {
480                                        /*
481                                         * we have no method to know the
482                                         * preference of this home agent.
483                                         * assume pref = 0.
484                                         */
485                                        mha->mha_pref = 0;
486                                        mip6_ha_list_reinsert(&mip6_ha_list, mha);
487                                }
488                                mip6_ha_update_lifetime(mha, 0);
```

_____mip6_icmp6.c

472–488 If the received address is already registered as a home agent of this mobile node, the mobile node updates the home agent information. If the existing entry has a preference value other than 0, the mobile node resets the value to 0 since the DHAAD reply message does not have any information about preference. The lifetime is assumed to be infinite for the same reason.

Listing 15-218

_____mip6_icmp6.c

```
489                        } else {
490                                /*
491                                 * create a new home agent entry and insert it
492                                 * to the internal home agent list
```

```
493                              * (mip6_ha_list).
494                              */
495                              mha = mip6_ha_create(&haddr_sa.sin6_addr,
496                                  ND_RA_FLAG_HOME_AGENT, 0, 0);
497                              if (mha == NULL) {
....
501                                      m_freem(m);
502                                      return (ENOMEM);
503                              }
504                              mip6_ha_list_insert(&mip6_ha_list, mha);
505                              mip6_dhaad_ha_list_insert(sc, mha);
506                      }
```
_____ mip6_icmp6.c

495–505 If the mobile node does not have a home agent entry for the received home agent address, a new `mip6_ha{}` instance is created by the `mip6_ha_create()` function and inserted into the list of home agents. The preference is assumed to be 0 and the lifetime is assumed to be infinite. The `mip6_dhaad_ha_list_insert()` function updates the home agent information of the virtual home network interface specified as its first parameter.

Listing 15-219
_____ mip6_icmp6.c

```
507                      if (mha_preferred == NULL) {
508                              /*
509                              * the home agent listed at the top of the
510                              * DHAAD reply packet is the most preferable
511                              * one.
512                              */
513                              mha_preferred = mha;
514                      }
515              }
```
_____ mip6_icmp6.c

507–514 The specification says that the address that is located at the head of the address list of the DHAAD reply message is the most preferred. The `mha_preferred` pointer is set to the `mip6_ha{}` instance processed first, which is the head of the list.

Listing 15-220
_____ mip6_icmp6.c

```
517              /*
518              * search bu_list and do home registration pending.  each
519              * binding update entry which can't proceed because of no home
520              * agent has an field of a home agent address equals to an
521              * unspecified address.
522              */
523              for (mbu = LIST_FIRST(&sc->hif_bu_list); mbu;
524                  mbu = LIST_NEXT(mbu, mbu_entry)) {
525                      if ((mbu->mbu_flags & IP6MU_HOME)
526                          && IN6_IS_ADDR_UNSPECIFIED(&mbu->mbu_paddr)) {
527                              /* home registration. */
528                              mbu->mbu_paddr = mha_preferred->mha_addr;
529                              if (!MIP6_IS_BU_BOUND_STATE(mbu)) {
530                                      if (mip6_bu_send_bu(mbu)) {
....
539                                      }
540                              }
541                      }
542              }
```

```
543
544              return (0);
545      }
```
——— mip6_icmp6.c

523–542 The `mip6_bu_send_bu()` function is called to send a Binding Update message with the home agent address received by the incoming DHAAD reply message for each waiting entry if the mobile node has binding update list entries that are waiting for the DHAAD reply message to determine the correct home agent address.

15.17.1   Update Prefix Information Entries

The `mip6_dhaad_ha_list_insert()` function inserts a home agent entry received via the Dynamic Home Agent Address Discovery mechanism into the home agent list.

Listing 15-221
——— mip6_icmp6.c
```
547      static int
548      mip6_dhaad_ha_list_insert(hif, mha)
549              struct hif_softc *hif;
550              struct mip6_ha *mha;
551      {
552              struct hif_prefix *hpfx;
553
554              for (hpfx = LIST_FIRST(&hif->hif_prefix_list_home); hpfx;
555                  hpfx = LIST_NEXT(hpfx, hpfx_entry)) {
556                      mip6_prefix_ha_list_insert(&hpfx->hpfx_mpfx->mpfx_ha_list,
557                          mha);
558              }
559
560              return (0);
561      }
```
——— mip6_icmp6.c

548–550 The `mip6_dhaad_ha_list_insert()` function has two parameters. The `hif` parameter is a pointer to the virtual home network interface of which the home agent information is going to be updated. The `mha` parameter is a pointer to the newly created instance of the `mip6_ha{}` structure, which is inserted into the virtual home interface.

554–558 The home agent information is referenced from each prefix information entry. In a virtual home network, the `hif_prefix_list_home` variable keeps all home prefix information entries. The `mip6_prefix_ha_list_insert()` function is called for all prefix information entries stored in the variable with the newly created home agent information. The home agent information is added to each prefix information entry.

15.18   Receiving ICMPv6 Error Messages

A mobile node sometimes receives an ICMPv6 error message from nodes with which the mobile node is communicating or from routers between the mobile node and communicating nodes. Some of these error messages need to be processed by the Mobile IPv6 stack.

Listing 15-222

_____ mip6_icmp6.c
```
129     int
130     mip6_icmp6_input(m, off, icmp6len)
131             struct mbuf *m;
132             int off;
133             int icmp6len;
134     {
....
195             case ICMP6_PARAM_PROB:
196                     if (!MIP6_IS_MN)
197                             break;
```
_____ mip6_icmp6.c

195–197 If a mobile node receives an ICMPv6 Parameter Problem message, the mobile node needs to see whether the message is related to Mobile IPv6 mobile node function. The error message is processed only in a mobile node.

Listing 15-223

_____ mip6_icmp6.c
```
199                     pptr = ntohl(icmp6->icmp6_pptr);
200                     if ((sizeof(*icmp6) + pptr + 1) > icmp6len) {
201                             /*
202                              * we can't get the detail of the
203                              * packet, ignore this...
204                              */
205                             break;
206                     }
```
_____ mip6_icmp6.c

199–205 The problem pointer (`icmp6_pptr`) points to the address where the error occurred. The pointer may point to an address that is larger than the end of the incoming ICMPv6 message. An ICMPv6 error message may not be able to contain all of the original packet because of the limitation on the packet size. In this case, the packet is ignored since it is impossible to know what the problem was.

Listing 15-224

_____ mip6_icmp6.c
```
208                     switch (icmp6->icmp6_code) {
209                     case ICMP6_PARAMPROB_OPTION:
210                             /*
211                              * XXX: TODO
212                              *
213                              * should we mcopydata??
214                              */
215                             origip6 = (caddr_t)(icmp6 + 1);
216                             switch (*(u_int8_t *)(origip6 + pptr)) {
217                             case IP6OPT_HOME_ADDRESS:
218                                     /*
219                                      * the peer doesn't recognize HAO.
220                                      */
....
223                                     IP6_EXTHDR_CHECK(m, off, icmp6len, EINVAL);
224                                     mip6_icmp6_find_addr(m, off, icmp6len,
225                                                     &laddr, &paddr);
....
229                                     /*
230                                      * if the peer doesn't support HAO, we
231                                      * must use bi-directional tunneling
232                                      * to continue communication.
```

```
233                                                     */
234                                    for (sc = LIST_FIRST(&hif_softc_list); sc;
235                                         sc = LIST_NEXT(sc, hif_entry)) {
236                                              mbu = mip6_bu_list_find_withpaddr
    (&sc->hif_bu_list, &paddr, &laddr);
237                                                  mip6_bu_fsm
    (mbu, MIP6_BU_PRI_FSM_EVENT_ICMP_PARAMPROB, NULL);
238                                          }
239                                          break;
240                                   }
241                            break;
```
——— mip6_icmp6.c

Note: Lines 236 and 237 are broken here for layout reasons. However, they are each a single line of code.

208–217 A mobile node may receive an ICMPv6 Parameter Problem message against the Home
Address option if the node with which the mobile node is communicating does not support
the Home Address option.

224–241 The `mip6_icmp6_find_addr()` function is called to get the source and
destination addresses of the original packet. With these addresses, the corresponding
binding update list entry is found by the `mip6_bu_list_find_withpaddr()`
function. If the problem pointer points to one of the destination options and the
option type is the Home Address option, the mobile node sends an
`MIP6_BU_PRI_FSM_EVENT_ICMP_PARAMPROB` event to the state machine of the binding update list entry of the remote node. The event will mark the entry that the node does
not support Mobile IPv6.

Listing 15-225
——— mip6_icmp6.c
```
243                     case ICMP6_PARAMPROB_NEXTHEADER:
244                          origip6 = (caddr_t)(icmp6 + 1);
245                          switch (*(u_int8_t *)(origip6 + pptr)) {
246                          case IPPROTO_MH:
247                                  /*
248                                   * the peer doesn't recognize mobility header.
249                                   */
250                                  mip6stat.mip6s_paramprobmh++;
251
252                                  IP6_EXTHDR_CHECK(m, off, icmp6len, EINVAL);
253                                  mip6_icmp6_find_addr(m, off, icmp6len,
254                                                    &laddr, &paddr);
....
258                                  for (sc = LIST_FIRST(&hif_softc_list); sc;
259                                       sc = LIST_NEXT(sc, hif_entry)) {
260                                            mbu = mip6_bu_list_find_withpaddr
    (&sc->hif_bu_list, &paddr, &laddr);
261                                                mip6_bu_fsm
    (mbu, MIP6_BU_PRI_FSM_EVENT_ICMP_PARAMPROB, NULL);
262                                        }
263                                        break;
264                                }
265                                break;
266                        }
267                  break;
....
269             }
270
271             return (0);
272     }
```
——— mip6_icmp6.c

Note: Lines 260 and 261 are broken here for layout reasons. However, they are each a single line of code.

243–267 A mobile node may receive an ICMPv6 Parameter Problem message against the next header value of the Mobility Header sent before if the peer node with which the mobile node is communicating does not recognize the extension header. In this case, the mobile node will send an `MIP6_BU_PRI_FSM_EVENT_ICMP_PARAMPROB` event to the state machine as well.

15.19   State Machine

Each binding update list entry has a state machine to hold the registration state and react to incoming events properly. Table 10-12 in Chapter 10 shows the list of the states of a state machine. Table 15-2 shows the list of the events that are sent to the state machine. Table 15-3

TABLE 15-2

Events (for primary state)	Description
`MIP6_BU_PRI_FSM_EVENT_MOVEMENT`	Moved from one foreign network to another foreign network
`MIP6_BU_PRI_FSM_EVENT_RETURNING_HOME`	Returned home
`MIP6_BU_PRI_FSM_EVENT_REVERSE_PACKET`	Received a bidirectional packet
`MIP6_BU_PRI_FSM_EVENT_RR_DONE`	Return routability procedure has been completed
`MIP6_BU_PRI_FSM_EVENT_RR_DONE`	Return routability procedure has been completed
`MIP6_BU_PRI_FSM_EVENT_RR_FAILED`	Return routability procedure failed
`MIP6_BU_PRI_FSM_EVENT_BRR`	Received a Binding Refresh Request message
`MIP6_BU_PRI_FSM_EVENT_BA`	Received a Binding Acknowledgment message
`MIP6_BU_PRI_FSM_EVENT_NO_BINDING`	(not used)
`MIP6_BU_PRI_FSM_EVENT_UNVERIFIED_HAO`	Received a Binding Error with UNKNOWN_HAO status
`MIP6_BU_PRI_FSM_EVENT_UNKNOWN_MH_TYPE`	Received a Binding Error with UNKNOWN_MH status
`MIP6_BU_PRI_FSM_EVENT_ICMP_PARAMPROB`	Received an ICMPv6 Parameter Problem message
`MIP6_BU_PRI_FSM_EVENT_RETRANS_TIMER`	Retransmission timer expired
`MIP6_BU_PRI_FSM_EVENT_REFRESH_TIMER`	Refresh timer expired
`MIP6_BU_PRI_FSM_EVENT_FAILURE_TIMER`	(not used)

Events (for secondary state)	Description
`MIP6_BU_SEC_FSM_EVENT_START_RR`	Return routability procedure is initiated
`MIP6_BU_SEC_FSM_EVENT_START_HOME_RR`	Return routability procedure for returning home is initiated
`MIP6_BU_SEC_FSM_EVENT_STOP_RR`	Return routability procedure needs to be stopped

(Continued)

TABLE 15-2 (*Cont.*)

Events (for secondary state)	Description
MIP6_BU_SEC_FSM_EVENT_HOT	Received a Home Test message
MIP6_BU_SEC_FSM_EVENT_COT	Received a Care-of Test message
MIP6_BU_SEC_FSM_EVENT_RETRANS_TIMER	Retransmission timer expired

Events for the state machine.

TABLE 15-3

Name	Description
MIP6_BU_IS_PRI_FSM_EVENT(*ev*)	True if ev is an event for the primary state machine
MIP6_BU_IS_SEC_FSM_EVENT(*ev*)	True if ev is an event for the secondary state machine

Macros to determine if the specified event is for the primary state machine or secondary state machine.

FIGURE 15-4

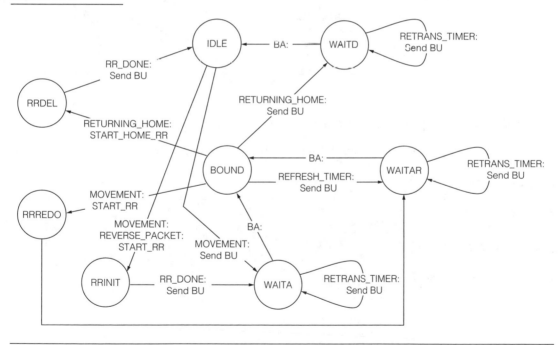

The state transition graph of the primary state machine.

shows macros to judge that the specified event is for the primary state machine or the secondary state machine.

Figures 15-4 and 15-5 show the basic state transition graph of both the primary and secondary state machines, respectively. The figures do not describe any error conditions. The error handling is discussed in Sections 15.20 and 15.21.

FIGURE 15-5

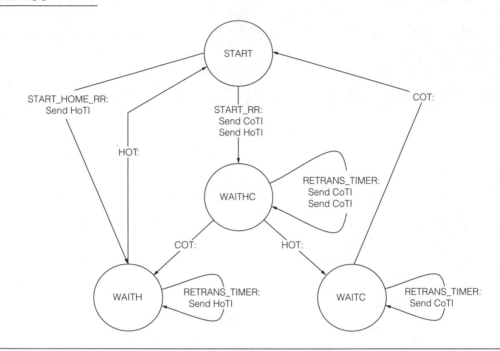

The state transition graph of the secondary state machine.

15.20   Primary State Machine

The `mip6_bu_fsm()` function is an entry point to the state machine.

Listing 15-226

_____ mip6_fsm.c

```
74      int
75      mip6_bu_fsm(mbu, event, data)
76              struct mip6_bu *mbu;
77              int event;
78              void *data;
79      {
80              /* sanity check. */
81              if (mbu == NULL)
82                      return (EINVAL);
83
84              if (MIP6_BU_IS_PRI_FSM_EVENT(event))
85                      return (mip6_bu_pri_fsm(mbu, event, data));
86              if (MIP6_BU_IS_SEC_FSM_EVENT(event))
87                      return (mip6_bu_sec_fsm(mbu, event, data));
88
89              /* invalid event. */
90              return (EINVAL);
91      }
```

_____ mip6_fsm.c

74–78 The `mip6_bu_fsm()` function has three parameters: the `mbu` parameter is a pointer
to the binding update list entry; the `event` parameter is one of the events listed in

Table 15-2; and the `data` parameter is a pointer to additional data depending on the event.

84–87 A binding update list entry has two kinds of states. Based on the type of input event, the `mip6_bu_pri_fsm()` function or the `mip6_bu_sec_fsm()` function is called. The former is for the primary state machine, which maintains the registration status of the binding update list entry. The latter is for the secondary state machine, which maintains the status of the return routability procedure.

Listing 15-227

```
                                                                          mip6_fsm.c
93        int
94        mip6_bu_pri_fsm(mbu, event, data)
95                struct mip6_bu *mbu;
96                int event;
97                void *data;
98        {
99                u_int8_t *mbu_pri_fsm_state;
100               int error;
....
102               struct timeval mono_time;
....
104               struct ip6_mh_binding_request *ip6mr;
105               struct ip6_mh_binding_ack *ip6ma;
106               struct ip6_mh_binding_error *ip6me;
107               struct icmp6_hdr *icmp6;
108               struct hif_softc *hif;
                                                                          mip6_fsm.c
```

93–97 The `mip6_bu_pri_fsm()` function has three parameters, which are the same as the values passed to the `mip6_bu_fsm()` function.

Listing 15-228

```
                                                                          mip6_fsm.c
118               mbu_pri_fsm_state = &mbu->mbu_pri_fsm_state;
119
120               /* set pointers. */
121               ip6mr = (struct ip6_mh_binding_request *)data;
122               ip6ma = (struct ip6_mh_binding_ack *)data;
123               ip6me = (struct ip6_mh_binding_error *)data;
124               icmp6 = (struct icmp6_hdr *)data;
125               hif = (struct hif_softc *)data;
                                                                          mip6_fsm.c
```

118 The `mbu_pri_fsm_state` pointer points to the address of the `mbu_pri_fsm_state` member variable. This is just for providing an easy way to access the member.

121–125 The `data` parameter points to various structures based on the event input to the function. `ip6mr`, `ip6ma`, `ip6me`, `icmp6`, and `hif` point to the same address as the `data` parameter providing access to each structure. Based on the event, zero or one of the variables has the real meaning.

15.20.1 `MIP6_BU_PRI_FSM_STATE_IDLE` State

Lines 130–214 process events when the primary state machine is in the `MIP6_BU_PRI_FSM_STATE_IDLE` state.

Listing 15-229

```
_____mip6_fsm.c
127             error = 0;
128
129             switch (*mbu_pri_fsm_state) {
130             case MIP6_BU_PRI_FSM_STATE_IDLE:
131                     switch (event) {
132                     case MIP6_BU_PRI_FSM_EVENT_MOVEMENT:
133                             if ((mbu->mbu_flags & IP6MU_HOME) != 0) {
134                                     /*
135                                      * Send BU,
136                                      * Reset retransmission counter,
137                                      * Start retransmission timer,
138                                      * XXX Start failure timer.
139                                      */
140                                     mbu->mbu_retrans_count = 0;
141
142                                     error = mip6_bu_pri_fsm_home_registration(mbu);
143                                     if (error) {
....
149                                             /* continue and try again. */
150                                     }
151
152                                     *mbu_pri_fsm_state
153                                         = MIP6_BU_PRI_FSM_STATE_WAITA;
_____mip6_fsm.c
```

132–153 If the `MIP6_BU_PRI_FSM_EVENT_MOVEMENT` event is sent to the binding update
list entry for home registration, the `mip6_bu_pri_fsm_home_registration()` function is called to perform the home registration procedure. The state is changed to
`MIP6_BU_PRI_FSM_STATE_WAITA`.

Listing 15-230

```
_____mip6_fsm.c
154                             } else {
155                                     /*
156                                      * Start RR.
157                                      */
158                                     error = mip6_bu_sec_fsm(mbu,
159                                         MIP6_BU_SEC_FSM_EVENT_START_RR,
160                                         data);
161                                     if (error) {
....
167                                             return (error);
168                                     }
169                                     *mbu_pri_fsm_state
170                                         = MIP6_BU_PRI_FSM_STATE_RRINIT;
171                             }
172                             break;
_____mip6_fsm.c
```

155–171 If the entry is not for home registration, the `mip6_bu_sec_fsm()` function is called
with the `MIP6_BU_SEC_FSM_EVENT_START_RR` event to initiate the return routability
procedure. The state is changed to `MIP6_BU_PRI_FSM_STATE_RRINIT`.

Listing 15-231

```
_____mip6_fsm.c
174                     case MIP6_BU_PRI_FSM_EVENT_REVERSE_PACKET:
175                             /*
176                              * Start RR.
```

```
177                                  */
178                                  if ((mbu->mbu_state & MIP6_BU_STATE_NEEDTUNNEL)
179                                      != 0) {
180                                      /*
181                                       * if the peer doesn't support MIP6,
182                                       * keep IDLE state.
183                                       */
184                                      break;
185                                  }
186                                  error = mip6_bu_sec_fsm(mbu,
187                                      MIP6_BU_SEC_FSM_EVENT_START_RR,
188                                      data);
189                                  if (error) {
....
194                                      return (error);
195                                  }
196
197                                  *mbu_pri_fsm_state = MIP6_BU_PRI_FSM_STATE_RRINIT;
198
199                                  break;
```
——— mip6_fsm.c

174–199 If the MIP6_BU_PRI_FSM_EVENT_REVERSE_PACKET event is input, the mip6_bu_sec_fsm() function is called with the MIP6_BU_PRI_FSM_EVENT_REVERSE_PACKET event, unless the binding update list entry does not have the MIP6_BU_STATE_NEEDTUNNEL flag set. The MIP6_BU_STATE_NEEDTUNNEL flag means that the peer node does not support Mobile IPv6 and there is no need to perform the return routability procedure.

The state is changed to MIP6_BU_PRI_FSM_STATE_RRINIT.

Listing 15-232
——— mip6_fsm.c
```
201                          case MIP6_BU_PRI_FSM_EVENT_ICMP_PARAMPROB:
202                              /*
203                               * Stop timers.
204                               */
205                              mip6_bu_stop_timers(mbu);
206
207                              *mbu_pri_fsm_state = MIP6_BU_PRI_FSM_STATE_IDLE;
208
209                              mbu->mbu_state |= MIP6_BU_STATE_DISABLE;
210
211                              break;
212                          }
213                      break;
```
——— mip6_fsm.c

201–212 If the MIP6_BU_PRI_FSM_EVENT_ICMP_PARAMPROB event is input, the mip6_bu_stop_timers() function is called to stop all timer functions related to this binding update list entry. The state remains MIP6_BU_PRI_FSM_STATE_IDLE. The MIP6_BU_STATE_DISABLE flag is set since the input of an ICMPv6 Parameter Problem message indicates that the peer node does not recognize the Home Address option or the Mobility Header.

15.20.2 MIP6_BU_PRI_FSM_STATE_RRINIT State

Lines 215–376 process events when the primary state machine is in the MIP6_BU_PRI_FSM_STATE_RRINIT state.

Listing 15-233

_____mip6_fsm.c
```
215              case MIP6_BU_PRI_FSM_STATE_RRINIT:
216                  switch (event) {
217                  case MIP6_BU_PRI_FSM_EVENT_RR_DONE:
218                      if ((mbu->mbu_flags & IP6MU_ACK) != 0) {
219                          /*
220                           * if A flag is set,
221                           *    Send BU,
222                           *    Reset retransmission counter,
223                           *    Start retransmission timer,
224                           *    Start failure timer.
225                           */
226
227                          /* XXX no code yet. */
228
229                          *mbu_pri_fsm_state
230                              = MIP6_BU_PRI_FSM_STATE_WAITA;
```
_____mip6_fsm.c

215–230 This part is intended to send a Binding Update message to a correspondent node with the Acknowledgment (A) flag set. However, the current KAME implementation never sets the A flag in the Binding Update message sent to correspondent nodes. This part is not implemented.

Listing 15-234

_____mip6_fsm.c
```
231                      } else {
232                          /*
233                           * if A flag is not set,
234                           *    Send BU,
235                           *    Start refresh timer.
236                           */
237                          error = mip6_bu_send_cbu(mbu);
238                          if (error) {
....
243                              return (error);
244                          }
245
246                          mbu->mbu_retrans
247                              = mono_time.tv_sec + mbu->mbu_lifetime;
248
249                          *mbu_pri_fsm_state
250                              = MIP6_BU_PRI_FSM_STATE_BOUND;
251                      }
252                      break;
```
_____mip6_fsm.c

231–251 If the entry does not have the A flag set, the `mip6_bu_send_cbu()` function is called to send a Binding Update message to the correspondent node related to the binding update list entry. The `mbu_retrans` variable is set to the lifetime of the binding update list entry to remove the entry when the lifetime of the entry expires. The state is changed to `MIP6_BU_PRI_FSM_STATE_BOUND`.

Listing 15-235

_____mip6_fsm.c
```
254                  case MIP6_BU_PRI_FSM_EVENT_UNKNOWN_MH_TYPE:
255                      /*
256                       * Stop timers,
```

```
257                          * Stop RR.
258                          */
259                         mip6_bu_stop_timers(mbu);
260
261                         error = mip6_bu_sec_fsm(mbu,
262                             MIP6_BU_SEC_FSM_EVENT_STOP_RR,
263                             data);
264                         if (error) {
....
269                                 return (error);
270                         }
271
272                         *mbu_pri_fsm_state = MIP6_BU_PRI_FSM_STATE_IDLE;
273
274                         mbu->mbu_state |= MIP6_BU_STATE_DISABLE;
275
276                         break;
```
_____ mip6_fsm.c

254–274 If the `MIP6_BU_PRI_FSM_EVENT_UNKNOWN_MH_TYPE` event is input,
the `mip6_bu_sec_fsm()` function is called with the
`MIP6_BU_SEC_FSM_EVENT_STOP_RR` event to stop the running return routability
procedure since the peer node does not recognize the Mobility Header. The timers
are stopped by the `mip6_bu_stop_timers()` function and the state is changed to
`MIP6_BU_PRI_FSM_STATE_IDLE`. Also, the `MIP6_BU_STATE_DISABLE` flag is set
to indicate that the peer node does not support Mobile IPv6.

Listing 15-236
_____ mip6_fsm.c

```
278                     case MIP6_BU_PRI_FSM_EVENT_MOVEMENT:
279                             /*
280                              * Stop timers,
281                              * Stop RR,
282                              * Start RR.
283                              */
284                             mip6_bu_stop_timers(mbu);
285
286                             error = mip6_bu_sec_fsm(mbu,
287                                 MIP6_BU_SEC_FSM_EVENT_STOP_RR,
288                                 data);
289
290                             if (error == 0) {
291                                     error = mip6_bu_sec_fsm(mbu,
292                                         MIP6_BU_SEC_FSM_EVENT_START_RR,
293                                         data);
294                             }
295                             if (error) {
....
300                                     return (error);
301                             }
302
303                             *mbu_pri_fsm_state = MIP6_BU_PRI_FSM_STATE_RRINIT;
304
305                             break;
```
_____ mip6_fsm.c

278–305 If the `MIP6_BU_PRI_FSM_EVENT_MOVEMENT` event is input when the current
state is `MIP6_BU_PRI_FSM_STATE_RRINIT`, the mobile node needs to stop the running return routability procedure and needs to start a new return routability procedure
using the new care-of address. The `mip6_bu_sec_fsm()` function is called with the

MIP6_BU_SEC_FSM_EVENT_STOP_RR event to stop the current return routability procedure and the same function is called with the MIP6_BU_SEC_FSM_EVENT_START_RR event immediately to start a new procedure. The state remains MIP6_BU_PRI_FSM_STATE_RRINIT.

Listing 15-237

```
                                                                       mip6_fsm.c
307                     case MIP6_BU_PRI_FSM_EVENT_RETURNING_HOME:
308                         /*
309                          * Stop timers,
310                          * Stop RR.
311                          */
312                         mip6_bu_stop_timers(mbu);
313
314                         error = mip6_bu_sec_fsm(mbu,
315                             MIP6_BU_SEC_FSM_EVENT_STOP_RR,
316                             data);
317                         if (error) {
....
322                             return (error);
323                         }
324
325                         *mbu_pri_fsm_state = MIP6_BU_PRI_FSM_STATE_IDLE;
326
327                         /* free mbu */
328                         mbu->mbu_lifetime = 0;
329                         mbu->mbu_expire = mono_time.tv_sec + mbu->mbu_lifetime;
330
331                         break;
                                                                       mip6_fsm.c
```

307–329 If the mobile node returns home, the MIP6_BU_PRI_FSM_EVENT_RETURNING_HOME event is input. In this case, the mip6_bu_stop_timers() function is called to stop all timer functions related to this entry and the mip6_bu_sec_fsm() function is called with the MIP6_BU_SEC_FSM_EVENT_STOP_RR event. The state is changed to MIP6_BU_PRI_FSM_STATE_IDLE. The lifetime of the binding update list entry is set to 0 and the mbu_expire variable is set to the current time. The entry will be removed when the mip6_bu_timer() function is called next time.

Listing 15-238

```
                                                                       mip6_fsm.c
333                     case MIP6_BU_PRI_FSM_EVENT_REVERSE_PACKET:
334                         /*
335                          * Start RR.
336                          */
337                         error = mip6_bu_sec_fsm(mbu,
338                             MIP6_BU_SEC_FSM_EVENT_START_RR,
339                             data);
340                         if (error) {
....
345                             return (error);
346                         }
347
348                         *mbu_pri_fsm_state = MIP6_BU_PRI_FSM_STATE_RRINIT;
349
350                         break;
                                                                       mip6_fsm.c
```

333–350 This code is almost the same as the `MIP6_BU_PRI_FSM_EVENT_REVERSE_PACKET` event in the `MIP6_BU_PRI_STATE_IDLE` state. The difference is this code does not check the `MIP6_BU_STATE_NEEDTUNNEL` flag of the binding update list entry. If the entry has the `MIP6_BU_STATE_NEEDTUNNEL` flag set, the return routability procedure has already been previously stopped. The mobile node never enters this part of the code in that case.

Listing 15-239

```
                                                                    mip6_fsm.c
352                  case MIP6_BU_PRI_FSM_EVENT_ICMP_PARAMPROB:
    . . . [See lines 255–274]
374                      break;
375                  }
376              break;
                                                                    mip6_fsm.c
```

352–374 A mobile node receives an ICMPv6 Parameter Problem message if the peer node does not support Mobile IPv6. The procedure when a mobile node receives the ICMPv6 message is the same as with receiving an unknown Mobility Header message described on lines 255–274.

15.20.3 `MIP6_BU_PRI_FSM_STATE_RRREDO` State

Lines 378–542 process events when the primary state machine is in the `MIP6_BU_PRI_FSM_STATE_RRREDO` state. Most of the code is the same as for the `MIP6_BU_PRI_FSM_STATE_RRREDO` state.

Listing 15-240

```
                                                                    mip6_fsm.c
378          case MIP6_BU_PRI_FSM_STATE_RRREDO:
379              switch (event) {
380              case MIP6_BU_PRI_FSM_EVENT_RR_DONE:
    . . . [See lines 218–251]
415                  break;
416
417              case MIP6_BU_PRI_FSM_EVENT_UNKNOWN_MH_TYPE:
    . . . [See lines 256–276]
439                  break;
440
441              case MIP6_BU_PRI_FSM_EVENT_MOVEMENT:
    . . . [See lines 279–304]
468                  break;
469
470              case MIP6_BU_PRI_FSM_EVENT_RETURNING_HOME:
    . . . [See lines 308–330]
497                  break;
    . . . .
518              case MIP6_BU_PRI_FSM_EVENT_ICMP_PARAMPROB:
    . . . [See lines 255–274]
540                  break;
                                                                    mip6_fsm.c
```

380–497, 518–540 The code when a state machine receives

the MIP6_BU_PRI_FSM_EVENT_RR_DONE,

the MIP6_BU_PRI_FSM_EVENT_UNKNOWN_MH_TYPE,

the MIP6_BU_PRI_FSM_EVENT_MOVEMENT,

the MIP6_BU_PRI_FSM_EVENT_RETURNING_HOME,

or the MIP6_BU_PRI_FSM_EVENT_ICMP_PARAMPROB event

is the same as for each event implemented in the MIP6_BU_PRI_FSM_STATE_RRINIT.

Listing 15-241
```
                                                               _____ mip6_fsm.c
499                     case MIP6_BU_PRI_FSM_EVENT_REVERSE_PACKET:
500                         /*
501                          * Start RR.
502                          */
503                         error = mip6_bu_sec_fsm(mbu,
504                             MIP6_BU_SEC_FSM_EVENT_START_RR,
505                             data);
506                         if (error) {
....
511                                 return (error);
512                         }
513
514                         *mbu_pri_fsm_state = MIP6_BU_PRI_FSM_STATE_RRREDO;
515
516                         break;
....
541                     }
542                     break;
                                                               _____ mip6_fsm.c
```

499–516 The code for the MIP6_BU_PRI_FSM_EVENT_REVERSE_PACKET event in the
MIP6_BU_PRI_FSM_STATE_RRREDO state is almost the same as the relevant code for
the MIP6_BU_PRI_FSM_STATE_RRREDO state. The only difference is that the next state
is set to the MIP6_BU_PRI_FSM_STATE_RRREDO state.

15.20.4 MIP6_BU_PRI_FSM_STATE_WAITA State

Lines 544–744 process events when the primary state machine is in the
MIP6_BU_PRI_FSM_STATE_WAITA state.

Listing 15-242
```
                                                               _____ mip6_fsm.c
544             case MIP6_BU_PRI_FSM_STATE_WAITA:
545                 switch (event) {
546                 case MIP6_BU_PRI_FSM_EVENT_BA:
547                     /* XXX */
548                     if ((mbu->mbu_flags & IP6MU_HOME) != 0) {
549                         /*
550                          * (Process BA,)
551                          * Stop timer,
552                          * Reset retransmission counter,
553                          * Start refresh timer.
554                          */
555
```

```
556                                    /* XXX home registration completed. */
557
558                                    mip6_bu_stop_timers(mbu);
559
560                                    mbu->mbu_retrans_count = 0;
561
562                                    mbu->mbu_retrans
563                                        = mono_time.tv_sec
564                                        + mbu->mbu_refresh;
565
566                                    *mbu_pri_fsm_state
567                                        = MIP6_BU_PRI_FSM_STATE_BOUND;
568                            } else {
569                                    /* XXX no code yet. */
570                            }
571                            break;
                                                                        mip6_fsm.c
```

546–567 Receiving the `MIP6_BU_PRI_FSM_EVENT_BA` event means that the registration message has been successfully accepted by the peer node. If the binding update list entry
related to this event is for home registration, the mobile node sets the `mbu_retrans`
variable to indicate the next refresh time so that the mobile node can perform
reregistration before the binding update entry expires. The state is changed to
`MIP6_BU_PRI_FSM_STATE_BOUND`.

568–570 This code is for a correspondent node. However, as we already discussed, the code
is empty since the current KAME implementation does not require correspondent nodes
to reply to a Binding Acknowledgment message.

Listing 15-243
                                                                        mip6_fsm.c
```
573                    case MIP6_BU_PRI_FSM_EVENT_RETRANS_TIMER:
574                            /*
575                             * Send BU,
576                             * Start retransmittion timer.
577                             */
578                            if ((mbu->mbu_flags & IP6MU_HOME) != 0) {
579                                    /*
580                                     * Send BU,
581                                     * Start retransmission timer.
582                                     */
583                                    error = mip6_bu_pri_fsm_home_registration(mbu);
584                                    if (error) {
....
590                                            /* continue and try again. */
591                                    }
592
593                                    *mbu_pri_fsm_state
594                                        = MIP6_BU_PRI_FSM_STATE_WAITA;
595                            } else {
....
615                            }
616                            break;
                                                                        mip6_fsm.c
```

573–594 If the `MIP6_BU_PRI_FSM_EVENT_RETRANS_TIMER` event is input while a mobile
node is in the `MIP6_BU_PRI_FSM_STATE_WAITA` state, a Binding Update message must
be resent. If the binding update list entry related to this event is for home registration,
the `mip6_bu_pri_fsm_home_registration()` function is called to send a Binding
Update message. The state is not changed.

595–615 This code is for a correspondent node and is never executed in the KAME implementation.

Listing 15-244

```
                                                                          mip6_fsm.c
618                    case MIP6_BU_PRI_FSM_EVENT_UNKNOWN_MH_TYPE:
619                        if ((mbu->mbu_flags & IP6MU_HOME) != 0) {
620                            /* XXX correct ? */
621                            break;
622                        }
623
624                        /*
625                         * Stop timers.
626                         */
627                        mip6_bu_stop_timers(mbu);
628
629                        *mbu_pri_fsm_state = MIP6_BU_PRI_FSM_STATE_IDLE;
630
631                        mbu->mbu_state |= MIP6_BU_STATE_DISABLE;
632
633                        break;
                                                                          mip6_fsm.c
```

618–633 If a mobile node receives the MIP6_BU_PRI_FSM_EVENT_UNKNOWN_MH_TYPE event while the node is in the MIP6_BU_PRI_FSM_STATE_WAITA state, the node stops all timer functions related to the binding update list entry and changes its state to MIP6_BU_PRI_FSM_STATE_IDLE. The MIP6_BU_STATE_DISABLE flag is also set in the entry. Note that we do not process this event for a home registration entry in order to always keep the entry active.

Listing 15-245

```
                                                                          mip6_fsm.c
635                    case MIP6_BU_PRI_FSM_EVENT_MOVEMENT:
636                        if ((mbu->mbu_flags & IP6MU_HOME) != 0) {
637                            /*
638                             * Send BU,
639                             * Reset retrans counter,
640                             * Start retransmission timer,
641                             * XXX Start failure timer.
642                             */
643                            mbu->mbu_retrans_count = 0;
644
645                            error = mip6_bu_pri_fsm_home_registration(mbu);
646                            if (error) {
....
652                                /* continue and try again. */
653                            }
654
655                            *mbu_pri_fsm_state
656                                = MIP6_BU_PRI_FSM_STATE_WAITA;
657                        } else {
658                            /*
659                             * Stop timers,
660                             * Start RR.
661                             */
662                            mip6_bu_stop_timers(mbu);
663
664                            error = mip6_bu_sec_fsm(mbu,
665                                MIP6_BU_SEC_FSM_EVENT_START_RR,
666                                data);
667                            if (error) {
```

```
        ....
673                                            return (error);
674                                      }
675                                      *mbu_pri_fsm_state
676                                          = MIP6_BU_PRI_FSM_STATE_RRINIT;
677                                  }
678                              break;
```
_____ mip6_fsm.c

635–678 The procedure when a mobile node receives the
`MIP6_BU_PRI_FSM_EVENT_MOVEMENT` event while it is in the
`MIP6_BU_PRI_FSM_STATE_WAITA` state is similar to the procedure that
is implemented for the `MIP6_BU_PRI_FSM_STATE_IDLE` state (on lines 132–172).
The difference is that the `mip6_bu_stop_timer()` function is called before
starting the return routability procedure for a correspondent node. In the
`MIP6_BU_PRI_FSM_STATE_IDLE` state, there is no active timer. However, in the
`MIP6_BU_PRI_FSM_STATE_WAITA` state, a retransmission timer is running, which
needs to be stopped before doing other state transitions.

Listing 15-246
_____ mip6_fsm.c

```
680                      case MIP6_BU_PRI_FSM_EVENT_RETURNING_HOME:
681                          if ((mbu->mbu_flags & IP6MU_HOME) != 0) {
682                                  /*
683                                   * Send BU,
684                                   * Reset retrans counter,
685                                   * Start retransmission timer.
686                                   */
687                                  mbu->mbu_retrans_count = 0;
688
689                                  error = mip6_bu_pri_fsm_home_registration(mbu);
690                                  if (error) {
        ....
696                                          /* continue and try again. */
697                                  }
698
699                                  *mbu_pri_fsm_state
700                                      = MIP6_BU_PRI_FSM_STATE_WAITD;
701                          } else {
702                                  /*
703                                   * Stop timers,
704                                   * Start Home RR.
705                                   */
706                                  mip6_bu_stop_timers(mbu);
707
708                                  error = mip6_bu_sec_fsm(mbu,
709                                      MIP6_BU_SEC_FSM_EVENT_START_HOME_RR,
710                                      data);
711                                  if (error) {
        ....
717                                          return (error);
718                                  }
719
720                                  *mbu_pri_fsm_state
721                                      = MIP6_BU_PRI_FSM_STATE_RRDEL;
722                          }
723                          break;
```
_____ mip6_fsm.c

681–700 If a mobile node returns home while it is in the `MIP6_BU_PRI_FSM_STATE_WAITA`
state, it starts the home deregistration procedure if the binding update list

entry is for home registration (the IP6MU_HOME flags are set). The mip6_bu_pri_fsm_home_registration() function is called to send a Binding Update message for deregistration, and the state is changed to MIP6_BU_PRI_FSM_STATE_WAITD.

If the entry is for a correspondent node, the mip6_bu_sec_fsm() function is called with the MIP6_BU_SEC_FSM_EVENT_START_HOME_RR event to perform the return routability procedure for deregistration. The state is changed to the MIP6_BU_PRI_FSM_STATE_RRDEL state.

Listing 15-247

```
                                                                      mip6_fsm.c
725                        case MIP6_BU_PRI_FSM_EVENT_REVERSE_PACKET:
.... [See lines 334–349]
742                            break;
743                        }
744                    break;
                                                                      mip6_fsm.c
```

725–742 If a mobile node receives the MIP6_BU_PRI_FSM_EVENT_REVERSE_PACKET event, it starts the return routability procedure. The code is the same as the code implemented for the MIP6_BU_PRI_FSM_STATE_RRINIT state on lines 334–349.

15.20.5 `MIP6_BU_PRI_FSM_STATE_WAITAR` State

Lines 746–948 process events when the primary state machine is in the MIP6_BU_PRI_FSM_STATE_WAITAR state. Most parts of the code are similar to the code for the MIP6_BU_PRI_FSM_STATE_WAITA state.

Listing 15-248

```
                                                                      mip6_fsm.c
746            case MIP6_BU_PRI_FSM_STATE_WAITAR:
747                switch (event) {
748                case MIP6_BU_PRI_FSM_EVENT_BA:
.... [See lines 547–571]
773                    break;
774
775                case MIP6_BU_PRI_FSM_EVENT_RETRANS_TIMER:
.... [See lines 574–592]
796                        *mbu_pri_fsm_state
797                            = MIP6_BU_PRI_FSM_STATE_WAITAR;
.... [See lines 596–615]
823                    break;
824
825                case MIP6_BU_PRI_FSM_EVENT_UNKNOWN_MH_TYPE:
.... [See lines 619–632]
840                    break;
841
842                case MIP6_BU_PRI_FSM_EVENT_MOVEMENT:
.... [See lines 636–677]
882                    break;
883
884                case MIP6_BU_PRI_FSM_EVENT_RETURNING_HOME:
.... [See lines 681–722]
```

```
927                              break;
928
929                         case MIP6_BU_PRI_FSM_EVENT_REVERSE_PACKET:
 . . . . [See lines 334–349]
946                              break;
947                         }
948                    break;
```
── mip6_fsm.c

746–948 The only difference from the code for the `MIP6_BU_PRI_FSM_STATE_WAITA` state is the next state for the `MIP6_BU_PRI_FSM_EVENT_RETRANS_TIMER` event. In the `MIP6_BU_PRI_FSM_STATE_WAITAR` state, the next state is kept unchanged when retransmitting a Binding Update message.

15.20.6   `MIP6_BU_PRI_FSM_STATE_WAITD` State

Lines 950–1075 process events when the primary state machine is in the `MIP6_BU_PRI_FSM_STATE_WAITD` state.

Listing 15-249
── mip6_fsm.c
```
950               case MIP6_BU_PRI_FSM_STATE_WAITD:
951                    switch (event) {
952                    case MIP6_BU_PRI_FSM_EVENT_BA:
953                         /* XXX */
954                         if ((mbu->mbu_flags & IP6MU_HOME) != 0) {
955                              /* XXX home de-registration completed. */
956                         } else {
957                              /* XXX no code yet. */
958                         }
959                    break;
```
── mip6_fsm.c

952–955 There is nothing to do for the `MIP6_BU_PRI_FSM_EVENT_BA` event while a mobile node is in the `MIP6_BU_PRI_FSM_STATE_WAITD` state, since all deregistration processing has already been done in the `mip6_process_hurbu()` function. The function has already been called before the state machine is called.

956–958 This part is for processing a Binding Acknowledgment from a correspondent node, which is not needed in the KAME implementation since KAME never requests a Binding Acknowledgment message to correspondent nodes.

Listing 15-250
── mip6_fsm.c
```
961                    case MIP6_BU_PRI_FSM_EVENT_RETRANS_TIMER:
962                         /* XXX */
963                         /*
964                          * Send BU,
965                          * Start retransmission timer.
966                          */
967                         if ((mbu->mbu_flags & IP6MU_HOME) != 0) {
968                              /*
969                               * Send BU,
970                               * Start retransmission timer.
971                               */
972                              error = mip6_bu_pri_fsm_home_registration(mbu);
973                              if (error) {
 . . . .
979                                   /* continue and try again. */
```

```
980                                                }
981
982                                       *mbu_pri_fsm_state
983                                          = MIP6_BU_PRI_FSM_STATE_WAITD;
984                          } else {
 ....
1008                                      }
1009                          break;
```
——— mip6_fsm.c

961–983 If the `MIP6_BU_PRI_FSM_EVENT_RETRANS_TIMER` event occurs while a mobile node is in the `MIP6_BU_PRI_FSM_STATE_WAITD` state, the `mip6_bu_pri_fsm_home_registration()` function is called to resend a Binding Update message for deregistration. The state is kept unchanged.

984–1008 The code for a correspondent node (the `IP6MU_HOME` flag is not set) is never executed since the KAME implementation does not retransmit the deregistration message because it does not set the A flag.

Listing 15-251
——— mip6_fsm.c
```
1011                     case MIP6_BU_PRI_FSM_EVENT_UNKNOWN_MH_TYPE:
 . . . . [See lines 619–632]
1026                          break;
```
——— mip6_fsm.c

1011–1026 If a mobile node receives the `MIP6_BU_PRI_FSM_EVENT_UNKNOWN_MH_TYPE` event while it is in the `MIP6_BU_PRI_FSM_STATE_WAITD` state, the binding update list entry is marked to indicate that the peer node does not support Mobile IPv6. The code is the same as the code for the `MIP6_BU_PRI_FSM_STATE_WAITA` state implemented on lines 619–632.

Listing 15-252
——— mip6_fsm.c
```
1028                     case MIP6_BU_PRI_FSM_EVENT_MOVEMENT:
 . . . . [See lines 636–677]
1073                          break;
1074                     }
1075                     break;
```
——— mip6_fsm.c

1028–1074 If a mobile node moves while it is in the `MIP6_BU_PRI_FSM_STATE_WAITD` state, it starts the registration procedure. The code is the same as the code for the `MIP6_BU_PRI_FSM_STATE_WAITA` state implemented on lines 635–678.

15.20.7 `MIP6_BU_PRI_FSM_STATE_RRDEL` State

Lines 1077–1198 process events when the primary state machine is in the `MIP6_BU_PRI_FSM_STATE_RRDEL` state.

Listing 15-253
——— mip6_fsm.c
```
1077            case MIP6_BU_PRI_FSM_STATE_RRDEL:
1078                switch (event) {
```

```
1079                           case MIP6_BU_PRI_FSM_EVENT_RR_DONE:
      .... [See lines 218–251]
1117                                  break;
1118
1119                           case MIP6_BU_PRI_FSM_EVENT_UNKNOWN_MH_TYPE:
      .... [See lines 256–276]
1143                                  break;
1144
1145                           case MIP6_BU_PRI_FSM_EVENT_MOVEMENT:
      .... [See lines 279–304]
1172                                  break;
1173
1174                           case MIP6_BU_PRI_FSM_EVENT_ICMP_PARAMPROB:
      .... [See lines 255–274]
1196                                  break;
1197                           }
1198                     break;
```
                                                                    —— mip6_fsm.c

1078–1198 The code for the MIP6_BU_PRI_FSM_STATE_RRDEL state is almost the same as
the code for the MIP6_BU_PRI_FSM_STATE_RRINIT state. In this state, the
MIP6_BU_PRI_FSM_EVENT_RETURNING_HOME event does not need to be processed
since the mobile node is already at home. For the other events, the mobile node performs
the same procedures as those in the MIP6_BU_PRI_FSM_STATE_RRINIT state.

15.20.8 `MIP6_BU_PRI_FSM STATE_BOUND` State

Lines 1201–1453 process events when the primary state machine is in the
MIP6_BU_PRI_FSM_STATE_BOUND state.

Listing 15-254

                                                                    —— mip6_fsm.c
```
1201                     case MIP6_BU_PRI_FSM_STATE_BOUND:
1202                             switch (event) {
1203                             case MIP6_BU_PRI_FSM_EVENT_BRR:
1204                                     if ((mbu->mbu_flags & IP6MU_HOME) != 0) {
1205                                             /*
1206                                              * Send BU,
1207                                              * Start retransmission timer.
1208                                              */
1209                                             error = mip6_bu_pri_fsm_home_registration(mbu);
1210                                             if (error) {
     ....
1216                                                     /* continue and try again. */
1217                                             }
1218
1219                                             *mbu_pri_fsm_state
1220                                                 = MIP6_BU_PRI_FSM_STATE_WAITAR;
1221                                     } else {
1222                                             /*
1223                                              * Stop timers,
1224                                              * Start RR.
1225                                              */
1226                                             mip6_bu_stop_timers(mbu);
1227
1228                                             error = mip6_bu_sec_fsm(mbu,
1229                                                 MIP6_BU_SEC_FSM_EVENT_START_RR,
1230                                                 data);
1231                                             if (error) {
```

```
....
1237                                              return (error);
1238                                      }
1239
1240                              *mbu_pri_fsm_state
1241                                  = MIP6_BU_PRI_FSM_STATE_RRREDO;
1242                          }
1243                      break;
```
——— mip6_fsm.c

1203–1243 If a mobile node receives the `MIP6_BU_PRI_FSM_EVENT_BRR` event while it is in the `MIP6_BU_PRI_FSM_STATE_BOUND` state, the node will resend a Binding Update message to the requesting peer node.

If the binding update list entry related to this event is for home registration, the `mip6_bu_pri_fsm_home_registration()` is called to perform the home registration procedure. The state is changed to the `MIP6_BU_PRI_FSM_STATE_WAITAR` state.

If the entry is for a correspondent node, the return routability procedure is started by calling the `mip6_bu_sec_fsm()` function with the `MIP6_BU_SEC_FSM_EVENT_START_RR` event. The state is changed to the `MIP6_BU_PRI_FSM_STATE_RRREDO` state.

Listing 15-255

——— mip6_fsm.c
```
1245                  case MIP6_BU_PRI_FSM_EVENT_MOVEMENT:
     .... [See lines 636–677]
1292                      break;
1293
1294                  case MIP6_BU_PRI_FSM_EVENT_RETURNING_HOME:
     .... [See lines 681–722]
1341                      break;
1342
1343                  case MIP6_BU_PRI_FSM_EVENT_REVERSE_PACKET:
     .... [See lines 500–516]
1360                      break;
```
——— mip6_fsm.c

1245–1341 The procedure when a mobile node receives the `MIP6_BU_PRI_FSM_EVENT_MOVEMENT` event or the `MIP6_BU_PRI_FSM_EVENT_RETURNING_HOME` event is same as the code for the `MIP6_BU_PRI_FSM_STATE_WAITA` state implemented on lines 636–722.

1343–1360 The procedure for the `MIP6_BU_PRI_FSM_EVENT_REVERSE_PACKET` event is the same procedure as for the `MIP6_BU_PRI_FSM_STATE_RRREDO` state implemented on lines 500–516.

Listing 15-256

——— mip6_fsm.c
```
1362                      case MIP6_BU_PRI_FSM_EVENT_REFRESH_TIMER:
     .... [See lines 1204–1242]
1402                      break;
```
——— mip6_fsm.c

1362–1402 The behavior when a mobile node receives the `MIP6_BU_PRI_FSM_EVENT_REFRESH_TIMER` event is the same as the procedure when the node receives the `MIP6_BU_PRI_FSM_EVENT_BRR` event implemented on lines 1204–1242. The mobile node performs the home registration procedure for the home registration entry or performs the return routability procedure for the entry of a correspondent node.

Listing 15-257

```
                                                                    mip6_fsm.c
1404                        case MIP6_BU_PRI_FSM_EVENT_UNVERIFIED_HAO:
  .... [See lines 636–677]
1451                            break;
1452                    }
1453                    break;
1454
1455            default:
1456                    panic("the state of the primary fsm is unknown.");
1457            }
1458
1459            return (0);
1460    }
                                                                    mip6_fsm.c
```

1404–1451 The `MIP6_BU_PRI_FSM_EVENT_UNVERIFIED_HAO` event indicates that a mobile node sends a packet with a Home Address option although the peer node does not have a binding cache entry for the mobile node. In this case, the mobile node needs to perform the registration procedure to create a proper binding cache entry on the peer node. The procedure is the same as the procedure for the `MIP6_BU_PRI_FSM_STATE_WAITA` state implemented on lines 636–677.

15.20.9   Initiate a Home Registration Procedure

The `mip6_bu_pri_fsm_home_registration()` is called when a mobile node needs to perform the home registration procedure.

Listing 15-258

```
                                                                    mip6_fsm.c
1462    static int
1463    mip6_bu_pri_fsm_home_registration(mbu)
1464            struct mip6_bu *mbu;
1465    {
1466            struct mip6_ha *mha;
1467            int error;
  ....
1469            struct timeval mono_time;
  ....
1471
1472            /* sanity check. */
1473            if (mbu == NULL)
1474                    return (EINVAL);
  ....
1477            microtime(&mono_time);
                                                                    mip6_fsm.c
```

1462–1464 This function calls the `mip6_home_registration2()`. The `mbu` parameter is a pointer to the instance of the `mip6_bu{}` structure, which needs home registration.

Listing 15-259

———mip6_fsm.c
```
1480                error = mip6_home_registration2(mbu);
1481                if (error) {
....
1485                        /* continue and try again. */
1486                }
1487
1488                if (mbu->mbu_retrans_count++ > MIP6_BU_MAX_BACKOFF) {
1489                        /*
1490                         * try another home agent.  if we have no alternative,
1491                         * set an unspecified address to trigger DHAAD
1492                         * procedure.
1493                         */
1494                        mha = hif_find_next_preferable_ha(mbu->mbu_hif,
1495                            &mbu->mbu_paddr);
1496                        if (mha != NULL)
1497                                mbu->mbu_paddr = mha->mha_addr;
1498                        else
1499                                mbu->mbu_paddr = in6addr_any;
1500                        mbu->mbu_retrans_count = 1;
1501                }
1502                mbu->mbu_retrans = mono_time.tv_sec + (1 << mbu->mbu_retrans_count);
1503
1504                return (error);
1505        }
```
———mip6_fsm.c

1480–1486 The `mip6_home_registration2()` function will send a Binding Update message to the home agent of the mobile node of the binding update list entry.

1488–1501 If the mobile node did not receive a Binding Acknowledgment message after `MIP6_BU_MAX_BACKOFF` times retries (currently seven retries), the node will try another home agent. The `hif_find_next_preferable_ha()` function returns the next candidate of a home agent. If there is no other candidate, the home agent address (`mbu_paddr`) is set to the unspecified address, which indicates the Dynamic Home Agent Address Discovery procedure is required.

1502 `mbu_retrans` is set to the time to send the next Binding Update message. The time is calculated using an exponential backoff algorithm.

15.20.10   Stop Timers of the Binding Update List Entry

The `mip6_bu_stop_timers()` function stops timers of the specified binding update list entry.

Listing 15-260

———mip6_fsm.c
```
1754    void
1755    mip6_bu_stop_timers(mbu)
1756            struct mip6_bu *mbu;
1757    {
1758            if (mbu == NULL)
1759                    return;
1760
1761            mbu->mbu_retrans = 0;
1762            mbu->mbu_failure = 0;
1763    }
```
———mip6_fsm.c

1754–1763 The function disables timer functions by setting the `mbu_retrans` and `mbu_failure` variables to 0. These member variables indicate the time for some tasks to be processed. These values are checked in the `mip6_bu_timer()` function, and the function does a proper task based on the state of the binding update list entry if these values are not 0.

15.21 Secondary State Machine

The `mip6_bu_sec_fsm()` function is the entry point of the secondary state machine of a binding update list entry.

Listing 15-261

```
                                                                    mip6_fsm.c
1507    int
1508    mip6_bu_sec_fsm(mbu, event, data)
1509            struct mip6_bu *mbu;
1510            int event;
1511            void *data;
1512    {
1513            u_int8_t *mbu_sec_fsm_state;
1514            int error;
....
1516            struct timeval mono_time;
....
1518            struct ip6_mh_home_test *ip6mh;
1519            struct ip6_mh_careof_test *ip6mc;
                                                                    mip6_fsm.c
```

1507–1511 The `mip6_bu_sec_fsm()` function has three parameters, which are as the same as the parameters for the `mip6_bu_pri_fsm()` function.

Listing 15-262

```
                                                                    mip6_fsm.c
1530            mbu_sec_fsm_state = &mbu->mbu_sec_fsm_state;
1531
1532            /* set pointers. */
1533            ip6mh = (struct ip6_mh_home_test *)data;
1534            ip6mc = (struct ip6_mh_careof_test *)data;
                                                                    mip6_fsm.c
```

1530 The `mbu_sec_fsm_state` pointer points to the `mbu_sec_fsm_state` variable, which indicates the status of the secondary state machine. This variable is used as a shortcut to the `mbu_sec_fsm_state` variable.

1533–1534 In the secondary state machine, the `data` pointer may point to a Home Test message or a Care-of Test message. The `ip6mh` and the `ip6mc` variables point to the `data` parameter.

15.21.1 `MIP6_BU_SEC_FSM_STATE_START` State

Lines 1539–1576 process events when the secondary state machine is in the `MIP6_BU_SEC_FSM_STATE_START` state.

Listing 15-263

```
                                                                    mip6_fsm.c
1536            error = 0;
1537
```

```
1538                 switch (*mbu_sec_fsm_state) {
1539                 case MIP6_BU_SEC_FSM_STATE_START:
1540                         switch (event) {
1541                         case MIP6_BU_SEC_FSM_EVENT_START_RR:
1542                                 /*
1543                                  * Send HoTI,
1544                                  * Send CoTI,
1545                                  * Start retransmission timer,
1546                                  * Start failure timer
1547                                  */
1548                                 if (mip6_bu_send_hoti(mbu) != 0)
1549                                         break;
1550                                 if (mip6_bu_send_coti(mbu) != 0)
1551                                         break;
1552                                 mbu->mbu_retrans
1553                                     = mono_time.tv_sec + MIP6_HOT_TIMEOUT;
1554                                 mbu->mbu_failure
1555                                     = mono_time.tv_sec + MIP6_HOT_TIMEOUT * 5;
 /* XXX */
1556                                 *mbu_sec_fsm_state = MIP6_BU_SEC_FSM_STATE_WAITHC;
1557
1558                                 break;
```
——mip6_fsm.c

Note: Line 1555 is broken here for layout reasons. However, it is a single line of code.

1541–1558 If a mobile node receives the MIP6_BU_SEC_FSM_EVENT_START_RR event while
it is in the MIP6_BU_SEC_FSM_STATE_START state, the node starts the return routability procedure. The mip6_bu_send_hoti() and the mip6_bu_send_coti() functions are called to send a Home Test Init message and a Care-of Test Init message. The
mbu_retrans variable is set to MIP6_HOT_TIMEOUT seconds (currently 5 s) to trigger
retransmission in case the message is lost. The mbu_failure variable indicates the failure timeout is set here; however, the variable is not used. The next state is changed to the
MIP6_BU_SEC_FSM_STATE_WAITHC state.

Listing 15-264

——mip6_fsm.c
```
1560                         case MIP6_BU_SEC_FSM_EVENT_START_HOME_RR:
1561                                 /*
1562                                  * Send HoTI,
1563                                  * Start retransmission timer,
1564                                  * Start failure timer
1565                                  */
1566                                 if (mip6_bu_send_hoti(mbu) != 0)
1567                                         break;
1568                                 mbu->mbu_retrans
1569                                     = mono_time.tv_sec + MIP6_HOT_TIMEOUT;
1570                                 mbu->mbu_failure
1571                                     = mono_time.tv_sec + MIP6_HOT_TIMEOUT * 5;
 /* XXX */
1572                                 *mbu_sec_fsm_state = MIP6_BU_SEC_FSM_STATE_WAITH;
1573
1574                                 break;
1575                         }
1576                 break;
```
——mip6_fsm.c

Note: Line 1571 is broken here for layout reasons. However, it is a single line of code.

1560–1574 If a mobile node receives the MIP6_BU_SEC_FSM_EVENT_START_HOME_RR
event while it is in the MIP6_BU_SEC_FSM_STATE_START state, the node starts the
returning home procedure. The node calls the mip6_bu_send_hoti() function to send

a Home Test Init message. A Care-of Test Init message does not need to be sent since the home address and the care-of address are the same at home. The `mbu_retrans` variable is set to 5 s after the current time for retransmission. The state is changed to the `MIP6_BU_SEC_FSM_STATE_WAITH` state.

15.21.2 `MIP6_BU_SEC_FSM_STATE_WAITHC` State

Lines 1578–1634 process events when the secondary state machine is in the `MIP6_BU_SEC_FSM_STATE_WAITHC` state.

Listing 15-265

mip6_fsm.c

```
1578            case MIP6_BU_SEC_FSM_STATE_WAITHC:
1579                    switch (event) {
1580                    case MIP6_BU_SEC_FSM_EVENT_HOT:
1581                            /*
1582                             * Store token, nonce index.
1583                             */
1584                            /* XXX */
1585                            mbu->mbu_home_nonce_index
1586                                = htons(ip6mh->ip6mhht_nonce_index);
1587                            bcopy(ip6mh->ip6mhht_keygen8, mbu->mbu_home_token,
1588                                sizeof(ip6mh->ip6mhht_keygen8));
1589
1590                            *mbu_sec_fsm_state = MIP6_BU_SEC_FSM_STATE_WAITC;
1591
1592                            break;
```

mip6_fsm.c

1578–1592 If a mobile node receives a Home Test message, the `MIP6_BU_SEC_FSM_EVENT_HOT` event is input. The home nonce index (`ip6mhht_nonce_index`) and the home keygen token (`ip6mhht_keygen8`) contained in the received message are stored in the `mbu_home_nouce_index` and the `mbu_home_token` variables. The state is changed to the `MIP6_BU_SEC_FSM_STATE_WAITC` state.

Listing 15-266

mip6_fsm.c

```
1594            case MIP6_BU_SEC_FSM_EVENT_COT:
1595                    /*
1596                     * Store token, nonce index.
1597                     */
1598                    /* XXX */
1599                    mbu->mbu_careof_nonce_index
1600                        = htons(ip6mc->ip6mhct_nonce_index);
1601                    bcopy(ip6mc->ip6mhct_keygen8, mbu->mbu_careof_token,
1602                        sizeof(ip6mc->ip6mhct_keygen8));
1603
1604                    *mbu_sec_fsm_state = MIP6_BU_SEC_FSM_STATE_WAITH;
1605                    break;
```

mip6_fsm.c

1594–1605 If a mobile node receives a Care-of Test message, the `MIP6_BU_SEC_FSM_EVENT_COT` is input. The care-of nonce index (`ip6mhct_nonce_index`) and the care-of keygen token (`ip6mhct_keygen8`) contained in the received message are stored in the `mbu_careof_nonce_index`

and the mbu_careof_token variables, respectively. The state is changed to the
MIP6_BU_SEC_FSM_STATE_WAITH state.

Listing 15-267

_____mip6_fsm.c

```
1607                    case MIP6_BU_SEC_FSM_EVENT_STOP_RR:
1608                        /*
1609                         * Stop timers.
1610                         */
1611                        mip6_bu_stop_timers(mbu);
1612
1613                        *mbu_sec_fsm_state = MIP6_BU_SEC_FSM_STATE_START;
1614
1615                        break;
```
_____mip6_fsm.c

1607–1615 If a mobile node cancels the running return routability procedure, the
MIP6_BU_SEC_FSM_EVENT_STOP_RR event is input. The mip6_bu_stop_timers()
function is called to stop all timers related to the binding update list entry, and the state
is changed to the MIP6_BU_SEC_FSM_STATE_START state.

Listing 15-268

_____mip6_fsm.c

```
1617                    case MIP6_BU_SEC_FSM_EVENT_RETRANS_TIMER:
1618                        /*

1620                         * Send CoTI,
1621                         * Start retransmission timer.
1622                         */
1623                        if (mip6_bu_send_hoti(mbu) != 0)
1624                                break;
1625                        if (mip6_bu_send_coti(mbu) != 0)
1626                                break;
1627                        mbu->mbu_retrans
1628                            = mono_time.tv_sec + MIP6_HOT_TIMEOUT;
1629
1630                        *mbu_sec_fsm_state = MIP6_BU_SEC_FSM_STATE_WAITHC;
1631
1632                        break;
1633                    }
1634                break;
```
_____mip6_fsm.c

1617–1630 If a mobile node does not receive either a Home Test message or a Care-
of Test message while it is in the MIP6_BU_SEC_FSM_STATE_WAITHC state, the
MIP6_BU_SEC_FSM_EVENT_RETRANS_TIMER event is input. The mobile node resends
a Home Test message or a Care-of Test message using the mip6_bu_send_hoti() and
the mip6_bu_send_coti() functions. The retransmission timer is set to 5 s
(MIP6_HOT_TIMEOUT) after the current time. Note that the retransmission interval is
not compliant with the specification. The specification says a mobile node needs to use
an exponential backoff when resending these messages.

15.21.3 MIP6_BU_SEC_FSM_STATE_WAITH State

Lines 1636–1690 process events when the secondary state machine is in the
MIP6_BU_SEC_FSM_STATE_WAITH state.

Listing 15-269

_____ mip6_fsm.c
```
1636            case MIP6_BU_SEC_FSM_STATE_WAITH:
1637                switch (event) {
1638                case MIP6_BU_SEC_FSM_EVENT_HOT:
1639                    /*
1640                     * Store token and nonce index,
1641                     * Stop timers,
1642                     * RR done.
1643                     */
1644                    mbu->mbu_home_nonce_index
1645                        = htons(ip6mh->ip6mhht_nonce_index);
1646                    bcopy(ip6mh->ip6mhht_keygen8, mbu->mbu_home_token,
1647                        sizeof(ip6mh->ip6mhht_keygen8));
1648
1649                    mip6_bu_stop_timers(mbu);
1650
1651                    error = mip6_bu_pri_fsm(mbu,
1652                        MIP6_BU_PRI_FSM_EVENT_RR_DONE,
1653                        data);
1654                    if (error) {
....
1659                        return (error);
1660                    }
1661
1662                    *mbu_sec_fsm_state = MIP6_BU_SEC_FSM_STATE_START;
1663
1664                    break;
```
_____ mip6_fsm.c

1636–1649 When a mobile node receives a Home Test message, the
`MIP6_BU_SEC_FSM_EVENT_HOT` event is input. The mobile node copies the home
nonce index (`ip6mhht_nonce_index`) and the home keygen token
(`ip6mhht_keygen8`) contained in the message to the `mbu_home_nonce_index`
and the `mbu_home_token` variables of the binding update list entry related to
the message. The mobile node stops all the timers and sends the
`MIP6_BU_PRI_FSM_EVENT_RR_DONE` event to the primary state machine of the
binding update list entry to notify it of the completion of the return routability procedure, since the node has received both care-of and home tokens from the peer
node. The primary state machine will initiate the registration procedure using the
tokens.

The state is reset to the `MIP6_BU_SEC_FSM_STATE_START` state.

Listing 15-270

_____ mip6_fsm.c
```
1666            case MIP6_BU_SEC_FSM_EVENT_STOP_RR:
1667                /*
1668                 * Stop timers.
1669                 */
1670                mip6_bu_stop_timers(mbu);
1671
1672                *mbu_sec_fsm_state = MIP6_BU_SEC_FSM_STATE_START;
1673
1674                break;
```
_____ mip6_fsm.c

1666–1674 When a mobile node receives the `MIP6_BU_SEC_FSM_EVENT_STOP_RR` event,
the running return routability procedure is stopped. The code is the same code as lines
1607–1615.

Listing 15-271

```
                                                                    mip6_fsm.c
1676                    case MIP6_BU_SEC_FSM_EVENT_RETRANS_TIMER:
1677                        /*
1678                         * Send HoTI,
1679                         * Start retransmission timer.
1680                         */
1681                        if (mip6_bu_send_hoti(mbu) != 0)
1682                            break;
1683                        mbu->mbu_retrans
1684                            = mono_time.tv_sec + MIP6_HOT_TIMEOUT;
1685
1686                        *mbu_sec_fsm_state = MIP6_BU_SEC_FSM_STATE_WAITH;
1687
1688                        break;
1689                    }
1690                    break;
                                                                    mip6_fsm.c
```

1676–1690 If a mobile node does not receive a Home Test message in 5 s after sending a Home Test Init message, the MIP6_BU_SEC_FSM_EVENT_RETRANS_TIMER event is input. The mobile node resends a Home Test message using the mip6_bu_send_hoti() function and resets the next retransmission timer. The state is unchanged.

15.21.4 MIP6_BU_SEC_FSM_STATE_WAITC State

Lines 1692–1746 process events when the secondary state machine is in the MIP6_BU_SEC_FSM_STATE_WAITC state.

Listing 15-272

```
                                                                    mip6_fsm.c
1692            case MIP6_BU_SEC_FSM_STATE_WAITC:
1693                switch (event) {
1694                case MIP6_BU_SEC_FSM_EVENT_COT:
1695                    /*
1696                     * Store token and nonce index,
1697                     * Stop timers,
1698                     * RR done.
1699                     */
1700                    mbu->mbu_careof_nonce_index
1701                        = htons(ip6mc->ip6mhct_nonce_index);
1702                    bcopy(ip6mc->ip6mhct_keygen8, mbu->mbu_careof_token,
1703                        sizeof(ip6mc->ip6mhct_keygen8));
1704
1705                    mip6_bu_stop_timers(mbu);
1706
1707                    error = mip6_bu_pri_fsm(mbu,
1708                        MIP6_BU_PRI_FSM_EVENT_RR_DONE,
1709                        data);
1710                    if (error) {
....
1715                        return (error);
1716                    }
1717
1718                    *mbu_sec_fsm_state = MIP6_BU_SEC_FSM_STATE_START;
1719
1720                    break;
                                                                    mip6_fsm.c
```

1692–1720 If a mobile node receives a Care-of Test message, the MIP6_BU_SEC_FSM_STATE_WAITC event is input. If the mobile node is in the MIP6_BU_SEC_FSM_STATE_WAITC state, the node copies the care-of nonce index (ip6mhct_nonce_index) and the care-of keygen token (ip6mhct_keygen8) to the mbu_careof_nonce_index and the mbu_careof_token variables, respectively. The node stops the timers related to the binding update list entry and sends the MIP6_BU_PRI_FSM_EVENT_RR_DONE event to notify the primary state machine that the return routability procedure has been completed. The state is changed to the MIP6_BU_SEC_FSM_STATE_START state.

Listing 15-273

_____ mip6_fsm.c

```
1722                    case MIP6_BU_SEC_FSM_EVENT_STOP_RR:
1723                            /*
1724                             * Stop timers.
1725                             */
1726                            mip6_bu_stop_timers(mbu);
1727
1728                            *mbu_sec_fsm_state = MIP6_BU_SEC_FSM_STATE_START;
1729
1730                            break;
```
_____ mip6_fsm.c

1722–1730 If a mobile node cancels the current running return routability procedure, the MIP6_BU_SEC_FSM_EVENT_STOP_RR event is input. The code is the same code as in lines 1607–1615.

Listing 15-274

_____ mip6_fsm.c

```
1732                    case MIP6_BU_SEC_FSM_EVENT_RETRANS_TIMER:
1733                            /*
1734                             * Send CoTI,
1735                             * Start retransmission timer.
1736                             */
1737                            if (mip6_bu_send_coti(mbu) != 0)
1738                                    break;
1739                            mbu->mbu_retrans
1740                                = mono_time.tv_sec + MIP6_HOT_TIMEOUT;
1741
1742                            *mbu_sec_fsm_state = MIP6_BU_SEC_FSM_STATE_WAITC;
1743
1744                            break;
1745                    }
1746            break;
1747
1748    default:
1749            panic("the state of the secondary fsm is unknown.");
1750    }
1751    return (0);
1752 }
```
_____ mip6_fsm.c

1732–1744 If a mobile node does not receive a Care-of Test message in 5 s after sending a Care-of Test Init message, the node resends a Care-of Test Init message. The code is almost the same as the code in lines 1676–1690.

15.22 Virtual Home Interface

A mobile node has a virtual interface, which represents its home network. The definition of the virtual interface was discussed in Section 10.32 of Chapter 10. The virtual home interface keeps the current location, the home address of the mobile node, and prefix information. The interface is also used as an output routine from the mobile node to correspondent nodes when the node is using the bidirectional tunneling mechanism.

15.22.1   Initialization of the Interface

The `hifattach()` function is the initialization function of the virtual home interface, which is defined as the `hif_softc{}` structure.

Listing 15-275
_____ if_hif.c

```
178     void
179     hifattach(dummy)
 ....
181             void *dummy;
 ....
185     {
186             struct hif_softc *sc;
187             int i;
188
189             LIST_INIT(&hif_softc_list);
190
191             sc = malloc(NHIF * sizeof(struct hif_softc), M_DEVBUF, M_WAIT);
192             bzero(sc, NHIF * sizeof(struct hif_softc));
193             for (i = 0 ; i < NHIF; sc++, i++) {
 ....
197                     sc->hif_if.if_name = "hif";
198                     sc->hif_if.if_unit = i;
 ....
200                     sc->hif_if.if_flags = IFF_MULTICAST | IFF_SIMPLEX;
201                     sc->hif_if.if_mtu = HIF_MTU;
202                     sc->hif_if.if_ioctl = hif_ioctl;
203                     sc->hif_if.if_output = hif_output;
204                     sc->hif_if.if_type = IFT_HIF;
 ....
209                     IFQ_SET_MAXLEN(&sc->hif_if.if_snd, ifqmaxlen);
210                     IFQ_SET_READY(&sc->hif_if.if_snd);
 ....
212                     if_attach(&sc->hif_if);
 ....
218                     bpfattach(&sc->hif_if, DLT_NULL, sizeof(u_int));
```
_____ if_hif.c

178–181 The basic procedure is no different from the initialization code of other network interfaces. We only discuss some specific parameters to the `hif_softc{}` structure here. The `HIF_MTU` macro is defined as 1280, which is the minimum MTU of IPv6 packets. This interface is used as an output function for packets that are tunneled from a mobile node to a home agent. To avoid the Path MTU discovery for this tunnel connection between a mobile node and a home agent, the implementation limits the MTU to the minimum size. The `IFT_HIF` macro is a new type number, which indicates the virtual home interface. The virtual home interface sometimes requires special care, since it is not a normal network interface. Assigning a new type number makes it easy to identify the virtual home interface when we need to perform specific tasks.

Listing 15-276
_____ if_hif.c

```
224             sc->hif_location = HIF_LOCATION_UNKNOWN;
225             sc->hif_coa_ifa = NULL;
```

```
    ....
230                     /* binding update list and home agent list. */
231                     LIST_INIT(&sc->hif_bu_list);
232                     LIST_INIT(&sc->hif_prefix_list_home);
233                     LIST_INIT(&sc->hif_prefix_list_foreign);
234
235                     /* DHAAD related. */
236                     sc->hif_dhaad_id = mip6_dhaad_id++;
237                     sc->hif_dhaad_lastsent = 0;
238                     sc->hif_dhaad_count = 0;
239
240                     /* Mobile Prefix Solicitation. */
241                     sc->hif_mps_id = mip6_mps_id++;
242                     sc->hif_mps_lastsent = 0;
243
244                     sc->hif_ifid = in6addr_any;
245
246                     /* create hif_softc list */
247                     LIST_INSERT_HEAD(&hif_softc_list, sc, hif_entry);
248             }
249     }
```
――― if_hif.c

224–225 The initial location is set to `HIF_LOCATION_UNKNOWN`. The location is determined when a mobile node receives the first Router Advertisement message. The care-of address is also undefined until the mobile node configures at least one address.

231–233 The list of binding update list entries (`hif_bu_list`) related to this home network, the list of home prefixes (`hif_prefix_list_home`), and the list of foreign prefixes (`hif_prefix_list_foreign`) are initialized.

236–242 The `mip6_dhaad_id` and the `mip6_mps_id` are unique identifiers that are used when sending a Dynamic Home Agent Address Discovery request message and a Mobile Prefix Solicitation message, respectively. These variables are managed as global variables and are incremented by 1 every time they are used to avoid duplication of identifiers.

244 The `hif_ifid` variable is used as the interface identifier part of home addresses of this virtual home network. The value is determined when it is first used.

15.22.2 I/O Control of the Virtual Home Interface

The `hif_ioctl()` function manages commands sent by the `ioctl()` system call.

Listing 15-277
――― if_hif.c

```
251     int
252     hif_ioctl(ifp, cmd, data)
253             struct ifnet *ifp;
254             u_long cmd;
255             caddr_t data;
256     {
257             int s;
258             struct hif_softc *sc = (struct hif_softc *)ifp;
259             struct hif_ifreq *hifr = (struct hif_ifreq *)data;
260             struct ifreq *ifr = (struct ifreq *)data;
261             int error = 0;
```
――― if_hif.c

251–255 The `hif_ioctl()` function has three parameters. The `ifp` parameter is a pointer to the virtual home interface to be controlled. The `cmd` and `data` parameters are the

TABLE 15-4

Name	Description
SIOCAHOMEPREFIX_HIF	Adds one home prefix information entry
SIOCAHOMEAGENT_HIF	Adds one home agent information entry

I/O control commands for hif_softc{}.

command number and pointer to the instance of the hif_ifreq{} structure that keeps related data to the command.

There are two commands currently used for I/O control of the hif_softc{} structure. The list of commands are shown in Table 15-4.

Listing 15-278

```
                                                                   if_hif.c
269              switch(cmd) {
    ....
294              case SIOCAHOMEPREFIX_HIF:
295                      error = hif_prefix_list_update_withprefix(sc, data);
296                      break;
    ....
337              case SIOCAHOMEAGENT_HIF:
338                      error = hif_prefix_list_update_withhaaddr(sc, data);
339                      break;
    ....
404
405              default:
406                      error = EINVAL;
407                      break;
408              }
409
410      hif_ioctl_done:
411
412              splx(s);
413
414              return (error);
415      }
                                                                   if_hif.c
```

294–339 The hif_prefix_list_update_withprefix() function adds the home prefix information passed from the user space. The hif_prefix_list_update_withhaaddr() function adds the home agent information passed from the user space.

15.22.3   Add Home Prefix Information

The hif_prefix_list_update_withprefix() function adds home prefix information to a virtual home network.

Listing 15-279

```
                                                                   if_hif.c
513      static int
514      hif_prefix_list_update_withprefix(sc, data)
515          struct hif_softc *sc;
516          caddr_t data;
```

```
517          {
518                  struct hif_ifreq *hifr = (struct hif_ifreq *)data;
519                  struct mip6_prefix *nmpfx, *mpfx;
520                  struct hif_softc *hif;
521                  int error = 0;
```
─── if_hif.c

513–516 The function adds home prefix information to the virtual home network specified as the sc parameter with the prefix information passed as the data parameter.

Listing 15-280
─── if_hif.c
```
526                  nmpfx = &hifr->ifr_ifru.ifr_mpfx;
527
528                  mpfx = mip6_prefix_list_find_withprefix(&nmpfx->mpfx_prefix,
529                      nmpfx->mpfx_prefixlen);
530                  if (mpfx == NULL) {
531                          mpfx = mip6_prefix_create(&nmpfx->mpfx_prefix,
532                              nmpfx->mpfx_prefixlen, nmpfx->mpfx_vltime,
533                              nmpfx->mpfx_pltime);
534                          if (mpfx == NULL) {
 ....
538                                  return (ENOMEM);
539                          }
540                          error = mip6_prefix_list_insert(&mip6_prefix_list, mpfx);
541                          if (error) {
542                                  return (error);
543                          }
```
─── if_hif.c

526–529 The prefix information is stored in the ifr_mpfx variable of the hif_ifreq{} structure. If a mobile node does not have the home prefix, it creates a new mip6_prefix{} instance using the mip6_prefix_create() function and inserts the new entry in the global prefix list by the mip6_prefix_list_insert() function.

Listing 15-281
─── if_hif.c
```
545                          for (hif = LIST_FIRST(&hif_softc_list); hif;
546                              hif = LIST_NEXT(hif, hif_entry)) {
547                                  if (hif == sc)
548                                          hif_prefix_list_insert_withmpfx(
549                                              &hif->hif_prefix_list_home, mpfx);
550                                  else
551                                          hif_prefix_list_insert_withmpfx(
552                                              &hif->hif_prefix_list_foreign, mpfx);
553                          }
554                  }
555
556                  mip6_prefix_update_lifetime(mpfx, nmpfx->mpfx_vltime,
557                      nmpfx->mpfx_pltime);
558
559                  return (0);
560          }
```
─── if_hif.c

545–553 The newly created prefix is a home prefix of the virtual home interface specified as the sc parameter. If a mobile node has more than two virtual home networks, the new prefix information can be considered foreign prefix information for the home interface not specified as the sc parameter. The new prefix is added to hif_prefix_list_home, which keeps all home prefix information of the specified virtual home interface by the `hif_prefix_list_insert_withmpfx()` function. For the rest of virtual home interfaces, the new prefix is added to `hif_prefix_list_foreign`, which is a list of foreign prefix information entries.

556–557 If the mobile node already has the same prefix information, the node updates the lifetime using the `mip6_prefix_update_lifetime()` function.

15.22.4   Add Home Agent Information

The `hif_prefix_list_update_withhaaddr()` function adds home agent information to a virtual home network.

Listing 15-282

—— `if_hif.c`

```
562     static int
563     hif_prefix_list_update_withhaaddr(sc, data)
564         struct hif_softc *sc;
565         caddr_t data;
566     {
567             struct hif_ifreq *hifr = (struct hif_ifreq *)data;
568             struct mip6_ha *nmha = (struct mip6_ha *)data;
569             struct mip6_ha *mha;
570             struct in6_addr prefix;
571             struct mip6_prefix *mpfx;
572             struct hif_softc *hif;
573             int error = 0;
....
575             struct timeval mono_time;
```
—— `if_hif.c`

562–565 The function adds the home agent information specified by the `data` parameter to the virtual home interface specified as the `sc` parameter. Apparently, the initialization on line 568 is wrong. `nmha` will be overwritten properly later.

Listing 15-283

—— `if_hif.c`

```
585             nmha = &hifr->ifr_ifru.ifr_mha;
586             if (IN6_IS_ADDR_UNSPECIFIED(&nmha->mha_addr)
587                 || IN6_IS_ADDR_LOOPBACK(&nmha->mha_addr)
588                 || IN6_IS_ADDR_LINKLOCAL(&nmha->mha_addr)
589                 || IN6_IS_ADDR_SITELOCAL(&nmha->mha_addr))
590                     return (EINVAL);
591
592             mha = mip6_ha_list_find_withaddr(&mip6_ha_list, &nmha->mha_addr);
593             if (mha == NULL) {
594                     mha = mip6_ha_create(&nmha->mha_addr, nmha->mha_flags,
595                         nmha->mha_pref, 0);
596                     if (mha == NULL) {
....
600                             return (ENOMEM);
601                     }
602                     mip6_ha_list_insert(&mip6_ha_list, mha);
603             }
604
605             mha->mha_addr = nmha->mha_addr;
606             mha->mha_flags = nmha->mha_flags;
607             mip6_ha_update_lifetime(mha, 0);
```
—— `if_hif.c`

585–590 The home agent information is stored in the `ifr_mha` variable of the `hif_ifreq{}` structure. If the specified address of the home agent is not suitable, the processing is aborted.

592–603 If a mobile node does not have the same home agent information as that passed from the user space, it creates a new `mip6_ha{}` instance using the `mip6_ha_create()` function and inserts the newly created entry in the global home agent information list.

604–607 If the mobile node already has the same home agent information, its flags and lifetime are updated. The lifetime is set to infinite when both creating a new entry or updating an existing entry. The actual lifetime is set when the mobile node receives a Router Advertisement message from the home agent. Until then, the mobile node assumes that the entry has an infinite lifetime.

Listing 15-284

```
                                                                        if_hif.c
609              /* add mip6_prefix, if needed. */
610              mpfx = mip6_prefix_list_find_withprefix(&mha->mha_addr, 64 /* XXX */);
611              if (mpfx == NULL) {
612                      bzero(&prefix, sizeof(prefix));
613                      prefix.s6_addr32[0] = mha->mha_addr.s6_addr32[0];
614                      prefix.s6_addr32[1] = mha->mha_addr.s6_addr32[1];
615                      mpfx = mip6_prefix_create(&prefix, 64 /* XXX */,
616                          65535 /* XXX */, 0);
617                      if (mpfx == NULL)
618                              return (ENOMEM);
619                      error = mip6_prefix_list_insert(&mip6_prefix_list, mpfx);
620                      if (error)
621                              return (error);
622                      for (hif = LIST_FIRST(&hif_softc_list); hif;
623                          hif = LIST_NEXT(hif, hif_entry)) {
624                              if (sc == hif)
625                                      hif_prefix_list_insert_withmpfx(
626                                          &sc->hif_prefix_list_home, mpfx);
627                              else
628                                      hif_prefix_list_insert_withmpfx(
629                                          &sc->hif_prefix_list_foreign, mpfx);
630                      }
631              }
632              mip6_prefix_ha_list_insert(&mpfx->mpfx_ha_list, mha);
633
634              return (0);
635      }
                                                                        if_hif.c
```

610–631 The prefix part of the home agent address means a home prefix. If the mobile node does not have the home prefix, it creates a new `mip6_prefix{}` instance based on the address of the home agent.

632 The home agent entry is considered an advertising router of the home prefix. The `mip6_prefix_ha_list_insert()` function adds a pointer to the home agent information from the home prefix information.

15.22.5   Create the `hif_prefix{}` Structure with the `mip6_prefix{}` Structure

The `hif_prefix_list_insert_withmpfx()` function inserts the specified prefix entry into the prefix list kept in a virtual home network structure.

Listing 15-285

if_hif.c

```
637     struct hif_prefix *
638     hif_prefix_list_insert_withmpfx(hif_prefix_list, mpfx)
639             struct hif_prefix_list *hif_prefix_list;
640             struct mip6_prefix *mpfx;
641     {
642             struct hif_prefix *hpfx;
643
644             if ((hif_prefix_list == NULL) || (mpfx == NULL))
645                     return (NULL);
646
647             hpfx = hif_prefix_list_find_withmpfx(hif_prefix_list, mpfx);
648             if (hpfx != NULL)
649                     return (hpfx);
650
651             MALLOC(hpfx, struct hif_prefix *, sizeof(struct hif_prefix), M_TEMP,
652                 M_NOWAIT);
653             if (hpfx == NULL) {
....
656                     return (NULL);
657             }
658             hpfx->hpfx_mpfx = mpfx;
659             LIST_INSERT_HEAD(hif_prefix_list, hpfx, hpfx_entry);
660
661             return (hpfx);
662     }
```

if_hif.c

637–640 The `hif_prefix_list_insert_withmpfx()` function creates a new `hif_prefix{}` instance, which points to the `mip6_prefix{}` instance specified by the `mpfx` parameter, and inserts the newly created entry into the prefix list specified by the `hif_prefix_list` parameter.

647–659 If `hif_prefix_list`, which is a list of the `hif_prefix{}` instances, does not contain the prefix information specified by the `mpfx` parameter, a mobile node allocates memory space for the new `hif_prefix{}` instance. If the allocation succeeds, the prefix information specified as the `mpfx` parameter is copied to the `hpfx_mpfx` variable. The new entry is inserted into the `hif_prefix_list` list.

Listing 15-286

if_hif.c

```
664     void
665     hif_prefix_list_remove(hpfx_list, hpfx)
666             struct hif_prefix_list *hpfx_list;
667             struct hif_prefix *hpfx;
668     {
669             if ((hpfx_list == NULL) || (hpfx == NULL))
670                     return;
671
672             LIST_REMOVE(hpfx, hpfx_entry);
673             FREE(hpfx, M_TEMP);
674     }
```

if_hif.c

664–674 The `hif_prefix_list_remove()` function removes the pointer to the `hif_prefix{}` instance from the list specified by the first parameter and releases the memory space allocated for the `hif_prefix{}` instance.

Listing 15-287

———————————————————————————————————— if_hif.c
```
676    struct hif_prefix *
677    hif_prefix_list_find_withprefix(hif_prefix_list, prefix, prefixlen)
678           struct hif_prefix_list *hif_prefix_list;
679           struct in6_addr *prefix;
680           int prefixlen;
681    {
682           struct hif_prefix *hpfx;
683           struct mip6_prefix *mpfx;
684
685           for (hpfx = LIST_FIRST(hif_prefix_list); hpfx;
686               hpfx = LIST_NEXT(hpfx, hpfx_entry)) {
687                   mpfx = hpfx->hpfx_mpfx;
688                   if (in6_are_prefix_equal(prefix, &mpfx->mpfx_prefix,
689                           prefixlen)
690                       && (prefixlen == mpfx->mpfx_prefixlen)) {
691                           /* found. */
692                           return (hpfx);
693                   }
694           }
695           /* not found. */
696           return (NULL);
697    }
```
———————————————————————————————————— if_hif.c

676–696 The `hif_prefix_list_find_withprefix()` function searches for a `hif_prefix{}` instance that has the prefix information specified by the second and third parameters from the list specified by the first parameter. Note that the `hif_prefix{}` structure itself does not include prefix information. It has a pointer to the related `mip6_prefix{}` instance as described in Figure 10-1 of Chapter 10. If the list specified as the first parameter contains the `hif_prefix{}` instance of which the `mip6_prefix{}` instance has the same prefix information specified by the second and third parameters, the function returns the pointer to the `hif_prefix{}` instance.

Listing 15-288

———————————————————————————————————— if_hif.c
```
699    struct hif_prefix *
700    hif_prefix_list_find_withhaaddr(hif_prefix_list, haaddr)
701           struct hif_prefix_list *hif_prefix_list;
702           struct in6_addr *haaddr;
703    {
704           struct hif_prefix *hpfx;
705           struct mip6_prefix *mpfx;
706           struct mip6_prefix_ha *mpfxha;
707           struct mip6_ha *mha;
708
709           for (hpfx = LIST_FIRST(hif_prefix_list); hpfx;
710               hpfx = LIST_NEXT(hpfx, hpfx_entry)) {
711                   mpfx = hpfx->hpfx_mpfx;
712                   for (mpfxha = LIST_FIRST(&mpfx->mpfx_ha_list); mpfxha;
713                       mpfxha = LIST_NEXT(mpfxha, mpfxha_entry)) {
714                           mha = mpfxha->mpfxha_mha;
715                           if (IN6_ARE_ADDR_EQUAL(&mha->mha_addr, haaddr))
716                                   return (hpfx);
717                   }
718           }
719           /* not found. */
720           return (NULL);
721    }
```
———————————————————————————————————— if_hif.c

699–720 The `hif_prefix_list_find_withhaaddr()` function performs tasks similar to the `hif_prefix_list_find_withprefix()` function. This function finds a `hif_prefix{}` instance that has the home agent address specified by the second parameter. Each `mip6_prefix{}` structure points to the home agent information as described in Figure 10-1 of Chapter 10. The function checks all `mip6_prefix{}` instances to which instances from `hif_prefix{}` point and compares the related home agent information to the address specified by the second parameter.

Listing 15-289

_____ if_hif.c

```
723     struct hif_prefix *
724     hif_prefix_list_find_withmpfx(hif_prefix_list, mpfx)
725             struct hif_prefix_list *hif_prefix_list;
726             struct mip6_prefix *mpfx;
727     {
728             struct hif_prefix *hpfx;
729
730             for (hpfx = LIST_FIRST(hif_prefix_list); hpfx;
731                 hpfx = LIST_NEXT(hpfx, hpfx_entry)) {
732                     if (hpfx->hpfx_mpfx == mpfx)
733                             return (hpfx);
734             }
735             /* not found. */
736             return (NULL);
737     }
```
_____ if_hif.c

723–737 The `hif_prefix_list_find_withmpfx()` function searches for an `hif_prefix{}` structure that has the `mip6_prefix{}` instance specified by the `mpfx` pointer.

Listing 15-290

_____ if_hif.c

```
739     static struct hif_prefix *
740     hif_prefix_list_find_withmha(hpfx_list, mha)
741             struct hif_prefix_list *hpfx_list;
742             struct mip6_ha *mha;
743     {
744             struct hif_prefix *hpfx;
745             struct mip6_prefix *mpfx;
746             struct mip6_prefix_ha *mpfxha;
747
748             for (hpfx = LIST_FIRST(hpfx_list); hpfx;
749                 hpfx = LIST_NEXT(hpfx, hpfx_entry)) {
750                     mpfx = hpfx->hpfx_mpfx;
751                     for (mpfxha = LIST_FIRST(&mpfx->mpfx_ha_list); mpfxha;
752                         mpfxha = LIST_NEXT(mpfxha, mpfxha_entry)) {
753                             if (mpfxha->mpfxha_mha == mha)
754                                     return (hpfx);
755                     }
756             }
757             /* not found. */
758             return (NULL);
759     }
```
_____ if_hif.c

739–759 The `hif_prefix_list_find_withmha()` function searches for an `hif_prefix{}` instance for which the `mip6_prefix{}` instance has a pointer to the home agent information that is the same as that specified by the `mha` pointer.

15.22.6 Find the Preferred Home Agent Information

The `hif_find_preferable_ha()` function returns a pointer to the `mip6_ha{}` instance of which the preference is highest.

Listing 15-291

```
435     struct mip6_ha *
436     hif_find_preferable_ha(hif)
437             struct hif_softc *hif;
438     {
439             struct mip6_ha *mha;
440
441             /*
442              * we assume mip6_ha_list is ordered by a preference value.
443              */
444             for (mha = TAILQ_FIRST(&mip6_ha_list); mha;
445                 mha = TAILQ_NEXT(mha, mha_entry)) {
446                     if (!hif_prefix_list_find_withmha(&hif->hif_prefix_list_home,
447                         mha))
448                             continue;
449                     if (IN6_IS_ADDR_LINKLOCAL(&mha->mha_addr))
450                             continue;
451                     /* return the entry we have found first. */
452                     return (mha);
453             }
454             /* not found. */
455             return (NULL);
456     }
```

435–437 The `hif` parameter is a pointer to the virtual home interface to which the home agent belongs.

444–453 The global list (`mip6_ha_list`) of `mip6_ha{}` instances is ordered by the preference value of each entry. This function checks each `mip6_ha{}` instance listed in the `mip6_ha_list` from the head of the list. If an `mip6_ha{}` instance is a home agent of a certain virtual home network, one of the home prefixes must have a pointer to the `mip6_ha{}` instance. The `hif_prefix_list_find_withmha()` function will return an `hif_prefix{}` instance if the `mip6_ha{}` instance specified by the second parameter is in the home prefix list specified by the first parameter.

Listing 15-292

```
462     struct mip6_ha *
463     hif_find_next_preferable_ha(hif, haaddr)
464             struct hif_softc *hif;
465             struct in6_addr *haaddr;
466     {
467             struct mip6_ha *curmha, *mha;
468
469             curmha = mip6_ha_list_find_withaddr(&mip6_ha_list, haaddr);
470             if (curmha == NULL)
471                     return (hif_find_preferable_ha(hif));
472
473             /*
474              * we assume mip6_ha_list is ordered by a preference value.
475              */
476             for (mha = TAILQ_NEXT(curmha, mha_entry); mha;
477                 mha = TAILQ_NEXT(mha, mha_entry)) {
478                     if (!hif_prefix_list_find_withmha(&hif->hif_prefix_list_home,
```

```
479                          mha))
480                              continue;
481                      /* return the entry we have found first. */
482                      return (mha);
483              }
484      /* not found. */
485      return (NULL);
486  }
```

462–465 The `hif_find_next_preferable_ha()` function returns a pointer to the `mip6_ha{}` instance, which is the next candidate for the home agent of the virtual home interface specified as the `hif` parameter. The `haaddr` parameter is the address of the current candidate home agent.

469–482 curmha is a pointer to the `mip6_ha{}` instance of the current candidate home agent searched by the `mip6_ha_list_find_withaddr()` function. In the loop from lines 476 to 483, all `mip6_ha{}` entries after the `curmha` pointer are checked. The first entry found by the `hif_prefix_list_find_withmha()` function is returned.

15.22.7 Find the Virtual Home Interface

The `hif_list_find_withhaddr()` function returns the virtual home network related to the specified home address.

Listing 15-293

```
492      struct hif_softc *
493      hif_list_find_withaddr(haddr)
494          struct in6_addr *haddr;
495      {
496              struct hif_softc *hif;
497              struct hif_prefix *hpfx;
498              struct mip6_prefix *mpfx;
499
500              for (hif = LIST_FIRST(&hif_softc_list); hif;
501                  hif = LIST_NEXT(hif, hif_entry)) {
502                      for (hpfx = LIST_FIRST(&hif->hif_prefix_list_home); hpfx;
503                          hpfx = LIST_NEXT(hpfx, hpfx_entry)) {
504                              mpfx = hpfx->hpfx_mpfx;
505                              if (IN6_ARE_ADDR_EQUAL(&mpfx->mpfx_haddr, haddr))
506                                      return (hif);
507                      }
508              }
509      /* not found. */
510      return (NULL);
511  }
```

492–510 The `hif_list_find_withhaddr()` function returns a pointer to the `hif_softc{}` instance, which has a home address specified by the `haddr` parameter. A home address is stored in each `mip6_prefix{}` structure. This function checks all the home prefix information of all virtual home interfaces. The home prefix information is stored in the `hif_prefix_list_home` member variable. The pointer to the `hif_softc{}` instance is returned if its home prefix information includes the home address specified by the `haddr` parameter.

15.22.8   Send a Packet in IPv6 in IPv6 Tunnel Format

The `hif_output()` function sends an IPv6 packet to the destination node using IPv6 in IPv6 tunneling between a mobile node and its home agent.

Listing 15-294
_____ if_hif.c
```
811     int
812     hif_output(ifp, m, dst, rt)
813          struct ifnet *ifp;
814          struct mbuf *m;
815          struct sockaddr *dst;
816          struct rtentry *rt;
817     {
818             struct mip6_bu *mbu;
819             struct hif_softc *hif = (struct hif_softc *)ifp;
820             struct ip6_hdr *ip6;
```
_____ if_hif.c

811–816 The `hif_output()` function has four parameters: the `ifp` parameter is a pointer to the virtual home interface; the `m` parameter is a pointer to the mbuf that contains the IPv6 packet; and the `dst` and `rt` parameters are the destination addresses of the packet in the `sockaddr_in6{}` format and a pointer to the routing entry, respectively.

Listing 15-295
_____ if_hif.c
```
822             /* This function is copied from looutput */
```
`.... [copied from looutput() function in if_loop.c]`
```
883
884             switch (dst->sa_family) {
885             case AF_INET6:
886                     break;
887             default:
888                     printf("hif_output: af=%d unexpected\n", dst->sa_family);
889                     m_freem(m);
890                     return (EAFNOSUPPORT);
891             }
```
_____ if_hif.c

884–891 The `hif_output()` function only supports IPv6. If the address family of the destination address is not IPv6, the packet is dropped.

Listing 15-296
_____ if_hif.c
```
897             ip6 = mtod(m, struct ip6_hdr *);
898             if (IN6_IS_ADDR_LINKLOCAL(&ip6->ip6_src)
899                 || IN6_IS_ADDR_LINKLOCAL(&ip6->ip6_dst)
900                 || IN6_IS_ADDR_SITELOCAL(&ip6->ip6_src)
901                 || IN6_IS_ADDR_SITELOCAL(&ip6->ip6_dst))
902                     goto done;
903
904             mbu = mip6_bu_list_find_home_registration(&hif->hif_bu_list,
905                 &ip6->ip6_src);
906             if (!mbu)
907                     goto done;
908
```

```
909                   if (IN6_IS_ADDR_UNSPECIFIED(&mbu->mbu_paddr))
910                           goto done;
```
                                                                                    if_hif.c

898–902 Addresses of the packet tunneled to the home agent must be global addresses.

904–907 The source address of the outer header is the care-of address of the mobile node. The `mip6_bu_list_find_home_registration()` function will return a home registration entry for the source address, which is the home address of the mobile node. The care-of address information is stored in the home registration entry.

909 If the mobile node has not gotten its home agent information, the peer address (`mbu_peer`) is the unspecified address. No packet can be sent in this case.

Listing 15-297
                                                                                    if_hif.c
```
912                   M_PREPEND(m, sizeof(struct ip6_hdr), M_DONTWAIT);
913                   if (m && m->m_len < sizeof(struct ip6_hdr))
914                           m = m_pullup(m, sizeof(struct ip6_hdr));
915                   if (m == NULL)
916                           return (0);
917
918                   ip6 = mtod(m, struct ip6_hdr *);
919                   ip6->ip6_flow = 0;
920                   ip6->ip6_vfc &= ~IPV6_VERSION_MASK;
921                   ip6->ip6_vfc |= IPV6_VERSION;
922                   ip6->ip6_plen = htons((u_short)m->m_pkthdr.len - sizeof(*ip6));
923                   ip6->ip6_nxt = IPPROTO_IPV6;
924                   ip6->ip6_hlim = ip6_deflim;
925                   ip6->ip6_src = mbu->mbu_coa;
926                   ip6->ip6_dst = mbu->mbu_paddr;
....
929                   /* XXX */
930                   return (ip6_output(m, 0, 0, IPV6_MINMTU, 0, &ifp
....
932                                           , NULL
....
934                                           ));
....
942     done:
943           m_freem(m);
944           return (0);
945     }
```
                                                                                    if_hif.c

912–926 An extra mbuf is prepended to the mbuf that contains the original packet to create a space for the outer IPv6 header. The care-of address of the mobile node (`mbu_coa`) is copied to the source address field of the outer header and the home agent address (`mbu_paddr`) is copied to the destination address.

930–934 The `ip6_output()` is called with the `IPV6_MINMTU` flag, which indicates the `ip6_output()` function is to send a packet with the minimum MTU size. This flag avoids the Path MTU Discovery procedure between the mobile node and its home agent.

15.23   Return Routability and Route Optimization

In this section, we discuss the detailed process of the return routability procedure.

15.23.1 Trigger Return Routability Procedure

A mobile node initiates the return routability procedure, described in Section 5.1 of Chapter 5, when the node receives a tunneled packet.

Listing 15-298

```
                                                                    ip6_input.c
377     void
378     ip6_input(m)
379             struct mbuf *m;
380     {
....
1092            while (nxt != IPPROTO_DONE) {
1146                    if ((nxt != IPPROTO_HOPOPTS) && (nxt != IPPROTO_DSTOPTS) &&
1147                        (nxt != IPPROTO_ROUTING) && (nxt != IPPROTO_FRAGMENT) &&
1148                        (nxt != IPPROTO_ESP) && (nxt != IPPROTO_AH) &&
1149                        (nxt != IPPROTO_MH) && (nxt != IPPROTO_NONE)) {
1150                            if (mip6_route_optimize(m))
1151                                    goto bad;
1152                    }
....
1156            }
                                                                    ip6_input.c
```

1092–1152 This `while` loop processes all the extension headers contained in an incoming IPv6 packet. The `mip6_route_optimize()` function is called during the process just before processing the upper layer protocol headers (e.g., Transmission Control Protocol [TCP] or User Datagram Protocol [UDP]).

Listing 15-299

```
                                                                    mip6_mncore.c
1341    int
1342    mip6_route_optimize(m)
1343            struct mbuf *m;
1344    {
1345            struct m_tag *mtag;
1346            struct in6_ifaddr *ia;
1347            struct ip6aux *ip6a;
1348            struct ip6_hdr *ip6;
1349            struct mip6_prefix *mpfx;
1350            struct mip6_bu *mbu;
1351            struct hif_softc *sc;
1352            struct in6_addr hif_coa;
1353            int error = 0;
                                                                    mip6_mncore.c
```

1341–1343 The `mip6_route_optimize()` function takes a pointer to the mbuf that contains the incoming packet. This function checks to see if the packet was route optimized and sends a Binding Update message if the packet was not route optimized.

A packet that is not route optimized is delivered as a tunnel packet to a mobile node. In the BSD network code, we cannot know if the packet being processed in the `ip6_input()` function is a tunneled packet or not, because the outer header of a tunnel packet is removed when it is processed and the inner packet does not have any clue about its outer header. This function checks to see if the packet includes a Type 2 Routing Header. If the header exists, that means the packet was sent with route optimization.

Listing 15-300

```
1355            if (!MIP6_IS_MN) {
1356                    /* only MN does the route optimization. */
1357                    return (0);
1358            }
1359
1360            ip6 = mtod(m, struct ip6_hdr *);
1361
1362            if (IN6_IS_ADDR_LINKLOCAL(&ip6->ip6_src) ||
1363                IN6_IS_ADDR_SITELOCAL(&ip6->ip6_src)) {      /* XXX */
1364                    return (0);
1365            }
1366            /* Quick check */
1367            if (IN6_IS_ADDR_LINKLOCAL(&ip6->ip6_dst) ||
1368                IN6_IS_ADDR_SITELOCAL(&ip6->ip6_dst) ||      /* XXX */
1369                IN6_IS_ADDR_MULTICAST(&ip6->ip6_dst)) {
1370                    return (0);
1371            }
1372
1373            for (ia = in6_ifaddr; ia; ia = ia->ia_next) {
1374                    if (IN6_ARE_ADDR_EQUAL(&ia->ia_addr.sin6_addr,
1375                        &ip6->ip6_src)) {
1376                            return (0);
1377                    }
1378            }
```

1362–1371 A simple sanity check is done. A packet of which the source address is not a global address will not be optimized. Also, a packet of which the destination address is not a global address or of which the address is a multicast address will not be optimized.

1373–1378 If the source address of the incoming packet is one of the addresses assigned to a mobile node, then there is no need to optimize the route.

Listing 15-301

```
1380            mtag = ip6_findaux(m);
1381            if (mtag) {
1382                    ip6a = (struct ip6aux *) (mtag + 1);
1383                    if (ip6a->ip6a_flags & IP6A_ROUTEOPTIMIZED) {
1384                            /* no need to optimize route. */
1385                            return (0);
1386                    }
1387            }
1388            /*
1389             * this packet has no rthdr or has a rthdr not related mip6
1390             * route optimization.
1391             */
```

1380–1391 The KAME Mobile IPv6 implementation sets the `IP6A_ROUTEOPTIMIZED` flag in an auxiliary mbuf when processing the Type 2 Routing Header. If the packet has the flag set, it is understood that the packet has been delivered with a Type 2 Routing Header, which means it is route optimized.

Listing 15-302

```
1394            sc = hif_list_find_withhaddr(&ip6->ip6_dst);
1395            if (sc == NULL) {
```

```
1396                         /* this dst addr is not one of our home addresses. */
1397                         return (0);
1398                 }
1399         if (sc->hif_location == HIF_LOCATION_HOME) {
1400                         /* we are home.  no route optimization is required. */
1401                         return (0);
1402                 }
```
————————————————————————————————————— mip6_mncore.c

1394–1402 The `hif_list_find_withhaddr()` function returns a pointer to the `hif_softc{}` instance that has the home address specified by the parameter. If the destination address of the packet is not the home address of the mobile node, the mobile node does not do anything. Also, if the mobile node is at home, no route optimization is needed.

Listing 15-303
————————————————————————————————————— mip6_mncore.c

```
1404                 /* get current CoA and recover its scope information. */
1405             if (sc->hif_coa_ifa == NULL) {
....
1409                         return (0);
1410                 }
1411             hif_coa = sc->hif_coa_ifa->ia_addr.sin6_addr;
1412
1413                 /*
1414                  * find a mip6_prefix which has a home address of received
1415                  * packet.
1416                  */
1417             mpfx = mip6_prefix_list_find_withhaddr(&mip6_prefix_list,
1418                 &ip6->ip6_dst);
1419             if (mpfx == NULL) {
1420                         /*
1421                          * no related prefix found.  this packet is
1422                          * destined to another address of this node
1423                          * that is not a home address.
1424                          */
1425                         return (0);
1426                 }
```
————————————————————————————————————— mip6_mncore.c

1404–1411 If the mobile node does not have a valid care-of address, the node cannot send a Binding Update message. If there is a valid care-of address, the address is set to the `hif_coa` variable.

1417–1426 Creating a new binding update list entry requires the prefix information of the home address of the mobile node. The `mip6_prefix_list_find_withhaddr()` function is called to search the prefix information related to the destination address of the incoming packet, which is the home address of the mobile node. If the mobile node does not have the prefix information, it cannot send a Binding Update message.

Listing 15-304
————————————————————————————————————— mip6_mncore.c

```
1428                 /*
1429                  * search all binding update entries with the address of the
1430                  * peer sending this un-optimized packet.
1431                  */
1432             mbu = mip6_bu_list_find_withpaddr(&sc->hif_bu_list, &ip6->ip6_src,
1433                 &ip6->ip6_dst);
```

```
1434                if (mbu == NULL) {
1435                        /*
1436                         * if no binding update entry is found, this is a
1437                         * first packet from the peer.  create a new binding
1438                         * update entry for this peer.
1439                         */
1440                        mbu = mip6_bu_create(&ip6->ip6_src, mpfx, &hif_coa, 0, sc);
1441                        if (mbu == NULL) {
1442                                error = ENOMEM;
1443                                goto bad;
1444                        }
1445                        mip6_bu_list_insert(&sc->hif_bu_list, mbu);
1446                } else {
....
1473                }
1474                mip6_bu_fsm(mbu, MIP6_BU_PRI_FSM_EVENT_REVERSE_PACKET, NULL);
1475
1476                return (0);
1477        bad:
1478                m_freem(m);
1479                return (error);
1480        }
```
———mip6_mncore.c

1432–1473 If the mobile node does not have a binding update list entry between its home
address (the destination address of the incoming packet) and the peer address (the
source address of the incoming packet), a new binding update list entry is created by
the `mip6_bu_create()` function. The new entry is inserted into the list of binding
update list entries of the virtual home interface to which the home address is assigned.

1446–1476 The `MIP6_BU_PRI_FSM_EVENT_REVERSE_PACKET` event is sent to the state
machine of the newly created binding update list entry, or to the existing entry if the
mobile node already has one. The event will initiate the return routability procedure.

15.23.2   Sending Home Test Init/Care-of Test Init Messages

A mobile node sends a Home Test Init and a Care-of Test Init message to initiate the
return routability procedure. The sending of these messages is implemented by the
`mip6_bu_send_hoti()` function and the `mip6_bu_send_coti()` functions, respectively.

Listing 15-305
———mip6_mncore.c
```
2061    int
2062    mip6_bu_send_hoti(mbu)
2063            struct mip6_bu *mbu;
2064    {
2065            struct mbuf *m;
2066            struct ip6_pktopts opt;
2067            int error = 0;
```
———mip6_mncore.c

2061–2063 The `mip6_bu_send_hoti()` function sends a Home Test Init message to the
correspondent node specified in the binding update list entry specified in the parameter.

Listing 15-306
———mip6_mncore.c
```
2069            ip6_initpktopts(&opt);
2070
```

```
2071                    m = mip6_create_ip6hdr(&mbu->mbu_haddr, &mbu->mbu_paddr,
2072                        IPPROTO_NONE, 0);
2073                    if (m == NULL) {
....
2077                            return (ENOMEM);
2078                    }
2079
2080                    error = mip6_ip6mhi_create(&opt.ip6po_mh, mbu);
2081                    if (error) {
....
2085                            m_freem(m);
2086                            goto free_ip6pktopts;
2087                    }
2088
....
2090                    error = ip6_output(m, &opt, NULL, 0, NULL, NULL
....
2092                                        , NULL
....
2094                                        );
2095                    if (error) {
....
2099                            goto free_ip6pktopts;
2100                    }
2101
2102          free_ip6pktopts:
2103                    if (opt.ip6po_mh)
2104                            FREE(opt.ip6po_mh, M_IP6OPT);
2105
2106                    return (0);
2107          }
```
                                                                          _____ mip6_mncore.c

2069–2087 A Home Test Init message is passed to the `ip6_output()` function in the form of
a packet option. The `ip6_initpacketopts()` function initializes the `ip6_pktopts`
variable, which will contain the created message. An IPv6 header is prepared by the
`mip6_create_ip6hdr()` function and a Home Test Init message is created by the
`mip6_ip6hi_create()` function, which is discussed later.

2090–2104 The created message is sent by the `ip6_outout()` function. The memory space
allocated for the message is released before returning.

Listing 15-307

                                                                          _____ mip6_mncore.c
```
2109    int
2110    mip6_bu_send_coti(mbu)
2111            struct mip6_bu *mbu;
2112    {
2113            struct mbuf *m;
2114            struct ip6_pktopts opt;
2115            int error = 0;
2116
2117            ip6_initpktopts(&opt);
2118            opt.ip6po_flags |= IP6PO_USECOA;
2119
2120            m = mip6_create_ip6hdr(&mbu->mbu_coa, &mbu->mbu_paddr,
2121                IPPROTO_NONE, 0);
2122            if (m == NULL) {
....
2126                    return (ENOMEM);
2127            }
2128
2129            error = mip6_ip6mci_create(&opt.ip6po_mh, mbu);
```

```
2130            if (error) {
....
2134                    m_freem(m);
2135                    goto free_ip6pktopts;
2136            }
....
2139            error = ip6_output(m, &opt, NULL, 0, NULL, NULL
....
2141                                    , NULL
....
2143                                    );
2144            if (error) {
....
2148                    goto free_ip6pktopts;
2149            }
2150
2151    free_ip6pktopts:
2152            if (opt.ip6po_mh)
2153                    FREE(opt.ip6po_mh, M_IP6OPT);
2154
2155            return (0);
2156    }
```
―― mip6_mncore.c

2109–2156 The `mip6_bu_send_coti()` function sends a Care-of Test Init message. This is almost the same code as the `mip6_bu_send_hoti()` function. There are two differences. One is that the function sets the `IP6PO_USECOA` flag in the packet option to cause the use of the care-of address of a mobile node when sending the message. If the mobile node does not specify the flag, then the packet will be sent with a Home Address option if there is a valid binding update list entry for the correspondent node. The Home Address option must not be used with the Care-of Test Init message. The other difference is that the `mip6_ip6mci_create()` function is called instead of the `mip6_ip6mhi_create()` function to create a Care-of Test Init message.

Listing 15-308
―― mip6_mncore.c
```
3508    int
3509    mip6_ip6mhi_create(pktopt_mobility, mbu)
3510            struct ip6_mh **pktopt_mobility;
3511            struct mip6_bu *mbu;
3512    {
3513            struct ip6_mh_home_test_init *ip6mhi;
3514            int ip6mhi_size;
```
―― mip6_mncore.c

3508–3511 The `mip6_ip6mhi_create()` function has two parameters. The `pktopt_mobility` parameter is a pointer to the memory space of the `ip6_mh{}` instance, which will be allocated in this function to hold a Home Test Init message. The `mbu` parameter is a pointer to the binding update list entry related to the Home Test Init message.

Listing 15-309
―― mip6_mncore.c
```
3520            *pktopt_mobility = NULL;
3521
3522            ip6mhi_size =
3523                    ((sizeof(struct ip6_mh_home_test_init) +7) >> 3) * 8;
```

```
3524
3525            MALLOC(ip6mhi, struct ip6_mh_home_test_init *,
3526                ip6mhi_size, M_IP6OPT, M_NOWAIT);
3527            if (ip6mhi == NULL)
3528                    return (ENOMEM);
3529
3530            bzero(ip6mhi, ip6mhi_size);
3531            ip6mhi->ip6mhhti_proto = IPPROTO_NONE;
3532            ip6mhi->ip6mhhti_len = (ip6mhi_size >> 3) - 1;
3533            ip6mhi->ip6mhhti_type = IP6_MH_TYPE_HOTI;
3534            bcopy(mbu->mbu_mobile_cookie, ip6mhi->ip6mhhti_cookie8,
3535                sizeof(ip6mhi->ip6mhhti_cookie8));
3536
3537            /* calculate checksum. */
3538            ip6mhi->ip6mhhti_cksum = mip6_cksum(&mbu->mbu_haddr, &mbu->mbu_paddr,
3539                ip6mhi_size, IPPROTO_MH, (char *)ip6mhi);
3540
3541            *pktopt_mobility = (struct ip6_mh *)ip6mhi;
3542
3543            return (0);
3544    }
```
── mip6_mncore.c

3520–3534 A memory space to hold the Home Test Init message is allocated and each header value is filled. The size of the message is calculated in the same manner as other extension headers. The size is in units of 8 bytes excluding the first 8 bytes. The type is set to `IP6_MH_TYPE_HOTI`. A cookie value, which is kept in the binding update list entry, is copied to the `ip6mhhti_cookie8` field.

`ip6mhhti_cksum` is filled with the checksum value of the message by the `mip6_cksum()` function.

Listing 15-310
── mip6_mncore.c

```
3546    int
3547    mip6_ip6mci_create(pktopt_mobility, mbu)
3548            struct ip6_mh **pktopt_mobility;
3549            struct mip6_bu *mbu;
3550    {
3551            struct ip6_mh_careof_test_init *ip6mci;
3552            int ip6mci_size;
....
3558            *pktopt_mobility = NULL;
3559
3560            ip6mci_size =
3561                ((sizeof(struct ip6_mh_careof_test_init) + 7) >> 3) * 8;
3562
3563            MALLOC(ip6mci, struct ip6_mh_careof_test_init *,
3564                ip6mci_size, M_IP6OPT, M_NOWAIT);
3565            if (ip6mci == NULL)
3566                    return (ENOMEM);
3567
3568            bzero(ip6mci, ip6mci_size);
3569            ip6mci->ip6mhcti_proto = IPPROTO_NONE;
3570            ip6mci->ip6mhcti_len = (ip6mci_size >> 3) - 1;
3571            ip6mci->ip6mhcti_type = IP6_MH_TYPE_COTI;
3572            bcopy(mbu->mbu_mobile_cookie, ip6mci->ip6mhcti_cookie8,
3573                sizeof(ip6mci->ip6mhcti_cookie8));
3574
3575            /* calculate checksum. */
3576            ip6mci->ip6mhcti_cksum = mip6_cksum(&mbu->mbu_coa, &mbu->mbu_paddr,
3577                ip6mci_size, IPPROTO_MH, (char *)ip6mci);
3578
3579            *pktopt_mobility = (struct ip6_mh *)ip6mci;
```

```
3580
3581            return (0);
3582    }
```

3546–3582 The `mip6_ip6mci_create()` function does almost the same thing as the `mip6_ip6mhi_create()` function to create a Care-of Test Init message. We skip the discussion.

15.23.3   Receiving Home Test Init/Care-of Test Init Messages and Replying Home Test/Care-of Test Messages

A correspondent node will reply with a Home Test message and a Care-of Test message in response to a Home Test Init message and a Care-of Test Init message, respectively, if the node supports route optimization.

Listing 15-311

```
1826    int
1827    mip6_ip6mhi_input(m0, ip6mhi, ip6mhilen)
1828            struct mbuf *m0;
1829            struct ip6_mh_home_test_init *ip6mhi;
1830            int ip6mhilen;
1831    {
1832            struct ip6_hdr *ip6;
1833            struct mbuf *m;
1834            struct m_tag *mtag;
1835            struct ip6aux *ip6a;
1836            struct ip6_pktopts opt;
1837            int error = 0;
```

1826-1830 The `mip6_ip6mhi_input()` function is called from the `mobility6_input()` function when a correspondent node receives a Home Test Init message. m0 is a pointer to the mbuf that contains the packet. The `ip6mhi` and the `ip6mhilen` variables are pointers to the received Home Test Init message and the size of the message, respectively.

Listing 15-312

```
1841            ip6 = mtod(m0, struct ip6_hdr *);
1842
1843            /* packet length check. */
1844            if (ip6mhilen < sizeof(struct ip6_mh_home_test_init)) {
....
1850                    /* send an ICMP parameter problem. */
1851                    icmp6_error(m0, ICMP6_PARAM_PROB, ICMP6_PARAMPROB_HEADER,
1852                        (caddr_t)&ip6mhi->ip6mhhti_len - (caddr_t)ip6);
1853                    return (EINVAL);
1854            }
```

1841–1854 If the length of the packet is smaller than the size of the `ip6_mh_home_test_init{}` structure, then the correspondent node replies with an ICMPv6 Parameter Problem message to the mobile node that sent the input message. The problem pointer of the ICMPv6 message is set to the length field of the incoming Home Test Init message.

Listing 15-313

```
1856              /* a home address destination option must not exist. */
1857              mtag = ip6_findaux(m0);
1858              if (mtag) {
1859                      ip6a = (struct ip6aux *) (mtag + 1);
1860                      if ((ip6a->ip6a_flags & IP6A_HASEEN) != 0) {
....
1866                              m_freem(m0);
1867                              /* stat? */
1868                              return (EINVAL);
1869                      }
1870              }
```

1856–1870 The Home Test Init message must not have the Home Address option. If the packet contains a Home Address option, that is, the auxiliary mbuf has the `IP6A_HASEEN` flag set, the packet is dropped.

Listing 15-314

```
1872              m = mip6_create_ip6hdr(&ip6->ip6_dst, &ip6->ip6_src, IPPROTO_NONE, 0);
1873              if (m == NULL) {
....
1877                      goto free_ip6pktopts;
1878              }
1879
1880              ip6_initpktopts(&opt);
1881              error = mip6_ip6mh_create(&opt.ip6po_mh, &ip6->ip6_dst, &ip6->ip6_src,
1882                      ip6mhi->ip6mhhti_cookie8);
1883              if (error) {
....
1887                      m_freem(m);
1888                      goto free_ip6pktopts;
1889              }
```

1872–1889 An IPv6 packet is prepared by the `mip6_create_ip6hdr()` function and a Home Test message, which is a reply message to the incoming Home Test Init message, is created by the `mip6_ip6mh_create()` function. As with other Mobility Headers, the Home Test message is prepared as an `ip6_pktopts{}` instance and passed to the `ip6_output()` function.

Listing 15-315

```
1892              error = ip6_output(m, &opt, NULL, 0, NULL, NULL
....
1894                                      , NULL
....
1896                                      );
1897              if (error) {
....
1901                      goto free_ip6pktopts;
1902              }
1903
1904      free_ip6pktopts:
1905              if (opt.ip6po_mh != NULL)
1906                      FREE(opt.ip6po_mh, M_IP6OPT);
1907
```

```
1908              return (0);
1909      }
```

1892–1906 The created Home Test message is sent by the `ip6_output()` function. The memory space allocated for the message is released before finishing this function.

Listing 15-316

```
1911    int
1912    mip6_ip6mci_input(m0, ip6mci, ip6mcilen)
1913              struct mbuf *m0;
1914              struct ip6_mh_careof_test_init *ip6mci;
1915              int ip6mcilen;
1916    {
1917              struct ip6_hdr *ip6;
1918              struct mbuf *m;
1919              struct m_tag *mtag;
1920              struct ip6aux *ip6a;
1921              struct ip6_pktopts opt;
1922              int error = 0;
```

1911–1915 The `mip6_ip6mci_input()` function does tasks similar to the `mip6_ip6mhi_input()` function for a Care-of Test Init message. `m0` is a pointer to the mbuf that contains the Care-of Test Init message. The `ip6mci` and the `ip6mclen` variables are the address of the head of the message and the size of the message, respectively. The difference from the `mip6_ip6mhi_input()` function is that this function has a validation code of the source address of the incoming Care-of Test Init message.

Listing 15-317

```
1926              ip6 = mtod(m0, struct ip6_hdr *);
1927
1928              if (IN6_IS_ADDR_UNSPECIFIED(&ip6->ip6_src) ||
1929                  IN6_IS_ADDR_LOOPBACK(&ip6->ip6_src)) {
1930                      m_freem(m0);
1931                      return (EINVAL);
1932              }
```

Listing 15-318

```
1934              /* packet length check. */
1935              if (ip6mcilen < sizeof(struct ip6_mh_careof_test_init)) {
....
1941                      /* send ICMP parameter problem. */
1942                      icmp6_error(m0, ICMP6_PARAM_PROB, ICMP6_PARAMPROB_HEADER,
1943                          (caddr_t)&ip6mci->ip6mhcti_len - (caddr_t)ip6);
1944                      return (EINVAL);
1945              }
....
1947              /* a home address destination option must not exist. */
1948              mtag = ip6_findaux(m0);
1949              if (mtag) {
1950                      ip6a = (struct ip6aux *) (mtag + 1);
1951                      if ((ip6a->ip6a_flags & IP6A_HASEEN) != 0) {
....
```

```
1957                          m_freem(m0);
1958                          /* stat? */
1959                          return (EINVAL);
1960                      }
1961              }
....
1963              m = mip6_create_ip6hdr(&ip6->ip6_dst, &ip6->ip6_src, IPPROTO_NONE, 0);
1964              if (m == NULL) {
....
1968                      goto free_ip6pktopts;
1969              }
1970
1971              ip6_initpktopts(&opt);
1972              error = mip6_ip6mc_create(&opt.ip6po_mh, &ip6->ip6_dst, &ip6->ip6_src,
1973                  ip6mci->ip6mhcti_cookie8);
1974              if (error) {
....
1978                      m_freem(m);
1979                      goto free_ip6pktopts;
1980              }
....
1983              error = ip6_output(m, &opt, NULL, 0, NULL, NULL
....
1985                                   , NULL
....
1987                                   );
1988              if (error) {
....
1992                      goto free_ip6pktopts;
1993              }
1994
1995      free_ip6pktopts:
1996              if (opt.ip6po_mh != NULL)
1997                      FREE(opt.ip6po_mh, M_IP6OPT);
1998
1999              return (0);
2000      }
```
———————————————————————————————— mip6_mncore.c

1926–1932 An inappropriate source address is rejected. The source address of the Care-of Test Init message must be the care-of address of a mobile node. The unspecified address and the loopback address cannot be care-of addresses.

1934–1999 The rest of the function performs the same tasks as the mip6_ip6mhi_input() function. We skip the discussion.

Listing 15-319
———————————————————————————————— mip6_cncore.c
```
2314    static int
2315    mip6_ip6mh_create(pktopt_mobility, src, dst, cookie)
2316            struct ip6_mh **pktopt_mobility;
2317            struct in6_addr *src, *dst;
2318            u_int8_t *cookie;                   /* home init cookie */
2319    {
2320            struct ip6_mh_home_test *ip6mh;
2321            int ip6mh_size;
2322            mip6_nodekey_t home_nodekey;
2323            mip6_nonce_t home_nonce;
```
———————————————————————————————— mip6_cncore.c

2314–2318 The mip6_ip6mh_create() function creates a Home Test message. pktopt_mobility is a pointer to the memory space for the message to be created. src and dst are the source and destination addresses of the message, respectively.

These addresses are used to compute a checksum value of the message. cookie is a pointer to the home cookie value.

Listing 15-320

```
2325            *pktopt_mobility = NULL;
2326
2327            ip6mh_size = sizeof(struct ip6_mh_home_test);
2328
2329            if ((mip6_get_nonce(nonce_index, &home_nonce) != 0) ||
2330                (mip6_get_nodekey(nonce_index, &home_nodekey) != 0))
2331                    return (EINVAL);
2332
2333            MALLOC(ip6mh, struct ip6_mh_home_test *, ip6mh_size,
2334                M_IP6OPT, M_NOWAIT);
2335            if (ip6mh == NULL)
2336                    return (ENOMEM);
```

2325–2330 The current nonce value and nodekey value, pointed to by the global variable nonce_index, are retrieved by the mip6_get_nonce() and the mip6_get_nodekey() functions, respectively.

2333–2336 A memory space to hold the created Home Test message is allocated.

Listing 15-321

```
2338            bzero(ip6mh, ip6mh_size);
2339            ip6mh->ip6mhht_proto = IPPROTO_NONE;
2340            ip6mh->ip6mhht_len = (ip6mh_size >> 3) - 1;
2341            ip6mh->ip6mhht_type = IP6_MH_TYPE_HOT;
2342            ip6mh->ip6mhht_nonce_index = htons(nonce_index);
2343            bcopy(cookie, ip6mh->ip6mhht_cookie8, sizeof(ip6mh->ip6mhht_cookie8));
2344            if (mip6_create_keygen_token(dst, &home_nodekey,
2345                &home_nonce, 0, ip6mh->ip6mhht_keygen8)) {
....
2349                    return (EINVAL);
2350            }
2351
2352            /* calculate checksum. */
2353            ip6mh->ip6mhht_cksum = mip6_cksum(src, dst, ip6mh_size, IPPROTO_MH,
2354                (char *)ip6mh);
2355
2356            *pktopt_mobility = (struct ip6_mh *)ip6mh;
2357
2358            return (0);
2359    }
```

2338–2354 The size of the message is calculated in the same manner as other extension headers. The size is in units of 8 bytes excluding the first 8 bytes. The type is set to IP6_MH_TYPE_HOT. The value of the nonce index field (ip6mhht_nonce_index) is copied from the global variable nonce_index. The cookie value sent from the mobile node is copied to the cookie field (ip6mhht_cookie8). The keygen token is computed by the mip6_create_keygen_token() function and is copied to the keygen field (ip6mhht_keygen8). Finally, the checksum value is computed by the mip6_cksum() function.

Listing 15-322

```
                                                              mip6_cncore.c
2361     static int
2362     mip6_ip6mc_create(pktopt_mobility, src, dst, cookie)
2363             struct ip6_mh **pktopt_mobility;
2364             struct in6_addr *src, *dst;
2365             u_int8_t *cookie;                    /* careof init cookie */
2366     {
2367             struct ip6_mh_careof_test *ip6mc;
2368             int ip6mc_size;
2369             mip6_nodekey_t careof_nodekey;
2370             mip6_nonce_t careof_nonce;
2371
2372             *pktopt_mobility = NULL;
2373
2374             ip6mc_size = sizeof(struct ip6_mh_careof_test);
2375
2376             if ((mip6_get_nonce(nonce_index, &careof_nonce) != 0) ||
2377                 (mip6_get_nodekey(nonce_index, &careof_nodekey) != 0))
2378                     return (EINVAL);
2379
2380             MALLOC(ip6mc, struct ip6_mh_careof_test *, ip6mc_size,
2381                 M_IP6OPT, M_NOWAIT);
2382             if (ip6mc == NULL)
2383                     return (ENOMEM);
2384
2385             bzero(ip6mc, ip6mc_size);
2386             ip6mc->ip6mhct_proto = IPPROTO_NONE;
2387             ip6mc->ip6mhct_len = (ip6mc_size >> 3) - 1;
2388             ip6mc->ip6mhct_type = IP6_MH_TYPE_COT;
2389             ip6mc->ip6mhct_nonce_index = htons(nonce_index);
2390             bcopy(cookie, ip6mc->ip6mhct_cookie8, sizeof(ip6mc->ip6mhct_cookie8));
2391             if (mip6_create_keygen_token(dst, &careof_nodekey, &careof_nonce, 1,
2392                         ip6mc->ip6mhct_keygen8)) {
....
2397                     return (EINVAL);
2398             }
2399
2400             /* calculate checksum. */
2401             ip6mc->ip6mhct_cksum = mip6_cksum(src, dst, ip6mc_size, IPPROTO_MH,
2402                 (char *)ip6mc);
2403
2404             *pktopt_mobility = (struct ip6_mh *)ip6mc;
2405
2406             return (0);
2407     }
                                                              mip6_cncore.c
```

2361–2407 The `mip6_ip6mc_create()` function performs almost the same task to create the Care-of Test message as the `mip6_ip6mh_create()` does for a Home Test message.

15.23.4   Receiving Home Test/Care-of Test Messages

When a mobile node receives a Home Test message or a Care-of Test message, it sends a related event to its state machine.

Listing 15-323

```
                                                              mip6_mncore.c
2697     int
2698     mip6_ip6mh_input(m, ip6mh, ip6mhlen)
2699             struct mbuf *m;
2700             struct ip6_mh_home_test *ip6mh;
2701             int ip6mhlen;
```

```
2702    {
2703            struct ip6_hdr *ip6;
2704            struct hif_softc *sc;
2705            struct mip6_bu *mbu;
2706            int error = 0;
```
── mip6_mncore.c

2697–2701 The `mip6_ip6mh_input()` function is called from the `mobility6_input()` function when a mobile node receives a Home Test message. m is a pointer to the mbuf that contains the received packet. The `ip6mh` and the `ip6mhlen` variables are pointers to the head of the Home Test message and its length, respectively.

Listing 15-324
── mip6_mncore.c
```
2710            ip6 = mtod(m, struct ip6_hdr *);
2711
2712            /* packet length check. */
2713            if (ip6mhlen < sizeof(struct ip6_mh_home_test)) {
....
2721                    /* send ICMP parameter problem. */
2722                    icmp6_error(m, ICMP6_PARAM_PROB, ICMP6_PARAMPROB_HEADER,
2723                        (caddr_t)&ip6mh->ip6mhht_len - (caddr_t)ip6);
2724                    return (EINVAL);
2725            }
```
── mip6_mncore.c

2713–2725 If the length of the received Home Test message is smaller than the size of the `ip6_mh_home_test{}` structure, the mobile node sends an ICMPv6 Parameter Problem message. The problem pointer points to the length field of the incoming message.

Listing 15-325
── mip6_mncore.c
```
2727            sc = hif_list_find_withhaddr(&ip6->ip6_dst);
2728            if (sc == NULL) {
....
2733                    m_freem(m);
....
2735                    return (EINVAL);
2736            }
2737            mbu = mip6_bu_list_find_withpaddr(&sc->hif_bu_list, &ip6->ip6_src,
2738                &ip6->ip6_dst);
2739            if (mbu == NULL) {
....
2744                    m_freem(m);
....
2746                    return (EINVAL);
2747            }
```
── mip6_mncore.c

2727–2736 The destination address of the incoming packet must be the home address of the mobile node. If there is not a virtual home interface that has the home address, the mobile node drops the packet.

2737–2747 The status of the return routability procedure is kept in a binding update list entry. If the mobile node does not have a binding update list entry related to the incoming packet, the packet is dropped.

Listing 15-326

```
                                                            ___ mip6_mncore.c
2749              /* check mobile cookie. */
2750              if (bcmp(&mbu->mbu_mobile_cookie, ip6mh->ip6mhht_cookie8,
2751                  sizeof(ip6mh->ip6mhht_cookie8)) != 0) {
 ....
2755                      m_freem(m);
 ....
2757                      return (EINVAL);
2758              }
2759
2760              error = mip6_bu_fsm(mbu, MIP6_BU_SEC_FSM_EVENT_HOT, ip6mh);
2761              if (error) {
 ....
2765                      m_freem(m);
2766                      return (error);
2767              }
2768
2769              mbu->mbu_home_nonce_index = ntohs(ip6mh->ip6mhht_nonce_index);
 ....
2774              return (0);
2775      }
                                                            ___ mip6_mncore.c
```

2750–2758 If the incoming Home Test message is not a response to the Home Test Init message sent from this mobile node, the packet is dropped. A Home Test message must include the same cookie value that is copied from the Home Test Init message sent previously.

2760–2769 The mobile node sends the `MIP6_BU_SEC_FSM_EVENT_HOT` event to its state machine to inform that a Home Test message is received. Line 2769 is redundant because the home nonce index is copied in the state machine as well.

Listing 15-327

```
                                                            ___ mip6_mncore.c
2777      int
2778      mip6_ip6mc_input(m, ip6mc, ip6mclen)
2779              struct mbuf *m;
2780              struct ip6_mh_careof_test *ip6mc;
2781              int ip6mclen;
2782      {
2783              struct ip6_hdr *ip6;
2784              struct hif_softc *sc;
2785              struct mip6_bu *mbu = NULL;
2786              int error = 0;
 ....
2790              ip6 = mtod(m, struct ip6_hdr *);
2791
2792              if (IN6_IS_ADDR_UNSPECIFIED(&ip6->ip6_src) ||
2793                  IN6_IS_ADDR_LOOPBACK(&ip6->ip6_src)) {
2794                      m_freem(m);
2795                      return (EINVAL);
2796              }
2797
2798              /* packet length check. */
2799              if (ip6mclen < sizeof(struct ip6_mh_careof_test)) {
 ....
2807                      /* send ICMP parameter problem. */
2808                      icmp6_error(m, ICMP6_PARAM_PROB, ICMP6_PARAMPROB_HEADER,
2809                          (caddr_t)&ip6mc->ip6mhct_len - (caddr_t)ip6);
2810                      return (EINVAL);
2811              }
                                                            ___ mip6_mncore.c
```

2777–2781 The `mip6_ip6mc_input()` function performs almost the same tasks for the Care-of Test message as the `mip6_ip6mh_input()` function does for the Home Test message.

2792–2796 The intention of this part is to validate the care-of address of the received message. Apparently, the code is wrong. It should check the destination address instead of the source address. The destination address must be the care-of address of the mobile node. A care-of address cannot be the loopback address or the unspecified address.

Listing 15-328

```
                                                          mip6_mncore.c
2813            /* too ugly... */
2814            for (sc = LIST_FIRST(&hif_softc_list); sc;
2815                sc = LIST_NEXT(sc, hif_entry)) {
2816                for (mbu = LIST_FIRST(&sc->hif_bu_list); mbu;
2817                    mbu = LIST_NEXT(mbu, mbu_entry)) {
2818                        if (IN6_ARE_ADDR_EQUAL(&ip6->ip6_dst, &mbu->mbu_coa) &&
2819                            IN6_ARE_ADDR_EQUAL(&ip6->ip6_src, &mbu->mbu_paddr))
2820                                break;
2821                }
2822                if (mbu != NULL)
2823                        break;
2824            }
2825            if (mbu == NULL) {
....
2830                m_freem(m);
....
2832                return (EINVAL);
2833            }
                                                          mip6_mncore.c
```

2813–2833 To get a binding update list entry related to the incoming Care-of Test message, the mobile node checks all the binding update list entries it has. If there is no binding update list entry of which the care-of address is the same as the destination address of the incoming packet and of which the peer address is the same as the source address of the incoming packet, the packet is dropped.

Listing 15-329

```
                                                          mip6_mncore.c
2835            /* check mobile cookie. */
2836            if (bcmp(&mbu->mbu_mobile_cookie, ip6mc->ip6mhct_cookie8,
2837                sizeof(ip6mc->ip6mhct_cookie8)) != 0) {
....
2841                m_freem(m);
....
2843                return (EINVAL);
2844            }
2845
2846            error = mip6_bu_fsm(mbu, MIP6_BU_SEC_FSM_EVENT_COT, ip6mc);
2847            if (error) {
....
2851                m_freem(m);
2852                return (error);
2853            }
2854
2855            mbu->mbu_careof_nonce_index = ntohs(ip6mc->ip6mhct_nonce_index);
....
2860            return (0);
2861    }
                                                          mip6_mncore.c
```

2835–2855 After checking the cookie value, the mobile node sends the `MIP6_BU_SEC_FSM_EVENT_COT` event to its state machine.

15.23.5 Sending a Binding Update Message to Correspondent Node

As we have already discussed in Section 15.19, a mobile node will call the `mip6_bu_send_cbu()` function to send a Binding Update message after receiving a Home Test and a Care-of Test message from a correspondent node. The route optimization preparation has been completed at this point. The KAME implementation does not require a Binding Acknowledgment message from a correspondent node.

15.24 Route-Optimized Communication

If both a mobile node and a correspondent node support route optimization, these nodes can communicate with each other directly using extension headers as described in Chapter 5.

15.24.1 Sending a Route-Optimized Packet

The extension headers used by the route optimization procedure are inserted during the output processing of a packet.

Listing 15-330

── ip6_output.c

```
372              if (opt) {
373                      /* Hop-by-Hop options header */
374                      MAKE_EXTHDR(opt->ip6po_hbh, &exthdrs.ip6e_hbh);
375                      /* Destination options header(1st part) */
376                      if (opt->ip6po_rthdr
....
378                          || opt->ip6po_rthdr2
....
380                          ) {
381                              /*
382                               * Destination options header(1st part)
383                               * This only makes sence with a routing header.
384                               * See section 9.2 of [RFC3776].
385                               * Disabling this part just for MIP6 convenience is
386                               * a bad idea.  We need to think carefully about a
387                               * way to make the advanced API coexist with MIP6
388                               * options, which might automatically be inserted in
389                               * the kernel.
390                               */
391                              MAKE_EXTHDR(opt->ip6po_dest1, &exthdrs.ip6e_dest1);
392                      }
393                      /* Routing header */
394                      MAKE_EXTHDR(opt->ip6po_rthdr, &exthdrs.ip6e_rthdr);
....
396                      /* Type 2 Routing header for MIP6 route optimization */
397                      MAKE_EXTHDR(opt->ip6po_rthdr2, &exthdrs.ip6e_rthdr2);
....
399                      /* Destination options header(2nd part) */
400                      MAKE_EXTHDR(opt->ip6po_dest2, &exthdrs.ip6e_dest2);
....
402                      MAKE_EXTHDR(opt->ip6po_mh, &exthdrs.ip6e_mh);
....
404              }
```

── ip6_output.c

372–404 The code is a part of the `ip6_output()` function. We can specify extension headers when calling the `ip6_output()` function using the `ip6_pktopts{}` structure. With regard to Mobile IPv6, the Type 2 Routing Header (line 397) and the Mobility Header (line 402) may be specified. A Home Address option is not usually specified by the `ip6_pktopts{}` structure at this point. The option is usually inserted automatically in the `mip6_exthdr_create()` function, discussed later.

Listing 15-331

```
                                                                    ip6_output.c
406              bzero((caddr_t)&mip6opt, sizeof(mip6opt));
407              if ((flags & IPV6_FORWARDING) == 0) {
408                      struct m_tag *n;
409                      struct ip6aux *ip6a = NULL;
410                      /*
411                       * XXX: reconsider the following routine.
412                       */
413                      /*
414                       * MIP6 extension headers handling.
415                       * insert HA, BU, BA, BR options if necessary.
416                       */
417                      n = ip6_findaux(m);
418                      if (n)
419                              ip6a = (struct ip6aux *) (n + 1);
420                      if (!(ip6a && (ip6a->ip6a_flags & IP6A_NOTUSEBC)))
421                              if (mip6_exthdr_create(m, opt, &mip6opt))
422                                      goto freehdrs;
423
424                      if ((exthdrs.ip6e_rthdr2 == NULL)
425                          && (mip6opt.mip6po_rthdr2 != NULL)) {
426                              /*
427                               * if a type 2 routing header is not specified
428                               * when ip6_output() is called and
429                               * mip6_exthdr_create() creates a type 2
430                               * routing header for route optimization,
431                               * insert it.
432                               */
433                              MAKE_EXTHDR(mip6opt.mip6po_rthdr2,
  &exthdrs.ip6e_rthdr2);
434                              /*
435                               * if a routing header exists dest1 must be
436                               * inserted if it exists.
437                               */
438                              if ((opt != NULL) && (opt->ip6po_dest1) &&
439                                  (exthdrs.ip6e_dest1 == NULL)) {
440                                      m_freem(exthdrs.ip6e_dest1);
441                                      MAKE_EXTHDR(opt->ip6po_dest1,
442                                          &exthdrs.ip6e_dest1);
443                              }
444                      }
445                      /* Home Address Destination Option. */
446                      if (mip6opt.mip6po_haddr != NULL)
447                              have_hao = 1;
448                      MAKE_EXTHDR(mip6opt.mip6po_haddr, &exthdrs.ip6e_haddr);
449              } else {
450                      /*
451                       * this is a forwarded packet.  do not modify any
452                       * extension headers.
453                       */
454              }
455
456              if (exthdrs.ip6e_mh) {
  ....
458                      if (ip6->ip6_nxt != IPPROTO_NONE || m->m_next != NULL)
459                              panic("not supported piggyback");
```

```
460                         exthdrs.ip6e_mh->m_next = m->m_next;
461                         m->m_next = exthdrs.ip6e_mh;
462                         *mtod(exthdrs.ip6e_mh, u_char *) = ip6->ip6_nxt;
463                         ip6->ip6_nxt = IPPROTO_MH;
464                         m->m_pkthdr.len += exthdrs.ip6e_mh->m_len;
465                         exthdrs.ip6e_mh = NULL;
466                 }
```
—— ip6_output.c

Note: Line 433 is broken here for layout reasons. However, it is a single line of code.

406–422 If the packet being sent requires any extension headers for route optimization, the node will insert those necessary headers by calling the mip6_exthdr_create() function. The IP6A_NOTUSEBC flag is used when a home agent needs to ignore a binding cache entry when sending a Binding Acknowledgment message since the Type 2 Routing Header is already inserted, if needed, when sending a Binding Acknowledgment message, as discussed in Section 14.13 of Chapter 14.

424–444 The mip6_exthdr_create() function creates a Destination Options Header, which includes a Home Address option and a Type 2 Routing Header. The created headers will be stored in the mip6_pktopts{} structure specified as the mip6opt variable. If a Type 2 Routing Header is not specified in the ip6_pktopts{} instance and the mip6_exthdr_create() function generates a Type 2 Routing Header, the node will insert the header using the MAKE_EXTHDR() macro. If there is a first Destination Options Header specified as ip6po_dest1, the node needs to insert it since the node has a Routing Header at this point.

445–448 If the mip6_exthdr_create() function generates a Destination Options Header, which includes a Home Address option, the node inserts the header.

456–466 When a caller of the ip6_output() function specifies a Mobility Header as the ip6_pktopts{} structure, the node rearranges the packet. Lines 458–459 is a simple validation of the mbuf that contains the outgoing packet. The Mobility Header must not have any following headers. That is, the next header field (ip6_nxt) must be IPPROTO_NONE and the IPv6 packet must not have a following mbuf.

The mbuf that contains the Mobility Header is rearranged as the next mbuf of the IPv6 header and the next header value is set to IPPROTO_MH.

Listing 15-332
—— ip6_output.c

```
824             if ((flags & IPV6_FORWARDING) == 0) {
825                     /*
826                      * after the IPsec processing, the IPv6 header source
827                      * address (this is the homeaddress of this node) and
828                      * the address currently stored in the Home Address
829                      * destination option (this is the coa of this node)
830                      * must be swapped.
831                      */
832                     if ((error = mip6_addr_exchange(m, exthdrs.ip6e_haddr)) != 0) {
....
836                             goto bad;
837                     }
838             } else {
839                     /*
840                      * this is a forwarded packet.  The typical (and
841                      * only ?) case is multicast packet forwarding.  The
842                      * swapping has been already done before (if
```

```
843                         * necessary).  we must not touch any extension
844                         * headers at all.
845                         */
846              }
```
_____ ip6_output.c

832–837 If the node is acting as a mobile node and the outgoing packet has a Home Address
option, the node needs to exchange the source address (which at this point is the care-of
address of the mobile node) and the home address stored in the Home Address option
before routing the packet.

Listing 15-333
_____ ip6_output.c
```
857              if (exthdrs.ip6e_rthdr)
858                      rh = (struct ip6_rthdr *)(mtod(exthdrs.ip6e_rthdr,
859                          struct ip6_rthdr *));
    ....
861              else if (exthdrs.ip6e_rthdr2)
862                      rh = (struct ip6_rthdr *)(mtod(exthdrs.ip6e_rthdr2,
863                          struct ip6_rthdr *));
    ....
865              if (rh) {
866                      struct ip6_rthdr0 *rh0;
867                      struct in6_addr *addr;
868                      struct sockaddr_in6 sa;
869
870                      finaldst = ip6->ip6_dst;
871                      switch (rh->ip6r_type) {
    ....
873                      case IPV6_RTHDR_TYPE_2:
    ....
875                      case IPV6_RTHDR_TYPE_0:
876                          rh0 = (struct ip6_rthdr0 *)rh;
877                          addr = (struct in6_addr *)(rh0 + 1);
```
.... [Address swapping procedure for a Routing Header]
```
909                      }
910              }
```
_____ ip6_output.c

857–910 If a correspondent node has a Type 2 Routing Header, the destination address and
the address included in the Routing Header needs to be exchanged, as well as the Home
Address option. The exchange procedure is almost the same as that of the Type 1 Routing
Header.

15.24.2   Receiving a Route-Optimized Packet

A route-optimized packet sent from a mobile node has a Home Address option. The option
processing is done while the receiving node is processing the Destination Options header. This
procedure was discussed in Section 14.15 of Chapter 14. A route-optimized packet sent from
a correspondent node has a Type 2 Routing Header. This processing code was discussed in
Section 15.7.

15.24.3   Creating Extension Headers

Extension headers for route-optimized packets are created by the `mip6_exthdr_create()`
function.

Listing 15-334

```
                                                      _____ mip6_cncore.c
513     int
514     mip6_exthdr_create(m, opt, mip6opt)
515             struct mbuf *m;                   /* ip datagram */
516             struct ip6_pktopts *opt;          /* pktopt passed to ip6_output */
517             struct mip6_pktopts *mip6opt;
518     {
519             struct ip6_hdr *ip6;
520             int s, error = 0;
....
522             struct hif_softc *sc;
523             struct mip6_bu *mbu;
524             int need_hao = 0;
                                                      _____ mip6_cncore.c
```

513–516 The mip6_exthdr_create() function has three parameters: The m parameter is a pointer to the mbuf that contains a packet to be sent; the opt parameter is a pointer to the ip6_pktopts{} instance, which has been passed to the ip6_output() function; and the mip6opt parameter is a pointer to the mip6_pktopts{} instance. The function will create all necessary header information and store it in the mip6opt variable.

Listing 15-335

```
                                                      _____ mip6_cncore.c
538             /*
539              * From section 6.1: "Mobility Header messages MUST NOT be
540              * sent with a type 2 routing header, except as described in
541              * Section 9.5.4 for Binding Acknowledgment".
542              */
543             if ((opt != NULL)
544                 && (opt->ip6po_mh != NULL)
545                 && (opt->ip6po_mh->ip6mh_type != IP6_MH_TYPE_BACK)) {
546                     goto skip_rthdr2;
547             }
                                                      _____ mip6_cncore.c
```

543–547 A Mobility Header cannot have a Type 2 Routing Header except for the Binding Acknowledgment message. If ip6po_mh, which contains a Mobility Header, is specified and the type is not a Binding Acknowledgment, the function skips the creation code of a Type 2 Routing Header.

Listing 15-336

```
                                                      _____ mip6_cncore.c
549             /*
550              * create rthdr2 only if the caller of ip6_output() doesn't
551              * specify rthdr2 already.
552              */
553             if ((opt != NULL) &&
554                 (opt->ip6po_rthdr2 != NULL))
555                     goto skip_rthdr2;
556
557             /*
558              * add the routing header for the route optimization if there
559              * exists a valid binding cache entry for this destination
560              * node.
561              */
562             error = mip6_rthdr_create_withdst(&mip6opt->mip6po_rthdr2,
563                 &ip6->ip6_dst, opt);
```

```
564                if (error) {
  ....
568                        goto bad;
569                }
570        skip_rthdr2:
```

553–555 If the caller of the `ip6_output()` function explicitly specifies a Type 2 Routing Header (that is, `ip6po_rthdr2` is specified), the function respects the intention of the caller.

562–569 A Type 2 Routing Header is created by the `mip6_rthdr_create_withdst()` function and stored in the `mip6po_rthdr2` pointer.

Listing 15-337

```
573                /* the following stuff is applied only for a mobile node. */
574                if (!MIP6_IS_MN) {
575                        goto skip_hao;
576                }
577
578                /*
579                 * find hif that has a home address that is the same
580                 * to the source address of this sending ip packet
581                 */
582                sc = hif_list_find_withhaddr(&ip6->ip6_src);
583                if (sc == NULL) {
584                        /*
585                         * this source address is not one of our home addresses.
586                         * we don't need any special care about this packet.
587                         */
588                        goto skip_hao;
589                }
```

574–576 A Home Address option is inserted only if the node is acting as a mobile node.

582–589 If the source address of the outgoing packet is one of the home addresses of the mobile node, a pointer to the virtual home interface related to the home address is searched for. If the source address is not a home address, then no related virtual home interface will be found. In this case, there is no need to insert a Home Address option.

Listing 15-338

```
591                /*
592                 * check home registration status for this destination
593                 * address.
594                 */
595                mbu = mip6_bu_list_find_withpaddr(&sc->hif_bu_list, &ip6->ip6_dst,
596                    &ip6->ip6_src);
597                if (mbu == NULL) {
598                        /* no registration action has been started yet. */
599                        goto skip_hao;
600                }
```

595–600 If a mobile node does not have a binding update list entry to the peer node, the node cannot perform a route-optimized communication.

Listing 15-339

——— mip6_cncore.c

```
602                 if ((opt != NULL) && (opt->ip6po_mh != NULL)) {
603                         if (opt->ip6po_mh->ip6mh_type == IP6_MH_TYPE_BU)
604                                 need_hao = 1;
605                         else {
606                                 /*
607                                  * From 6.1 Mobility Header: "Mobility Header
608                                  * messages also MUST NOT be used with a Home
609                                  * Address destination option, except as
610                                  * described in Section 11.7.1 and Section
611                                  * 11.7.2 for Binding Update."
612                                  */
613                                 goto skip_hao;
614                         }
615                 }
```

——— mip6_cncore.c

602–615 A Home Address option must not be used with the Mobility Header except with a Binding Update message. If the packet has a Mobility Header (that is, `ip6po_mh` is specified) and the type is not Binding Update, the creation of a Home Address option is skipped.

Listing 15-340

——— mip6_cncore.c

```
616                 if ((mbu->mbu_flags & IP6MU_HOME) != 0) {
617                         /* to my home agent. */
618                         if (!need_hao &&
619                             (mbu->mbu_pri_fsm_state == MIP6_BU_PRI_FSM_STATE_IDLE ||
620                              mbu->mbu_pri_fsm_state == MIP6_BU_PRI_FSM_STATE_WAITD))
621                                 goto skip_hao;
622                 } else {
623                         /* to any of correspondent nodes. */
624                         if (!need_hao && !MIP6_IS_BU_BOUND_STATE(mbu))
625                                 goto skip_hao;
626                 }
```

——— mip6_cncore.c

616–621 The outgoing packet is destined for the home agent of the mobile node. A Home Address option must be used only when the mobile node is registered. If the registration state is `MIP6_BU_PRI_FSM_STATE_IDLE` or `MIP6_BU_PRI_FSM_STATE_WAITD`, the packet must not have a Home Address option.

622–626 The same conditions are checked for a correspondent node.

Listing 15-341

——— mip6_cncore.c

```
627                 /* create a home address destination option. */
628                 error = mip6_haddr_destopt_create(&mip6opt->mip6po_haddr,
629                     &ip6->ip6_src, &ip6->ip6_dst, sc);
630                 if (error) {
....
634                         goto bad;
635                 }
636         skip_hao:
637                 error = 0; /* normal exit. */
....
639
640         bad:
641                 splx(s);
```

```
642                return (error);
643        }
```
_____ mip6_cncore.c

628–629 The `mip6_haddr_destopt_create()` function is called, and a Destination Options Header that contains the Home Address option is created.

15.24.4   Creating a Type 2 Routing Header with a Destination Address

The `mip6_rthdr_create_withdst()` function creates a Type 2 Routing Header for a mobile node with the specified home address of the mobile node.

Listing 15-342
_____ mip6_cncore.c

```
681     static int
682     mip6_rthdr_create_withdst(pktopt_rthdr, dst, opt)
683             struct ip6_rthdr **pktopt_rthdr;
684             struct in6_addr *dst;
685             struct ip6_pktopts *opt;
686     {
687             struct mip6_bc *mbc;
688             int error = 0;
689
690             mbc = mip6_bc_list_find_withphaddr(&mip6_bc_list, dst);
691             if (mbc == NULL) {
692                     /* no binding cache entry for this dst is found. */
693                     return (0);
694             }
695
696             error = mip6_rthdr_create(pktopt_rthdr, &mbc->mbc_pcoa, opt);
697             if (error) {
698                     return (error);
699             }
700
701             return (0);
702     }
```
_____ mip6_cncore.c

681–685 The `dst` parameter is the home address of a mobile node.

690–699 The `mip6_bc_list_find_withphaddr()` function is called to find a binding cache entry for the mobile node whose home address is `dst`. If a correspondent node has such a binding cache entry, the `mip6_rthdr_create()` function is called with the care-of address of the mobile node to create a Type 2 Routing Header that contains the care-of address of the mobile node. The address in the Routing Header will be exchanged with the destination address of an IPv6 header in the `ip6_output()` function later.

15.24.5   Creating a Home Address Option

The `mip6_haddr_destopt_create()` function creates a Destination Options Header that contains a Home Address option.

Listing 15-343
_____ mip6_mncore.c

```
2420    int
2421    mip6_haddr_destopt_create(pktopt_haddr, src, dst, sc)
2422            struct ip6_dest **pktopt_haddr;
```

```
2423                struct in6_addr *src;
2424                struct in6_addr *dst;
2425                struct hif_softc *sc;
2426      {
2427                struct in6_addr hif_coa;
2428                struct ip6_opt_home_address haddr_opt;
2429                struct mip6_buffer optbuf;
2430                struct mip6_bu *mbu;
2431                struct in6_addr *coa;
```
── mip6_mncore.c

2420–2425 The `pktopt_haddr` parameter is a pointer where the created header is placed. The `src` and `dst` parameters are the source address and the destination address of a mobile node, respectively. The `sc` parameter is a pointer to the virtual home interface of the mobile node.

Listing 15-344
── mip6_mncore.c
```
2433                if (*pktopt_haddr) {
2434                        /* already allocated ? */
2435                        return (0);
2436                }
2437
2438                /* get current CoA and recover its scope information. */
2439                if (sc->hif_coa_ifa == NULL) {
 ....
2443                        return (0);
2444                }
2445                hif_coa = sc->hif_coa_ifa->ia_addr.sin6_addr;
```
── mip6_mncore.c

2433–2435 If the caller of the `ip6_output()` function has already specified a Home Address option, the caller's intention is respected.

2439–2445 The care-of address of the mobile node is taken from the virtual home interface. If the mobile node does not have a valid care-of address, a Home Address option cannot be created.

Listing 15-345
── mip6_mncore.c
```
2447                bzero(&haddr_opt, sizeof(struct ip6_opt_home_address));
2448                haddr_opt.ip6oh_type = IP6OPT_HOME_ADDRESS;
2449                haddr_opt.ip6oh_len = IP6OPT_HALEN;
2450
2451                mbu = mip6_bu_list_find_withpaddr(&sc->hif_bu_list, dst, src);
2452                if (mbu && ((mbu->mbu_state & MIP6_BU_STATE_NEEDTUNNEL) != 0)) {
2453                        return (0);
2454                }
2455                if (mbu)
2456                        coa = &mbu->mbu_coa;
2457                else
2458                        coa = &hif_coa;
2459                bcopy((caddr_t)coa, haddr_opt.ip6oh_addr, sizeof(struct in6_addr));
```
── mip6_mncore.c

2451–2454 If the destination node does not support Mobile IPv6 (the `MIP6_BU_STATE_NEEDTUNNEL` flag is set), a Home Address option will not be created.

2452–2459 The care-of address of the mobile node is copied to the option. The address will be swapped with the source address of the IPv6 packet later in the `ip6_output()` function.

> Note that the variable mbu will never be a NULL pointer. The code from lines 2457 to 2458 is old and was never executed. The code was used in the older Mobile IPv6 specification, which allowed using a Home Address option without a correct binding between a mobile node and a correspondent node.

Listing 15-346

---
mip6_mncore.c

```
2461             MALLOC(optbuf.buf, u_int8_t *, MIP6_BUFFER_SIZE, M_IP6OPT, M_NOWAIT);
2462             if (optbuf.buf == NULL) {
2463                     return (ENOMEM);
2464             }
2465             bzero((caddr_t)optbuf.buf, MIP6_BUFFER_SIZE);
2466             optbuf.off = 2;
2467
2468             /* Add Home Address option */
2469             mip6_add_opt2dh((u_int8_t *)&haddr_opt, &optbuf);
2470             mip6_align_destopt(&optbuf);
2471
2472             *pktopt_haddr = (struct ip6_dest *)optbuf.buf;
....
2476             return (0);
2477     }
```
---
mip6_mncore.c

2461–2472 Memory space to hold a Destination Options Header is allocated, and the Home Address option created just before is inserted on line 2469.

15.25   Tunnel Control

15.25.1   Adding/Removing a Tunnel

A mobile node and a home agent create a tunnel interface between them to send/receive packets between the mobile node and correspondent nodes. The tunnel is created/destroyed by the generic tunnel mechanism provided by the `mip6_tunnel_control()` function.

Listing 15-347

---
mip6_cncore.c

```
2869    int
2870    mip6_tunnel_control(action, entry, func, ep)
2871            int action;
2872            void *entry;
2873            int (*func)(const struct mbuf *, int, int, void *);
2874            const struct encaptab **ep;
2875    {
....
2877    #ifdef MIP6_MOBILE_NODE
2878            struct mip6_bu *mbu;
2879            struct sockaddr_in6 haddr_sa, coa_sa, paddr_sa;
2880    #endif
2881    #ifdef MIP6_HOME_AGENT
2882            struct mip6_bc *mbc;
2883            struct sockaddr_in6 phaddr_sa, pcoa_sa, addr_sa;
2884    #endif
```
---
mip6_cncore.c

2869–2874 The `mip6_tunnel_control()` function has four parameters. The `action` parameter is the type of operation. The `MIP6_TUNNEL_ADD`, `MIP6_TUNNEL_CHANGE`, or
`MIP6_TUNNEL_DELETE` command can be specified. The `entry` parameter is a pointer
to a binding update list entry or a binding cache entry based on the type of the node. The
`func` parameter is a pointer to the function that decides whether the node should receive
a tunneled packet. `ep` is a pointer to the tunnel information provided by the generic tunnel
mechanism.

Listing 15-348

```
                                                               mip6_cncore.c
2886            if ((entry == NULL) && (ep == NULL)) {
2887                    return (EINVAL);
2888            }
        .... [IPsec processing]
2959            /* before doing anything, remove an existing encap entry. */
2960            switch (action) {
2961            case MIP6_TUNNEL_ADD:
2962            case MIP6_TUNNEL_CHANGE:
2963            case MIP6_TUNNEL_DELETE:
2964                    if (*ep != NULL) {
2965                            encap_detach(*ep);
2966                            *ep = NULL;
2967                    }
2968            }
                                                               mip6_cncore.c
```

2960–2968 Existing tunnel information is removed by the `encap_detach()` function,
which removes the tunnel interface entry from the kernel, before adding/changing the
information.

Listing 15-349

```
                                                               mip6_cncore.c
2970            switch (action) {
2971            case MIP6_TUNNEL_ADD:
2972            case MIP6_TUNNEL_CHANGE:
2973                    *ep = encap_attach_func(AF_INET6, IPPROTO_IPV6,
2974                                            func,
2975                                            (struct protosw *)&mip6_tunnel_protosw,
2976                                            (void *)entry);
2977                    if (*ep == NULL) {
....
2982                            return (EINVAL);
2983                    }
2984                    break;
2985            }
2986
2987            return (0);
2988    }
                                                               mip6_cncore.c
```

2970–2984 If the operation is removing, the process has been done already. If the operation is adding or changing, a new tunnel information entry is created by the
`encap_attach_func()` function, which inserts the tunnel information into the kernel.

15.25.2 Validation of Packets Received from a Tunnel

The `mip6_bu_encapcheck()` function is called when a mobile node receives an IPv6 in IPv6
packet.

Listing 15-350
_____ mip6_mncore.c

```
1859    int
1860    mip6_bu_encapcheck(m, off, proto, arg)
1861            const struct mbuf *m;
1862            int off;
1863            int proto;
1864            void *arg;
1865    {
1866            struct ip6_hdr *ip6;
1867            struct mip6_bu *mbu = (struct mip6_bu *)arg;
1868            struct hif_softc *sc;
1869            struct mip6_prefix *mpfx;
1870            struct in6_addr *haaddr, *myaddr, *mycoa;
```
_____ mip6_mncore.c

1859–1864 The function must return a positive value (from 1 to 128) if the packet is acceptable and must return 0 if it is not acceptable. m is a pointer to the incoming packet. off and proto are the offset to the inner packet header and a protocol number of the inner packet, respectively. These two parameters are not used in this function. arg is a pointer that is the pointer passed when the tunnel information is registered by the encap_attach_func() function on line 2973 of mip6_cncore.c (Listing 15–349). In a mobile node case, a pointer to a binding update entry is set.

Listing 15-351
_____ mip6_mncore.c

```
1879            ip6 = mtod(m, struct ip6_hdr*);
1880
1881            haaddr = &mbu->mbu_paddr;
1882            myaddr = &mbu->mbu_haddr;
1883            mycoa = &mbu->mbu_coa;
1884
1885            /*
1886             * check whether this packet is from the correct sender (that
1887             * is, our home agent) to the CoA or the HoA the mobile node
1888             * has registered before.
1889             */
1890            if (!IN6_ARE_ADDR_EQUAL(&ip6->ip6_src, haaddr) ||
1891                !(IN6_ARE_ADDR_EQUAL(&ip6->ip6_dst, mycoa) ||
1892                  IN6_ARE_ADDR_EQUAL(&ip6->ip6_dst, myaddr))) {
1893                    return (0);
1894            }
```
_____ mip6_mncore.c

1881–1894 The source address of the tunnel packet must be the address of the home agent of the mobile node. The destination address of the tunnel packet must be either the home address or the care-of address of the mobile node. Otherwise, the packet is dropped.

Listing 15-352
_____ mip6_mncore.c

```
1896            /*
1897             * XXX: should we compare the ifid of the inner dstaddr of the
1898             * incoming packet and the ifid of the mobile node's?  these
1899             * check will be done in the ip6_input and later.
1900             */
1901
1902            /* check mn prefix */
1903            for (mpfx = LIST_FIRST(&mip6_prefix_list); mpfx;
1904                 mpfx = LIST_NEXT(mpfx, mpfx_entry)) {
```

```
1905                             if (!in6_are_prefix_equal(myaddr, &mpfx->mpfx_prefix,
1906                                 mpfx->mpfx_prefixlen)) {
1907                                     /* this prefix doesn't match my prefix.
1908                                         check next. */
1909                                     continue;
1910                             }
1911                             goto match;
1912                     }
1913             return (0);
1914     match:
1915             return (128);
1916     }
```
—— mip6_mncore.c

1903–1912 If the prefix part of the home address is not one of the prefixes that the mobile node is managing, the packet is dropped. This code is redundant and can be removed, since in the KAME implementation the prefix of a home address is always one of the prefixes that a mobile node is managing.

Listing 15-353
—— mip6_hacore.c

```
662     int
663     mip6_bc_encapcheck(m, off, proto, arg)
664             const struct mbuf *m;
665             int off;
666             int proto;
667             void *arg;
668     {
669             struct ip6_hdr *ip6;
670             struct mip6_bc *mbc = (struct mip6_bc *)arg;
671             struct in6_addr *mnaddr;
672
673             if (mbc == NULL) {
674                     return (0);
675             }
676
677             ip6 = mtod(m, struct ip6_hdr*);
678
679             mnaddr = &mbc->mbc_pcoa;
680
681             /* check mn addr */
682             if (!IN6_ARE_ADDR_EQUAL(&ip6->ip6_src, mnaddr)) {
683                     return (0);
684             }
685
686             /* check my addr */
687             /* XXX */
688
689             return (128);
690     }
```
—— mip6_hacore.c

662–667 The `mip6_bc_encapcheck()` does the validation check for the incoming tunnel packets on a home agent. The meaning of the `m`, `off`, and `proto` parameters are the same as the parameters of the `mip6_bu_encapcheck()` function. `arg` is a pointer to a binding cache entry, which is registered by the `encap_attach_func()` function on line 2973 in Listing 15-349 of `mip6_cncore.c`.

679–684 The function returns 128 if the source address of the incoming tunnel packet is the care-of address of a mobile node. The destination address of the tunnel packet is not checked at this moment, but in theory it should also be checked.

15.26   Receiving Packets from a Tunnel

If a mobile node or a home agent accepts a tunnel packet, the packet is input to the
`mip6_tunnel_input()` function through the protocol switch mechanism.

Listing 15-354

in6_proto.c

```
229     struct ip6protosw inet6sw[] = {
....
530     struct ip6protosw mip6_tunnel_protosw =
531     { SOCK_RAW,      &inet6domain,    IPPROTO_IPV6,    PR_ATOMIC|PR_ADDR,
532       mip6_tunnel_input, rip6_output,       0,         rip6_ctloutput,
....
534       0,
....
538       0,              0,               0,               0,
....
540       &rip6_usrreqs
....
542     };
```

in6_proto.c

530–542 The `inet6sw[]` array specifies an input function based on the protocol number.
The `mip6_tunnel_protosw` variable declares the `mip6_tunnel_input()` function
as the IPv6 in IPv6 protocol handing function. The information is passed when the
`encap_attach_func()` function is called on line 2973 in Listing 15-349 of
`mip6_cncore.c`.

Listing 15-355

mip6_cncore.c

```
2803    int
2804    mip6_tunnel_input(mp, offp, proto)
2805            struct mbuf **mp;
2806            int *offp, proto;
2807    {
2808            struct mbuf *m = *mp;
2809            struct ip6_hdr *ip6;
....
2811            int s;
```

mip6_cncore.c

2803–2806 The `mip6_tunnel_input()` function is called when a mobile node or a home
agent receives an IPv6 in IPv6 packet from its home agent or mobile nodes,
respectively.

Listing 15-356

mip6_cncore.c

```
2814            m_adj(m, *offp);
2815
2816            switch (proto) {
2817            case IPPROTO_IPV6:
2818                    if (m->m_len < sizeof(*ip6)) {
2819                            m = m_pullup(m, sizeof(*ip6));
2820                            if (!m)
2821                                    return (IPPROTO_DONE);
2822                    }
```

```
2823
2824                        ip6 = mtod(m, struct ip6_hdr *);
....
2831                        s = splimp();
....
2838                        if (IF_QFULL(&ip6intrq)) {
2839                                IF_DROP(&ip6intrq);      /* update statistics */
2840                                splx(s);
2841                                goto bad;
2842                        }
2843                        IF_ENQUEUE(&ip6intrq, m);
....
2851                        splx(s);
....
2853                        break;
2854                default:
....
2859                        goto bad;
2860                }
2861
2862                return (IPPROTO_DONE);
2863
2864        bad:
2865                m_freem(m);
2866                return (IPPROTO_DONE);
2867        }
```
── mip6_cncore.c

2814 The outer header is stripped.

2816–2867 The `mip6_tunnel_input()` function accepts only an IPv6 packet as an inner packet. If the inner packet is an IPv6 packet, the packet is enqueued to the IPv6 input packet queue by the `IF_ENQUEUE()` macro, as long as the queue is not full. If the queue is full, then the packet is dropped. All other packets of which the protocol is not IPv6 are also dropped.

15.27 I/O Control

An input/output (I/O) control interface is provided to manage the Mobile IPv6 features from a user space program. The I/O control is implemented as the `mip6_ioctl()` function. Table 15-5 shows the list of I/O control commands.

Listing 15-357
── mip6_cncore.c
```
264     int
265     mip6_ioctl(cmd, data)
266             u_long cmd;
267             caddr_t data;
268     {
269             int subcmd;
270             struct mip6_req *mr = (struct mip6_req *)data;
271             int s;
```
── mip6_cncore.c

264–267 The `mip6_ioctl()` function has two parameters. The `cmd` parameter is one of the control commands listed in Table 15-5, and the `data` parameter is a pointer to the `mip6_req{}` structure that contains data related to the command.

TABLE 15-5

Name	Description
SIOCSMIP6CFG	Configure features; the following subcommands are required
SIOCSMIP6CFG_ENABLEMN	Enable a mobile node feature
SIOCSMIP6CFG_DISABLEMN	Disable a mobile node feature
SIOCSMIP6CFG_ENABLEHA	Enable a home agent feature
SIOCSMIP6CFG_ENABLEIPSEC	Enable IPsec signal packets protection
SIOCSMIP6CFG_DISABLEIPSEC	Disable IPsec signal packets protection
SIOCSMIP6CFG_ENABLEDEBUG	Enable debug messages
SIOCSMIP6CFG_DISABLEDEBUG	Disable debug message
SIOCDBC	Remove binding cache entries
SIOCSPREFERREDIFNAMES	Set preferences of interface names when a care-of address is chosen

I/O control commands for Mobile IPv6.

Listing 15-358

_____ mip6_cncore.c

```
279             switch (cmd) {
280             case SIOCSMIP6CFG:
281                     subcmd = *(int *)data;
282                     switch (subcmd) {
```
_____ mip6_cncore.c

280–282 The SIOCSMIP6CFG command configures various Mobile IPv6 features, and it has several subcommands.

Listing 15-359

_____ mip6_cncore.c

```
284                 case SIOCSMIP6CFG_ENABLEMN:
285                 {
286                         int error;
287                         error = mip6_mobile_node_start();
288                         if (error) {
289                                 splx(s);
290                                 return (error);
291                         }
292                 }
293                         break;
294
295                 case SIOCSMIP6CFG_DISABLEMN:
296                         mip6_mobile_node_stop();
297                         break;
....
301                 case SIOCSMIP6CFG_ENABLEHA:
....
305                         mip6ctl_nodetype = MIP6_NODETYPE_HOME_AGENT;
306                         break;
```
_____ mip6_cncore.c

284–306 The SIOCSMIP6CFG_ENABLEMN subcommand activates the mobile node feature. The activation code is implemented in the mip6_mobile_node_start()

function. The `SIOCSMIP6CFG_DISABLEMN` subcommand stops the mobile node function. The `mip6_mobile_node_stop()` function does the actual work. `SIOCSMIP6CFG_ENABLEHA` activates the home agent function. Currently, no deactivation mechanism is provided to stop the home agent function.

Listing 15-360

```
                                                              mip6_cncore.c
309                    case SIOCSMIP6CFG_ENABLEIPSEC:
....
313                            mip6ctl_use_ipsec = 1;
314                            break;
315
316                    case SIOCSMIP6CFG_DISABLEIPSEC:
....
320                            mip6ctl_use_ipsec = 0;
321                            break;
322
323                    case SIOCSMIP6CFG_ENABLEDEBUG:
....
327                            mip6ctl_debug = 1;
328                            break;
329
330                    case SIOCSMIP6CFG_DISABLEDEBUG:
....
334                            mip6ctl_debug = 0;
335                            break;
336
337                    default:
338                            splx(s);
339                            return (EINVAL);
340                    }
341                    break;
                                                              mip6_cncore.c
```

309–341 The `SIOCSMIP6CFG_ENABLEIPSEC`, the `SIOCSMIP6CFG_DISABLEIPSEC`, the `SIOCSMIP6CFG_ENABLEDEBUG`, and the `SIOCSMIP6CFG_DISABLEDEBUG` set Mobile IPv6-related global variables. These global variables can be modified via the sysctl mechanism in addition to the I/O control mechanism.

Listing 15-361

```
                                                              mip6_cncore.c
363            case SIOCDBC:
364                    if (IN6_IS_ADDR_UNSPECIFIED(&mr->mip6r_ru.mip6r_in6)) {
365                            struct mip6_bc *mbc;
366
367                            /* remove all binding cache entries. */
368                            while ((mbc = LIST_FIRST(&mip6_bc_list)) != NULL) {
369                                    (void)mip6_bc_list_remove(&mip6_bc_list, mbc);
370                            }
371                    } else {
372                            struct mip6_bc *mbc;
373
374                            /* remove a specified binding cache entry. */
375                            mbc = mip6_bc_list_find_withphaddr(&mip6_bc_list,
376                                &mr->mip6r_ru.mip6r_in6);
377                            if (mbc != NULL) {
378                                    (void)mip6_bc_list_remove(&mip6_bc_list, mbc);
379                            }
380                    }
381                    break;
                                                              mip6_cncore.c
```

363 The SIOCDBC command removes a specified binding cache entry if the IPv6 address to be removed is specified. Otherwise, it removes all binding cache entries kept in a node.

364–370 If the specified IPv6 address is the unspecified address, the mip6_bc_list_remove() function is called for all binding cache entries listed in the global variable mip6_bc_list. All cache entries are removed.

371–380 If an IPv6 address is specified, the related binding cache entry found by the mip6_bc_list_find_withphaddr() function is removed from the cache list.

Listing 15-362

--- mip6_cncore.c

```
439             case SIOCSPREFERREDIFNAMES:
440             {
441                     /*
442                      * set preferrable ifps for selecting CoA.  we must
443                      * keep the name as a string because other information
444                      * (such as a pointer, interface index) may be changed
445                      * when removing the devices.
446                      */
447                     bcopy(&mr->mip6r_ru.mip6r_ifnames, &mip6_preferred_ifnames,
448                         sizeof(mr->mip6r_ru.mip6r_ifnames));
449                     mip6_process_movement();
450             }
451
452                     break;
    ....
454             }
455
456             splx(s);
457
458             return (0);
459     }
```

--- mip6_cncore.c

439–450 The SIOCSPREFERREDIFNAMES command specifies the order in which network interface names are referenced when a mobile node chooses a care-of address from multiple network interfaces. The array of the interface names is copied to the mip6_preferred_ifnames variable. The mip6_process_movement() function is called to choose the most preferable network interface.

Mobile IPv6 Operation

In this chapter, we discuss the configuration of the Mobile Internet Protocol version 6 (IPv6) function using the KAME implementation. We assume the readers know the basic installation of the KAME protocol stack. If you are not familiar with the KAME kit, please read the instruction document that is placed in the top of the KAME distribution directory.

16.1 Rebuilding a Kernel with Mobile IPv6 Extension

The Mobile IPv6 features are not enabled by default. You must prepare a new kernel configuration file and rebuild your kernel to be able to support the Mobile IPv6 protocol. Some user-space programs also need to be rebuilt.

16.1.1 Kernel Options for Mobile Node

To enable mobile node features, the following kernel options need to be added to your kernel configuration file.

```
options MIP6
options MIP6_MOBILE_NODE
pseudo-device hif 1
```

`hif` indicates a virtual home interface. If you have more than one home network, you need to specify the number of home networks you use.

16.1.2   Kernel Options for Home Agent

To enable home agent features, the following kernel options need to be added to your kernel configuration file.

```
options MIP6
options MIP6_HOME_AGENT
```

16.1.3   Kernel Options for Correspondent Node

To enable correspondent node features, the following option needs to be added to your kernel configuration file.

```
options MIP6
```

16.2   Rebuilding User Space Programs

There are four user space programs related to Mobile IPv6:

(1) **rtadvd**: the router advertisement daemon

(2) **had**: the daemon program that provides the Dynamic Home Agent Address Discovery and Mobile Prefix Solicitation/Advertisement mechanisms

(3) **mip6control**: the control program of the KAME Mobile IPv6 functions

(4) **mip6stat**: the program that displays statistics of packets related to Mobile IPv6

16.2.1   Rebuilding rtadvd

You need to add a compiler option to the Makefile of the **rtadvd** daemon. You will find the Makefile for `rtadvd` in `${KAME}/freebsd4/sbin/rtadvd/` directory. The following line needs to be added to the Makefile.

```
CFLAGS+=-DMIP6
```

The **rtadvd** daemon compiled with the above option will include a Home Agent Information option in Router Advertisement messages when the daemon is launched with the `-m` switch. Also, the option will relax the restriction of the advertisement interval. In the basic IPv6 specification, a router must wait a minimum of 3 s when sending Router Advertisement messages periodically. With the option, a router can send Router Advertisement messages every 50 ms at minimum.

16.2.2   Building Other Programs

Other user space programs (**had**, **mip6control**, **mip6stat**) require no additional configuration. These commands will be installed to the `/usr/local/v6/sbin/` directory with other KAME programs.

16.3   IPsec Signal Protection

The Mobile IPv6 specification requires that the signaling packets be protected by the IPsec mechanism. KAME Mobile IPv6 users must set up by themselves properly IPsec configuration between mobile nodes and its home agent; however, it would be tough work especially for those who are not familiar with IPsec and Mobile IPv6. The KAME implementation provides scripts to generate configuration files to set up IPsec configuration between a mobile node and a home agent.

- **mip6makeconfig.sh**: generate necessary configuration files
- **mip6seccontrol.sh**: install IPsec configuration using configuration files generated by the **mip6makeconfig.sh**: script

These scripts are located in ${KAME}/kame/kame/mip6control/ directory.

16.3.1   Configuration Directory

The default configuration directory is /usr/local/v6/etc/mobileip6/. The configuration files for each node are created in a separate directory under this directory. In this example, we use the directory named mobile_node_0. The names of the directories that contain configuration files are arbitrary.

```
# mkdir /usr/local/v6/etc/mobileip6
# mkdir /usr/local/v6/etc/mobileip6/mobile_node_0
```

16.3.2   Preparing a Base Configuration File

The next step is to create a base configuration file for a mobile node. The name of the configuration file must be config. The file is placed under the configuration directory for each node. In this case, the file should be in the /usr/local/v6/etc/mobileip6/mobile_node_0/ directory.

The configuration file provides the following parameters:

- mobile_node: the address of a mobile node
- home_agent: the address of a home agent
- transport_spi_mn_to_ha: the SPI value for the transport mode IPsec packets from a mobile node to a home agent
- transport_spi_mn_to_ha: the SPI value for the transport mode IPsec packets from a home agent to a mobile node
- transport_protocol: the name of IPsec transport mode protocol esp or ah can be specified
- transport_esp_algorithm: the name of the encryption algorithm used by IPsec transport mode
- transport_esp_secret: the secret value for the encryption algorithm

- `transport_auth_algorithm`: the name of the authentication algorithm used by IPsec transport mode

- `transport_auth_secret`: the secret value for the authentication algorithm

- `tunnel_spi_mn_to_ha`: the SPI value for the tunnel mode IPsec packets from a mobile node to a home agent

- `tunnel_spi_ha_to_mn`: the SPI value for the tunnel mode IPsec packets from a home agent to a mobile node

- `tunnel_uid_mn_to_ha`: the identifier to bind a security association and security policy for the tunnel mode connection from a mobile node to a home agent

- `tunnel_uid_ha_to_mn`: the identifier to bind a security association and security policy for the tunnel mode connection from a home agent to a mobile node

- `tunnel_esp_algorithm`: the name of the encryption algorithm used by IPsec tunnel mode

- `tunnel_esp_secret`: the secret value for the encryption algorithm

- `tunnel_auth_algorithm`: the name of the authentication algorithm used by IPsec tunnel mode

- `tunnel_auth_secret`: the secret value for the authentication algorithm

When esp is set to `transport_protocol` parameter, both the encryption algorithm/secret and the authentication algorithm/secret have to be set to `transport_esp_algorithm`, `transport_esp_secret` and `transport_auth_algorithm`, `transport_auth_ secret` parameters.

When ah is set to `transport_protocol` parameter, `transport_esp_algorithm` and `transport_esp_secret` parameters can be omitted.

Figure 16-1 is a sample configuration file.

The algorithms can be selected from all algorithms, which are supported by the **setkey** command.

16.3.3   Generating Configuration Files

The **mip6makeconfig.sh** script generates several configuration files from the base configuration file. Launching the script with the name of the configuration directory will generate these files.

```
# mip6makeconfig.sh mobile_node_0
```

The above command will generate the following files under the configuration directory.

- `add`: add security associations

- `delete`: delete security associations

- `spdadd_mobile_node`: add security policies for a mobile node

- `spddelete_mobile_node`: delete security policies for a mobile node

FIGURE 16-1

```
mobile_node=2001:db8:0:0:201:11ff:fe54:4fde
home_agent=2001:db8:0:0:201:11ff:fe54:5ffc
transport_spi_mn_to_ha=2000
transport_spi_ha_to_mn=2001
transport_protocol=esp
transport_esp_algorithm=blowfish-cbc
transport_esp_secret='"THIS_IS_ESP_SECRET!!"'
transport_auth_algorithm=hmac-sha1
transport_auth_secret='"THIS_IS_AUTH_SECRET"'
tunnel_spi_mn_to_ha=2002
tunnel_spi_ha_to_mn=2003
tunnel_uid_mn_to_ha=2002
tunnel_uid_ha_to_mn=2003
tunnel_esp_algorithm=blowfish-cbc
tunnel_esp_secret='"THIS_IS_ESP_SECRET!!"'
tunnel_auth_algorithm=hmac-sha1
tunnel_auth_secret='"THIS_IS_AUTH_SECRET"'
```

A sample configuration file for IPsec.

- `spdadd_home_agent`: add security policies for a home agent
- `spddelete_home_agent`: delete security policies for a home agent

The contents of the above files are actually a command list for the **setkey** command.

16.3.4   Installing IPsec Configuration

Basically, you do not need to install the configuration by hand. The startup script discussed in the next section will handle installation of the configuration file at boot time. If you need to manage the configuration, you can use the **mip6seccontrol.sh** script.

The **mip6seccontrol.sh** script installs/uninstalls IPsec configuration based on the files generated by the **mip6makeconfig.sh** script. The usage format of the script is shown in Figure 16-2.

On a mobile node, the -m switch must be specified, and on a home agent, -g must be specified. `installall` and `deinstallall` operations install or deinstall all configuration

FIGURE 16-2

```
mip6seccontrol.sh[-m|-g][installall|install|deinstallall|deinstall]
[config_directory]
```

mip6seccontrol.sh *format.*

parameters placed in the configuration directory. If you want to manipulate configuration parameters for a single node, you need to use `install` and `deinstall` operations specifying the directory name of the node as the last argument.

16.4   Configuring Node

To provide an easy way to configure a Mobile IPv6 node, the KAME implementation provides a startup script. You will find the `rc` and `rc.mobileip6` scripts in the `${KAME}/freebsd4/etc/` directory. The scripts should be copied to the `/etc/` directory.

The script provides the following configuration parameters.

- `ipv6_mobile_enable`: set to YES if you use Mobile IPv6

- `ipv6_mobile_config_dir`: set the directory that contains configuration files for IPsec between a home agent and a mobile node

- `ipv6_mobile_nodetype`: define a type of node, either `mobile_node` or `home_agent` can be set

- `ipv6_mobile_home_prefixes`: set the home prefixes in a form of *prefix/prefixlen*; multiple prefixes can be defined by separating them with a space character

- `ipv6_mobile_home_link` [home agent only]: set the physical interface name of a home network

- `ipv6_mobile_debug_enable` setting to YES will print debug messages to console

- `ipv6_mobile_security_enable` setting to NO will disable IPsec check against the signaling messages used by Mobile IPv6

16.4.1   Configuring Mobile Node

Figure 16-3 shows the sample network used in this configuration example. In this example, the home network is `2001:db8:0:0::/64`. The `rc.conf` file will include the following lines:

```
ipv6_mobile_enable="YES"
ipv6_mobile_config_dir="/usr/local/v6/etc/mobileip6"
ipv6_mobile_nodetype="mobile_node"
ipv6_mobile_home_prefixes="2001:db8:0:0::/64"
```

You need to generate all the necessary configuration files for IPsec signal protection by the **mip6makeconfig.sh** script and put them in the directory specified by the `ip6_mobile_config_dir` parameter.

If you do not want to use IPsec signal protection, you need to remove all configuration files from the `/usr/local/v6/etc/mobileip6/` directory and add the following line to your `rc.conf` file.

```
ipv6_mobile_security_enable="NO"
```

FIGURE 16-3

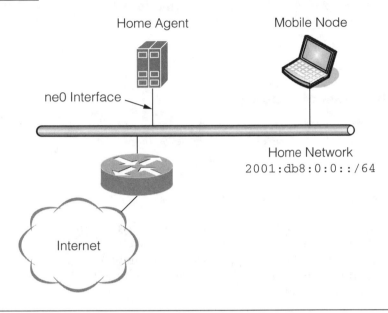

Sample operation network.

Note that the specification says the signaling message must be protected by the IPsec mechanism. Turning the protection off must be used only for testing purposes.

After restarting, your node will start acting as a mobile node.

16.4.2 Configuring a Home Agent

As illustrated in Figure 16-3, the home network is `2001:db8:0:0::/64` and the name of the network interface with which it is used as a home network is `ne0`.

```
ipv6_mobile_enable="YES"
ipv6_mobile_config_dir="/usr/local/v6/etc/mobileip6"
ipv6_mobile_nodetype="home_agent"
ipv6_mobile_home_prefixes="2001:db8:0:0::/64"
ipv6_mobile_home_link="ne0"
```

The configuration files for IPsec signal protection, generated by the **mip6makeconfig.sh** script in the same manner as you did for the mobile node, are stored in `/usr/local/v6/etc/mobileip6/`. You can use the same set of files generated for the mobile node since the **mip6makeconfig.sh** script generates files for both a home agent and a mobile node.

Similar to the configuration procedure for a mobile node, if you want to disable IPsec signal protection, you need to remove all configuration files in the `/usr/local/v6/etc/mobileip6/` directory and add the following line in your `rc.conf` file.

```
ipv6_mobile_security_enable="NO"
```

16.4.3   Configuring Correspondent Node

There is no special procedure to enable correspondent node features. A node that has a kernel with the `MIP6` option acts as a correspondent node by default.

16.5   Viewing Status Information

The `mip6control` command provides a way to get the current binding information on a node. To get the current binding update list entries, the `-bl` switch is used. The `-b` switch indicates getting the information about binding update list entries, and the `-l` switch prints the address of nodes in a long format. You can use the `-c` switch to get the binding cache entries instead of the binding update list entries.

The following is the sample output of `mip6control -bl`.

```
$ mip6control -bl
paddr         haddr        coa          lifetim ltexp refresh retexp seqno  flags pfsm  sfsm  state
ha.kame.net   mn.kame.net  2001:200:11  420     312   210     102    54345  AHL   BOUND INIT
www.kame.net  mn.kame.net  2001:200:11  420     130   420     130    34122        IDLE  INIT  D
cn.kame.net   mn.kame.net  2001:200:11  420     229   420     229    3423         BOUND INIT
```

In the example, there are three binding update list entries. The first entry is a home registration entry. The other two entries are to IPv6 nodes. The node shown in the middle entry (`www.kame.net`) does not support Mobile IPv6. The state field will show a `D` flag when the node does not support Mobile IPv6. The last entry is performing route-optimized communication with the mobile node.

The following is the sample output of `mip6control -cl`.

```
$ mip6control -cl
phaddr        pcoa        addr flags  seqno lifetim ltexp  state refcnt
mn.kame.net   2001:200:11 AHL    54345 420     312    BOUND 0
fe80::203:2   2001:200:11 AHL    23414 420     312    BOUND 1
mn.wide.ad.jp 2001:280:45        8473  420     123    BOUND 0
```

The above node has three entries. The first two are home registration entries for mn.kame.net. The upper entry is for the global address of the mobile node, and the lower entry is for the link-local address of the mobile node. Since the mobile node has sent a Binding Update message with the `L` flag, the home agent also needs to protect its link-local address. The last entry is for another mobile node that has registered to another home agent. This node and the last node (`mn.wide.ad.jp`) are performing route-optimized communication.

The complete usage of the `mip6control` command can be found in the manual page.

16.6   Viewing Statistics

The `mip6stat` command shows the statistics collected in a kernel. Figure 16-4 shows the sample output of the command. Each entry is related to the each member variable of the `mip6stat{}` structure described in Figure 16-4.

FIGURE 16-4

```
Input:
        82977 Mobility Headers
        1245 HoTI messages
        1245 CoTI messages
        0 HoT messages
        0 CoT messages
        82977 BU messages
        0 BA messages
        0 BR messages
        0 BE messages
        83439 Home Address Option
        12 unverified Home Address Option
        0 Routing Header type 2
        920861 reverse tunneled input
        0 bad MH checksum
        0 bad payload protocol
        0 unknown MH type
        0 not my home address
        0 no related binding update entry
        0 home init cookie mismatch
        0 careof init cookie mismatch
        29 unprotected binding signaling packets
        146 BUs discarded due to bad HAO
        0 RR authentication failed
        4 seqno mismatch
        0 parameter problem for HAO
        0 parameter problem for MH
        0 Invalid Care-of address
        0 Invalid mobility options
Output:
        82774 Mobility Headers
        1245 HoTI messages
        1245 CoTI messages
        0 HoT messages
        0 CoT messages
        0 BU messages
        82774 BA messages
        82774 binding update accepted
        0 BR messages
        12 BE messages
        0 Home Address Option
        83175 Routing Header type 2
        1223198 reverse tunneled output
```

Sample output of the `mip6stat` *command.*

Appendix: The Manual Page of `mip6control`

A.1 Name

mip6control — control KAME/MIP6 features

A.2 Synopsis

mip6control [-**i** *ifname*][-**abcghlmMnNw**][-**H** *home_prefix*][-**P** *prefixlen*]

[-**A** *home_agent_global_addr*][-**L** *home_agent_linklocal_addr*]

[-**C** *addr*][-**u** *address#port*][-**v** *address#port*]

[-**F** *ifp1[:ifp2[:ifp3]]*][-**S** *0|1*][-**D** *0|1*]

A.3 Description

mip6control sets/gets KAME/MIP6-related information. If no argument is specified,
mip6control shows the current status of the node.

-**i**   *ifname*
Specify home interface of the mobile node. The default value is hif( ).

-**H**   *home_prefix*
Set *home_prefix* as a home prefix of the mobile node. You must specify the prefix
length of *home_prefix* with **-P** option.

-**P**   *prefixlen*
Specify the length of the prefix to be assigned to the mobile node. Use with **-H** option.

-A *home_agent_global_address*
Specify the global address of the home agent of this mobile node. If your home agent supports Dynamic Home Agent Address Discovery (DHAAD), you need not use this switch. Use with **-L** option.

-L *home_agent_linklocal_address*
Specify the link-local address of the home agent of this mobile node. If your home agent supports DHAAD, you need not use this switch. Use with **-A** option.

-m   Start acting as a mobile node.

-M   Stop acting as a mobile node.

-n   Show network addresses as numbers.

-h   Show the home prefixes currently set to this mobile node.

-a   Show the home agents list.

-b   Show the binding update list.

-g   Start acting as a home agent.

-c   Show the binding cache entries.

-C *addr*
Remove the binding cache entry specified by `addr`. `addr` is a home address of the binding cache entry. If :: is specified, **mip6control** removes all binding cache entries.

-l   Show information in a long format.

-u *Address#Port*
Add a rule that MN doesn't add a Home Address option to the outgoing packet.

-v *Address#Port*
Delete the rule that specified one.

-w   Show the rule.

-F *ifp1[:ifp2[:ifp3]]*
Set preferable network interfaces for CoA selection. Specify nothing to remove the preferences.

-S *0|1*
When set to 0, the IPsec protection check of the incoming binding updates and binding acks will not be performed (always pass the check).

-D *0|1*
When set to 0, no debug messages are printed.

-N   Show the list of nonces that this host maintains as a correspondent node. The first column is the number of nonce index and the second is a nonce value.

A.4   Examples

To make a node act as a mobile node, issue the following commands as a root.

```
root# mip6control -i hif0 -H 2001:200:1:1:: -P 64
root# mip6control -m
```

Replace 2001:200:1:1:: with your home network prefix.

To make a node act as a home agent, issue the following commands as a root.

```
root\# mip6control -g
```

To set a rule to avoid adding home address option when querying DNS, issue the following commands as a root.

```
root\# mip6control -u ::\#53
```

A.5 History

The `mip6control` command first appeared in WIDE/KAME IPv6 protocol stack kit.

A.6 Bugs

Many :).

References

Most of the references for this book are RFCs. Some specifications are in the process of standardization or revision, for which Internet drafts are referred to. Both types of documents are freely available from the IETF web page: http://www.ietf.org. Note, however, that an Internet draft is a work-in-progress material, which may expire or may have become an RFC by the time this book is published. There are WWW or FTP sites on the Internet that provide a copy of old versions of Internet drafts when necessary. At the time of this writing, the KAME project's FTP server provides this service, which is located at ftp://ftp.kame.net/pub/internet-drafts/.

[CN-IPSEC]	F. Dupont and J.M. Combes, "Using IPsec between Mobile and Correspondent IPv6 Nodes," Internet draft: draft-ietf-mip6-cn-ipsec-03.txt, August 2006.
[IPV4-ADDRESS-REPORT]	G. Huston, "IPv4 Address Report," available at http://www.potaroo.net/tools/ipv4/index.html, December 2008.
[MEXT-NEMO-V4TRAVERSAL]	H. Soliman (ed.), "Mobile IPv6 Support for Dual Stack Hosts and Routers," Internet draft: draft-ietf-mext-nemo-v4traversal-10.txt, Internet Engineering Task Force, July 14, 2008.
[MIP6-NEMO-V4TRAVERSAL]	H. Soliman, et al., "Mobile IPv6 Support for Dual Stack Hosts and Routers (DSMIPv6)," Internet draft: draft-ietf-mip6-nemo-v4traversal-02.txt, June 2006.
[RFC2461]	T. Narten, et al., "Neighbor Discovery for IP Version 6 (IPv6)," RFC2461, December 1998.
[RFC2463]	A. Conta and S. Deering, "Internet Control Message Protocol (ICMPv6) for the Internet Protocol Version 6 (IPv6) Specification," RFC2463, December 1998.

[RFC2526]   D. Johnson and S. Deering, "Reserved IPv6 Subnet Anycast Addresses," RFC2526, March 1999.

[RFC2960]   R. Stewart, et al., "Stream Control Transmission Protocol," RFC2960, Internet Engineering Task, October 2000.

[RFC3041]   T. Narten and R. Draves, "Privacy Extensions for Stateless Address Autoconfiguration in IPv6," RFC3041, January 2001.

[RFC3261]   J. Rosenberg et al., "SIP: Session Initiation Protocol," RFC3261, Internet Engineering Task Force, June 2002.

[RFC3344]   B. Patil, P. Roberts, and C. E. Perkins, "IP Mobility Support for IPv4," RFC3344, Internet Engineering Task Force, August 2002.

[RFC3484]   R. Draves, "Default Address Selection for Internet Protocol version 6 (IPv6)," RFC3484, February 2003.

[RFC3775]   D. Johnson, C. E. Perkins, and J. Arkko, "Mobility Support in IPv6," RFC3775, Internet Engineering Task Force, June 2004.

[RFC3776]   W. Stevens et al., "Advanced Sockets Application Program Interface (API) for IPv6," RFC3776, May 2003.

[RFC4423]   R. Moskowitz and P. Nikander, "Host Identity Protocol (HIP) Architecture," RFC4423, Internet Engineering Task Force, May 2006.

[RFC4584]   S. Chakrabarti and E. Nordmark, "Extension to Sockets API for Mobile IPv6," RFC4584, July 2006.

[RFC4640]   A. Patel and G. Giaretta, "Problem Statement for Bootstrapping Mobile IPv6 (MIPv6)," RFC4640, September 2006.

Index

Home address (*continued*)
    ip6_opt_home_address{} structure
        representing, 96
    mip6_prefix{} structure and, 263, 269
    mobile node, care-of address v., 305
    mobile node configuring, 259–268
    prefix entry, finding with, 327
    prefix entry lifetime impacting, 347
    prohibited, 207
    registering, 307
    renumbering problems for, 73
    role of, 18
    source address swapped, in unverified packet,
        with, 204–208
    source address swapped with, 203–204
Home address field, 27–28, 35
Home Address option
    BE and, 128–129
    correspondent node receiving, 201–208
    defining, 25
    Destination Options Header, created for, 424–426
    Destination Options Header, search for, 206
    format of, 26, 26f
    IP6A_HASEEN flag and, 201
    ip6_opt_home_address{} structure size
        and length of, 202
    MIP6_BU_PRI_FSM_STATE_IDLE state and, 423
    MIP6_BU_PRI_FSM_STATE_WAITD state and, 423
    packet verification for, 202–203
    processing, 201–203
    using, 26–27
Home agent. *See also* Dynamic Home Agent Address
        Discovery (DHAAD) Reply messages; Dynamic
        Home Agent Address Discovery (DHAAD)
        Request messages
    adding/removing, 69
    bidirectional tunnel completed by mobile node
        and, 51–52, 52–53f
    capabilities of, 137
    configuration files for, 441–442, 441f
    creating, 309–311
    files used by, 138t
    finding, 315–316
    forwarding mechanism of, 18–19
    had program updating information of, 221–224
    home deregistration tasks performed by, 188
    home network generating list for, 67, 68f, 109f
    inserting, 312
    kernel options for, 436
    lifetime of, 223, 311, 314
    mip6_ha{} structure representing mobile node
        information of, 106–107
    Mobile IPv6 and registering, 15, 16f
    Mobile Prefix Advertisement messages sent by, 72
    Mobile Prefix Solicitation received by, 71–72
    next timeout set for, 316–317
    packet forwarding conditions for, 208–209

    packet intercepting and, 54, 54f
    periodical tasks of, 317–318
    prefix entry, finding with, 328–329
    prefix information option including global address
        for, 223–224
    prefix information structures' relationship with,
        108, 109f
    reinserting, 312–314
    security policy entries for, 76t
    sockaddr_in6{} instance created for, 184
    tunnel packet received by, 430–431
    updating, 311
    updating information of, 314–315
Home Agent Addresses field, 44
Home agent anycast address
    computation rules for, 352–353
    computing, 42, 43f
    creating, 352
Home Agent Information option. *See also* Preferred
        home agent information
    defining, 25
    format of, 42f
    mobile node and, 393
    nd_opt_homeagent_info{} structure
        representing, 99
    prefix entry inserting, 327–328
    structures, 218–220, 220–221f
    using, 41–42
    virtual home interface adding, 392–393
Home Agent Lifetime field, 42
Home agent list
    management of, 218–231
    mobile nodes and management of, 309
Home Agent Preference field, 41
Home deregistration, 54–55
    BU message for, 140
    home agent tasks performed for, 188
    mobile node and, 290
    procedure for, 188–190
    return routability procedure after, 64
Home Init Cookie, 59
    field, 30
Home interface. *See* Virtual home interface
Home Keygen Token, 32
    computing, 60
Home network, 18
    home agent list maintained on, 67, 68f, 109f
    Neighbor Advertisement message sent to, 291
Home Nonce Index, 31
Home prefix information, 390–392. *See also* Prefix
        information
Home registration, 308
    aborting, 272
    binding cache creation compared to, 170
    binding cache lifetime for, 145
    binding update list entry and, 289
    BU flags set for, 139